World Survey of Climatology Volume 16

FUTURE CLIMATES OF THE WORLD:
A MODELLING PERSPECTIVE

World Survey of Climatology

World Survey of Climatology Volume 16

Future climates of the world: a modelling perspective

edited by

A. HENDERSON-SELLERS

Macquarie University, Sydney (Australia)

1995

ELSEVIER
Amsterdam – Lausanne – New York – Oxford – Shannon – Tokyo

ELSEVIER SCIENCE B.V.
Sara Burgerhartstraat 25
P.O. Box 211, 1000 AE Amsterdam, The Netherlands

ISBN 0-444-89322-9

This book is printed on acid-free paper.

Printed in The Netherlands.

World Survey of Climatology

Editor in Chief: H.E. LANDSBERG

Volume 1 General Climatology, 1
Editor: H. FLOHN

Volume 2 General Climatology, 2
Editor: H. FLOHN

Volume 3 General Climatology, 3
Editor: H. FLOHN

Volume 4 Climate of the Free Atmosphere
Editor: D. F. REX

Volume 5 Climates of Northern and Western Europe
Editor: C. C. WALLÉN

Volume 6 Climates of Central and Southern Europe
Editor. C. C. WALLÉN

Volume 7 Climates of the Soviet Union
Editor: P. E. LYDOLPH

Volume 8 Climates of Northern and Eastern Asia
Editor: H. ARAKAWA

Volume 9 Climates of Southern and Western Asia
Editor: H. ARAKAWA

Volume 10 Climates of Africa
Editor: J. F GRIFFITHS

Volume 11 Climates of North America
Editor: R. A. BRYSON

Volume 12 Climates of Central and South America
Editor: W. SCHWERDTFEGER

Volume 13 Climates of Australia and New Zealand
Editor: J. GENTILLI

Volume 14 Climates of the Polar Regions
Editor: S. ORVIG

Volume 15 Climates of the Oceans
Editor: H. VAN LOON

Volume 16 Future Climates of the World: A Modelling Perspective
Editor: A. HENDERSON-SELLERS

Dedication for:

The Holly Grange Household:
Laura, Patrick, Philip, Stephen, Nicholas, Laurence, Alice and Eve

List of Contributors to this Volume

M. O. Andreae
Max Planck Institut für Chemie
Mainz (Germany)

E. J. Barron
Earth System Science Center
University Park, Pennsylvania (U.S.A.)

A. Berger
Université Catholique de Louvain
Institut d'Astronomie et de Geophysique
Louvain-la-Neuve (Belgium)

G. Brasseur
National Center for Atmospheric Research
Boulder, Colorado (U.S.A.)

H. Cleugh
Centre for Environmental Mechanics
CSIRO
Canberra (Australia)

H. Diaz
Climate Research Division, ARL
Environmental Research Laboratories
Boulder, Colorado (U.S.A.)

R. E. Dickinson
Institute of Atmospheric Physics
University of Arizona
Tucson, Arizona (U.S.A.)

M. P. Dudek
Atmospheric Sciences Research Center
University at Albany
Albany, New York (U.S.A.)

J. C. Gille
National Center for Atmospheric Research
Boulder, Colorado (U.S.A.)

A. Henderson-Sellers
Climatic Impacts Centre
Macquarie University
North Ryde (Australia)

G. J. Holland
BMRC
Melbourne
Victoria (Australia)

P. D. Jones
Climatic Research Unit
University of East Anglia
Norwich (U.K.)
England

G. N. Kiladis
Cooperative Institute for Research in Environmental Sciences
Boulder, Colorado (U.S.A.)

L. Kump
Earth System Science Center
University Park, Pennsylvania (U.S.A.)

X.-Z. Liang
Atmospheric Sciences Research Center
University at Albany
Albany, New York (U.S.A.)

J. E. Lovelock
Cornwall (U.K.)

Chapter 6. CLIMATIC VARIABILITY ON DECADAL TO CENTURY TIME-SCALES
by H. F. DIAZ and G. N. KILADIS

Introduction 191
Climate variations in the past 1,000 years 193
The Medieval Warm Period 193
The Little Ice Age 195
Possible causes of climatic variability on decadal to century time-scales 199
Solar variability 199
Volcanism and climate 205
El Niño and the Southern Oscillation 207
What is ENSO?, 208—Mean conditions, 208—ENSO perturbations, 210—Teleconnections, 214—Precipitation, 214—Temperature, 220—ENSO variability over time, 222—Future climate change and ENSO, 224
Other sources of internal climate variability 225
Climates of the future 230
Changes from anthropogenic sources 230
References 233

Chapter 7. SATELLITE SYSTEMS AND MODELS FOR FUTURE CLIMATE CHANGE
by R. E. DICKINSON

Introduction 245
Relationship of satellite observations to the climate system 245
Some of the key satellites and their instruments 245
Current climate models: limitations, outstanding questions and satellite paths to improvement 250
Global radiation balance and the role of clouds 252
Solar constant 252
Relating global temperature to global radiative balance 253
Earth Radiation Budget Satellite measurements 254
Satellite cloud climatology 255
Cloud feedbacks 256
Precipitation 260
Atmospheric dynamic and thermodynamic fields 262
Data assimilation issues 263
Inclusion of diabatic processes 264
Land surface properties for climate models 264
Radiative forcing 265
Latent heat atmospheric forcing 267
Land cover properties 269
Surface parameterizations and field programmes 270
Oceans and cryosphere 271
Oceans 271

Cryosphere 272
Satellites and future climate 274
Acknowledgements 275
References 275

Chapter 8. DYNAMICS OF FUTURE CLIMATES
by B. J. McAvaney and G. J. Holland

Introduction 281
Diagnostic methods 282
Diagnostic models 282
Climate models 282
Climate analysis 283
Large-scale dynamics 288
Atmospheric waves 288
Wave generation 289
Barotropic and baroclinic instability, 289—Mountain torques, 290—Diabatic heating, 290
Energetic constraints on the general circulation 291
Angular momentum, 291—Thermodynamic energy, 293—Water vapour, 296
Large scale interactions 297
Regional-scale interactions 300
Storm tracks 302
Summer monsoon circulation in the western North Pacific 304
Future climate dynamics 306
References 310

III. Human Factors Affecting Future Climate

Chapter 9. THE GREENHOUSE EFFECT OF TRACE GASES
by W.-C. Wang, M. P. Dudek and X.-Z. Liang

Introduction 317
The greenhouse effect 317
Radiative forcing 320
Approaches for studying climate change 320
Uniformly mixed gases 322
Emission and concentration of recent years 322
Future atmospheric concentrations 323
Emission scenarios, 323—Concentration level, 324
Climatic effects 325
Equilibrium experiment, 327—Transient response experiments, 331
Atmospheric ozone 334
Stratospheric ozone depletion, 335—Tropospheric ozone, 338

SA90	IPCC 1990 Business-as-Usual Emission Scenario
SAGE	Stratospheric Aerosol and Gas Experiment
SAR	Synthetic Aperture Radar
SHIM	surface heat island model
SNR	signal-to-noise ratio
SPM	suspended particulate matter
SPOT	Système Pour l'Observation de la Terre
SRB	Surface Radiation Budget
SSMI	Special Sensor Microwave sounder and Imager
SST	sea surface temperature
Sv	Sverdrup (= 1×10^6 m^3 s^{-1})
SVF	sky view factor
t	time
T_e	effective temperature
TOA	top of the atmosphere
TOMS	Total Ozone Mapping Spectrometer
TOPEX	Ocean Topography Experiment
TRMM	Tropical Rainfall Measuring Mission
T_s	surface temperature
TUTT	tropical upper tropospheric trough
UARS	Upper Atmosphere Research Satellite
UBL	urban boundary layer
UCL	urban canopy layer
UKMO	United Kingdom Meteorological Office
UN	United Nations
UNCOD	United Nations Conference on Desertification
UNEP	United Nations Environment Programme
USGS	United States Geological Survey
UV	ultraviolet
VI	vegetation index
W	width
WCRP	World Climate Research Programme
WMO	World Meteorological Organization
WSBW	Weddell Sea Bottom Water

Chemical Formulae and Nomenclature

C_2F_6	hexafluoroethane
CCl_2F_2	dichlorotrifluoromethane
CCl_3F	trichlorofluoromethane
CCl_4	carbon tetrachloride

CF_4	tetrafluoromethane
CH_3CCl_3	methyl chloroform
CH_4	methane
$CHClF_2$	chlorodifluoromethane
CO	carbon monoxide
CO_2	carbon dioxide
H_2O	water vapour
N_2O	nitrous oxide
NO_x	oxides of nitrogen (NO, NO_2, NO_3)
O_3	ozone
SF_6	sulphur hexafluoride

Chapter 1

Climates of the future

A. HENDERSON-SELLERS

Why consider the climate of the future?

Human beings are curious: we seek to understand and hence to predict. We are dependent upon the climate and so our desire to predict is not driven solely by curiosity but by a need to predict future weather, its variability and other characteristics: future climates.

The tools we employ in our attempts to predict future climates are the same as those employed by everyone attempting to understand other complexities of the world: observation, reconstruction of past events, interpretation, extrapolation and process studies. In the case of climate, observations comprise surface and satellite data retrievals; past events can be reconstructed from the geological record and also from the historical record (both of observations and of proxy information); extrapolation can take many forms but is increasingly being synthesized with simulations from computer-based models, themselves derived from interpretation of past events and studies of climatic and weather processes.

So far our predictive skills are rather poor. In addition, technology and the harnessing of natural resources appear to have decreased the need to predict the future climate of industrial, housing or even agricultural developments. However, although oil and gas pipelines can be laid across permafrost, indoor climates can be controlled independently of the outdoors, airports operate year-round and irrigated land can grow a wide variety of crops, there are still massive costs, in lives and revenue, associated with weather extremes and with climatic shifts.

Marginal changes in climate can be "managed" although this may be costly: financially in developed nations but in terms of human life in the less-developed nations. On the other hand, if the oceans were to "boil off" or if this planet were plunged into a global glaciation, it is difficult to see how technology could aid us. Thus it is arguable that global-scale climatic changes should be our greatest concern while regional to local changes might be viewed as of somewhat lesser importance. However, we inhabit "regions", called nations, so political responses are closely linked to regional climates and even local conditions. Overall, however, unless or until humans colonize other planets, our primary concern must remain the habitability and sustainability of the Earth as a whole.

Focus of this book

Like all its predecessors in this distinguished series, this text was designed to offer an up-to-date view of climatic science. Unlike its predecessors this text will be ephemeral. It is not a which is usually interpreted as indicating that globally averaged climatic conditions, such as temperature and precipitation, were not too different from the present day (see Chapter 15 by LOVELOCK and KUMP). Figure 1f shows mean temperatures rising slightly over the whole of Earth's history because we know that solar luminosity has increased by about 30% since the Earth was formed. It should also be noted that Fig. 1f is an interpretation of the probable record of the Earth's global mean temperatures based partly on observational evidence and partly on results from climate models which indicate that negative feedback effects can reduce large swings while "recovery" of habitable conditions from either a total planetary freeze or a "runaway greenhouse" is virtually impossible (KASTING, 1989; HENDERSON-SELLERS et al., 1991).

The format of this book is guided, in part, by our understanding of this record of past climates and, in part, by the increasing recognition that human activities have the capacity to be as significant an influence on future climate as the natural factors which have, so far, shaped climatic evolution. The first section of this book reviews the geological history of climatic change assessing astronomical factors, palæo-evidence and massively disruptive events. The second section reviews observations of climate including both the historical record and the exploitation of satellite data. In the third section, the possible impacts of human activities on future climates are considered including the trace gas greenhouse, increased aerosols, pollution, ozone depletion and land-use change. In this section the future climates of the Earth's cities are also reviewed. Finally, the last section of the book closes the loop back to the geological record by reviewing the potential range of future climates in the context of global geophysiology: the perspective which considers the Earth system holistically.

A potentially useful way of predicting patterns of future climate is to search for periods in the past when the global mean temperatures were similar to those we expect in the future, and then use the past spatial patterns as analogues of those which will arise in the future. For a good analogue, it is also necessary for the forcing factors (for example, greenhouse gases, orbital variations) and other conditions (for example, ice cover, topography) to be similar; direct comparisons with climate situations for which these conditions do not apply cannot be easily interpreted. HOUGHTON et al. (1990, p xxv) state that "analogues of future greenhouse-gas-changed climates have not been found" and therefore go on to say that, in the context of the IPCC scenarios for the next century, "we cannot therefore advocate the use of palæo-climates as predictions of regional climate change due to future increases in greenhouse gases. However, palæo-climatological information can provide useful insights into climate processes, and can assist in the validation of climate models." (HOUGHTON et al., 1990, p xxv).

In this book, palæo-climatological evidence will be used (i) to offer a more complete view of "natural variability" (Chapter 2 by BERGER) and the processes underlying this variability, (ii) to set the framework for the longer time-scales of interest here and to examine the validation of climate models (Chapter 3 by BARRON) and (iii) to permit a quantitative discussion of unlikely events (Chapter 4 by RAMPINO) affecting climate or non-linear responses causing rapid changes (Chapter 14 by PENG). However, all chapters draw on the results of numerical models of the Earth's climate.

Modelling the Earth's climate

The simplest possible way of constructing a model of the Earth's climate with which to predict its future is to consider the radiative balance of the globe as a whole. This is a zero dimensional model often written in the form

$$S(1 - A) = \sigma T_e^4 \tag{1a}$$

$$T_s = T_e + T_{greenhouse} \tag{1b}$$

Here, S, the solar input, has a value of about 1370 W m^{-2} divided by 4 to give about 342 W m^{-2}, the amount of solar radiation instantaneously incident at the planet per unit area of its (spherical) surface. Taking the Earth's albedo, A, as about 0.3, with σ, Stefan-Boltzmann constant of 5.67×10^{-8} W m^{-2} K^{-4}, the effective blackbody radiating temperature of the Earth, T_e, is found to be around 255 K. This is lower than the current global mean surface temperature of 288 K, the difference, about 33°C, largely being due to the greenhouse effect. In Fig. 2, the Earth's globally and annually averaged radiation balance can be seen graphically. The numbers in parentheses represent energy as a percentage of the average solar constant – about 342 W m^{-2} – at the top of the atmosphere. Note that nearly half the incoming solar radiation penetrates the clouds and greenhouse gases to the Earth's surface. These gases and clouds re-radiate most (i.e. 88 units) of the absorbed energy back down towards the surface. This is the basis of the mechanism of the greenhouse effect. The magnitude of the greenhouse effect is commonly measured as the difference between the blackbody emission at the surface temperature (a global average of 288 K gives 390 W m^{-2}) and the outgoing infrared radiation at the top of the atmosphere (here 70 units or 239 W m^{-2}), i.e. 151 W m^{-2} (see Fig. 9.1 in Chapter 9 by WANG et al.).

In this zero dimensional model, the climate can only be changed if the incoming solar radiation alters, if the albedo changes or if the greenhouse effect is modified. Thus, to first order,

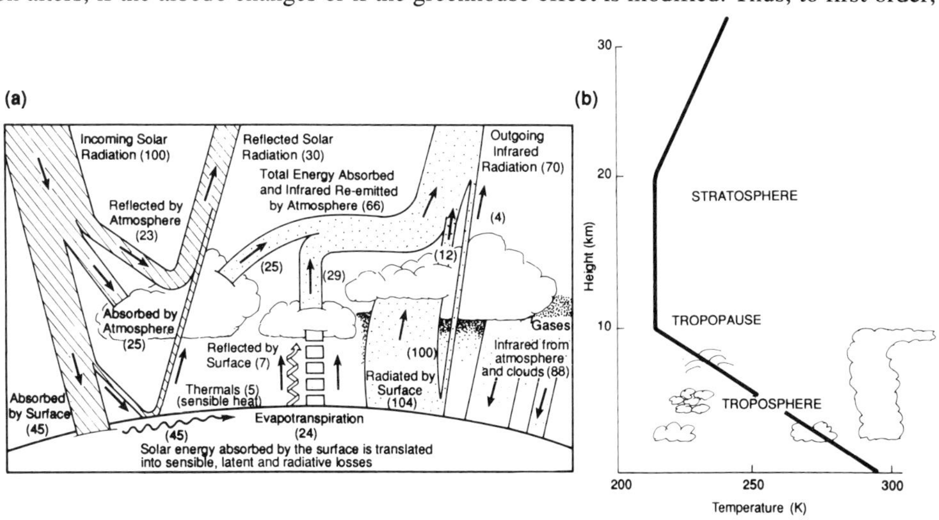

Fig. 2. Schematic of the Earth's energy budget (a) (modified from SCHNEIDER, 1992, reproduced with the permission of Cambridge University Press and the University Corporation for Atmospheric Research) and temperature profile (b) of the lower atmosphere. Units are percentages of the incident solar radiation, 342 W m^{-2} and temperature.

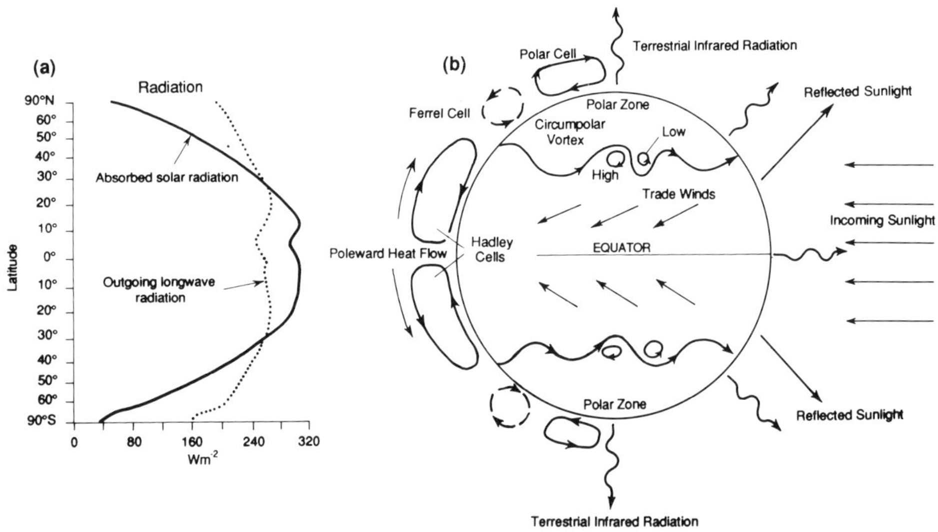

Fig. 3. Schematic of (a) the latitudinal energy balance of the Earth and (b) the dynamics of the atmosphere (modified from HENDERSON-SELLERS and MCGUFFIE, 1987).

the Earth's climate is controlled by incident solar radiation, by the amount of this which is absorbed by the planet and by the thermal absorptivity of the gases in the atmosphere (e.g. KIEHL, 1992).

Additional features control the climate. Of these, the most important are the latitudinal distribution of absorbed solar radiation (large at low latitudes and much less near the poles) as compared to the emitted thermal infrared radiation which varies much less with latitude (Fig. 3a). This latitudinal imbalance of net radiation for the surface-plus-atmosphere system as a whole (positive in low latitudes and negative in higher latitudes) combined with the effect of the Earth's rotation on its axis produces the dynamical circulation system of the atmosphere (Fig. 3b). Figures 2 and 3 describe schematically the most important processes controlling the atmospheric circulation and characteristics. The latitudinal radiative imbalance tends to warm air which rises in equatorial regions and would sink in polar regions were it not for the rotation of the Earth. The westerly waves in the upper troposphere in mid-latitudes and the associated high and low pressure systems are the product of planetary rotation affecting the thermally driven atmospheric circulation. The overall circulation pattern comprises thermally direct cells in low latitudes, strong waves in the mid-latitudes and weak direct cells in polar regions. This circulation combined with the vertical distribution of temperature (Fig. 2b) represent the major aspects of the atmospheric climate system (e.g. SCHNEIDER, 1992). It is possible that even these fundamental characteristics of atmospheric dynamics could change in the future (Chapter 8 by MCAVANEY and HOLLAND).

Processes in the atmosphere are strongly coupled to the land surface (Fig. 2a), to the oceans and to those parts of the Earth covered with ice and snow (the cryosphere). There is also strong coupling to the biosphere (the vegetation and other living systems on the land and in the ocean). These five components, the atmosphere, land, ocean, ice and biosphere (Fig. 4a), are now often considered to comprise the climate system (GARP, 1975). However, their time-scales differ quite significantly:

atmosphere	~minutes to days
oceans	~weeks to thousands of years
biosphere	~years to hundreds of years
ice	~thousands to millions (or even hundreds of millions) of years
land surface	~millions to hundreds of millions of years

The result of these different time-scales and of the complex interactions between these components of the climate system is a rich spectrum of climatic change (Fig. 4b). The largest

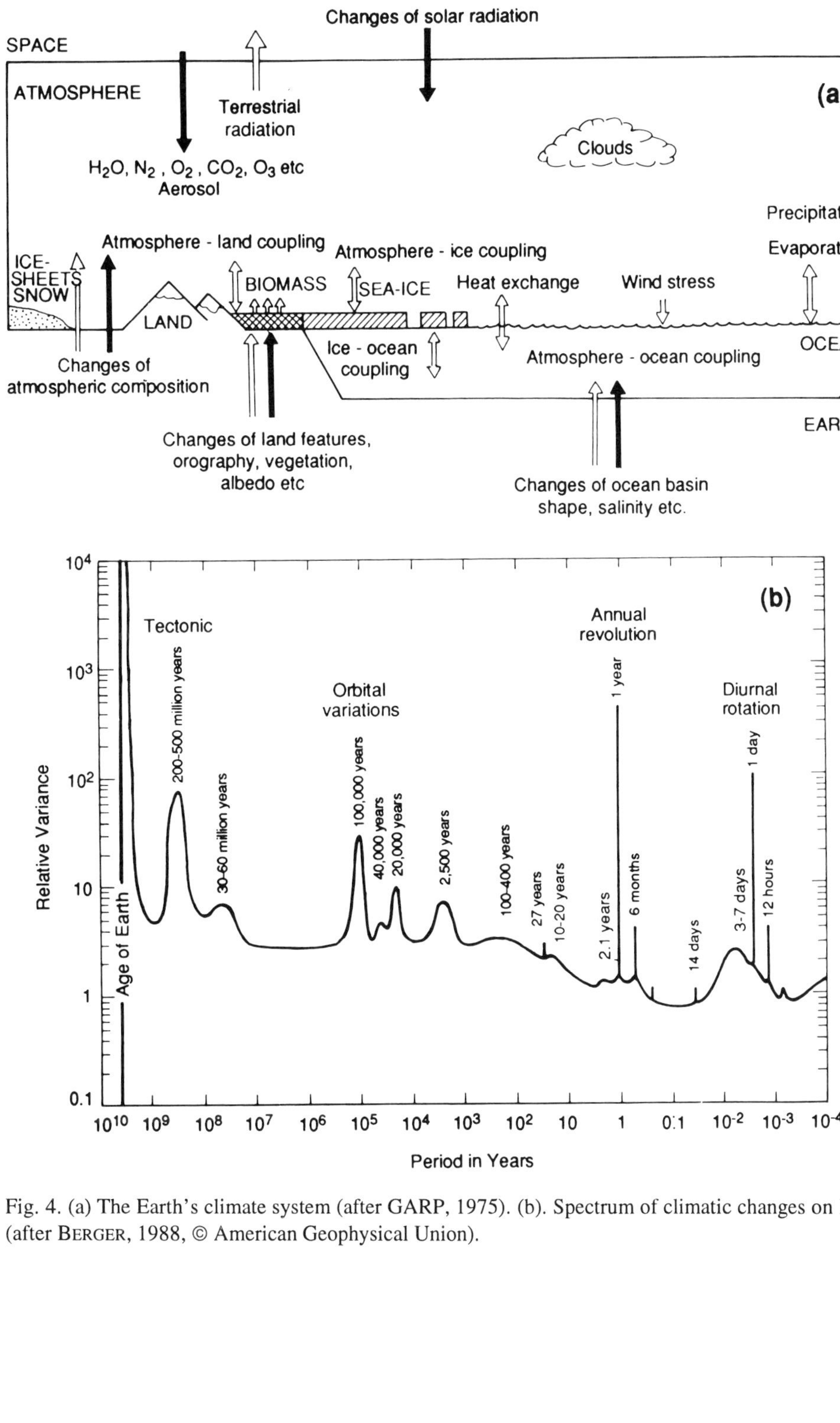

Fig. 4. (a) The Earth's climate system (after GARP, 1975). (b). Spectrum of climatic changes on Earth (after BERGER, 1988, © American Geophysical Union).

peaks in this spectrum relate to astronomical forcings: the Earth's rotation, its revolution around the Sun, variations in this orbit and the formation of the Solar System (see Chapter 2 by BERGER). Coupling amongst processes within the climate system components, such as the atmosphere with the oceans, the oceans and ice masses, is probably responsible for other peaks in this climate change spectrum.

In this book, the perspective is longer than that adopted by IPCC, viz. the next century. Whilst recognizing the potential importance of anthropogenic activities, particularly but not exclusively the release of greenhouse gases (see Chapter 9 by WANG et al.), for future climate these are placed in the context of astrophysical, geological and biological variations which may also modify future climates.

The most highly developed tool which we have to predict future climate is known as a general circulation model or global climate model, abbreviated to GCM. These models are based on the laws of physics and use descriptions in simplified physical terms (called parameterizations) of the smaller-scale processes such as those due to clouds and deep mixing in the ocean. In a climate model, an atmospheric component, essentially the same as a weather prediction model, is coupled to a model of the ocean, which can be equally complex (e.g. HENDERSON-SELLERS and MCGUFFIE, 1987; SCHLESINGER, 1988). The term GCM is nowadays taken to mean fully three-dimensional models of the atmosphere and oceans coupled together. If only the atmospheric component is represented the term AGCM (atmospheric GCM) is used. Such a model is generally linked to a simple representation of ocean temperatures such as offered by a mixed-layer ocean model. However, because the time-scales of, for example, the ice masses and the carbon cycle are very long they have not yet been incorporated into GCMs. For very long time-scale simulations of future, and past, climates simpler models are often used: energy balance models (EBMs) and two-dimensional statistical dynamical models (2-D SDs).

EBMs are called "one-dimensional", the dimension in which they vary being latitude. Vertical variations are ignored and the models are used with surface temperature as the dependent variable. Since the energy balance is allowed to vary from latitude to latitude, a horizontal energy transfer term must be introduced, so that the basic equation for the energy balance at each latitude, θ is

$$S[1 - A(\theta)] = \sigma T_e(\theta)^4 + \rho C \frac{\Delta T_s(\theta)}{\Delta T} + \text{transport out of zone } \theta \qquad (2)$$

where ρC is the heat capacity of the system and can be thought of as the system's "thermal inertia", $S[1 - A(\theta)]$ is the absorbed solar radiation, $\sigma T_e(\theta)^4$ the top-of-the-atmosphere net longwave radiative loss and T_s is the surface temperature varying with time, t. The radiation fluxes at the Earth's surface must be parameterized with care since conditions in the vertical are not considered. To a large extent, the effects of vertical temperature changes are treated explicitly. One possible use of EBMs is to employ them to obtain plausible distributions of ocean surface temperatures which can then be included in an AGCM simulation of a different climatic regime, such as a previous or future glacial epoch.

The general circulation of the atmosphere can be schematically represented as cellular circulations. These Hadley, Ferrel and polar cells are meridional features, i.e. they consist solely of latitudinally averaged movement between zones (Fig. 3b). Most two-dimensional statistical dynamical climate models are constructed to simulate these motions. The two dimensions they represent explicitly are height in the atmosphere and latitude. Atmospheric variations around latitude zones (i.e. longitudinal variations) are neither resolved nor described. These models, in common with GCMs, solve numerically the fundamental equations listed in Table 1 and produce simulations of the large scale two-dimensional flow. The fundamental difference between these models and full atmospheric GCMs is that all the variables of interest are zonally averaged values. This zonal averaging is identified below by angle brackets. The equations to be solved are:

Zonal momentum

$$\frac{\partial \langle u \rangle}{\partial t} - f\langle v \rangle + \frac{\partial \langle u' v' \rangle}{\partial y} = F \tag{3}$$

Meridional momentum (geostrophic balance)

$$f\langle u \rangle + R\langle T \rangle \frac{\partial}{\partial y}(\ln\langle p \rangle) = 0 \tag{4}$$

Hydrostatic balance (vertical component)

$$\frac{\partial (\ln\langle p \rangle)}{\partial z} = -\frac{g}{R\langle T \rangle} \tag{5}$$

Thermodynamic balance

$$\frac{\partial \langle T \rangle}{\partial t} + \frac{\partial \langle v' T' \rangle}{\partial y} + \frac{\partial \langle w' T' \rangle}{\partial z} + \langle w \rangle \left[\frac{g}{\langle \rho \rangle c_p} + \frac{\partial \langle T \rangle}{\partial z} \right] = \frac{Q}{\langle \rho \rangle c_p} \tag{6}$$

Continuity

$$\frac{\partial \langle \rho \rangle \langle v \rangle}{\partial y} + \frac{\partial \langle \rho \rangle \langle w \rangle}{\partial z} = 0 \tag{7}$$

where u, v and w are the zonal mean velocities in the eastward (x), northward (y) and vertical (z) directions, T is the temperature, Q is the diabatic heating and F the friction term, R the universal gas constant, c_p the specific heat at constant pressure, g the acceleration due to gravity, f the Coriolis parameter and $\langle \rho \rangle = \langle p \rangle / R\langle T \rangle$. The primed notation denotes a deviation from the zonal average of these parameters.

As these equations are essentially those which are solved in GCMs, although here they are written in a simpler form, they are worth considering in some detail. The momentum equations are themselves expressed for unit mass, so that the density terms cancel and the momentum (mass times velocity) is represented simply by the velocity component in the direction of interest. Zonal momentum changes (with time) are thus represented by the first term on the left hand side of equation (3). These temporal changes are balanced by the Coriolis term, $f\langle v \rangle$ and the rate of change in the poleward direction of the correlation term, $\langle u'v' \rangle$, i.e. the eddy transport of momentum in the poleward direction. Finally there is an additional frictional dissipation term, F, to be taken into consideration.

The meridional momentum equation is similarly constructed but the small temporal changes are neglected; if they were not the model might accumulate errors and predict non-zero momentum fluxes at the poles. Consequently the balance equation is simply between the horizontal (y component) perturbations (a barotropic instability) converts kinetic energy of the zonal wind to eddy energy through the flux of horizontal eddy momentum $\langle u'v' \rangle$.

The parameterization of momentum fluxes is considerably more complex when theoretically based than the parameterization of heat fluxes. It has been shown that potential vorticity is more suitable for the treatment of $\langle u'v' \rangle$ as the eddy momentum flux can be obtained once eddy potential vorticity and eddy heat fluxes have been derived. A further problem with eddy flux parameterizations is the existence in the atmosphere of large stationary eddies forced both by topography and by land/ocean temperature contrasts. Since the parameterization schemes described above represent only transient baroclinic eddies, it is possible that there will be an underestimation of the total eddy transport produced due to the neglect of these stationary eddies. However, observational data suggest that a compensatory mechanism may exist since total eddy flux seems to be correlated better with observed temperature gradients than is the transient eddy flux alone.

Full formulations of two-dimensional statistical dynamical models (2-D SDs) often also include vertical and horizontal eddy transports of water vapour as well as those of heat and momentum described above. Since 2-D models attempt to parameterize only the eddy transport, whilst the mean meridional transport terms are computed explicitly, it is hoped that inadequacies in the eddy transport parameterizations are compensated for in the explicit calculation of meridional transport. Some 2-D SDs also include quite complex representations of other components of the climate system not generally incorporated into GCMs. For example, the LLN 2-D SD (see Chapter 2 by BERGER) incorporates a submodel of ice masses and ice sheet dynamics.

All climate models employed for climate prediction comprise a set of non-linear differential equations which have to be solved by time-stepping forward into the future. All these equations represent only an approximation of the real climate system. These approximations are termed parameterizations. The solution of these sets of equations depend critically upon: (i) these parameterizations and how successfully they represent processes; (ii) the complexity of the system of equations itself; (iii) the values of internal parameters and initializations and (iv) the time periods and time steps involved in the solution. All climate models require extensive evaluation against observations before their predictions can be viewed with any confidence.

Climates of the future

This book places the predictions of a human-induced enhanced greenhouse warming in context by noting that this planet has undergone many natural large-scale climatic changes in the past. The climate in the past million years has been characterized by a series of quasi-cyclic glaciations. The response of the Earth's climate system to these natural variations is recorded in ocean sediments, peat bogs and polar ice. The results of analyzing the most recent record over the past 100,000 years indicate that Earth's climate does not respond to forcing in a smooth and gradual way. Instead, it responds with rapid, and sometimes discontinuous, changes, especially in the case of warm forcing. If this is correct, a lesson we might learn from the past is that the main responses of the Earth system to greenhouse gas buildup could come in "jumps" whose timing and magnitude are unpredictable.

On the other hand, the very long time-scale record of the Earth's climate seems to have been remarkably stable (Fig. 1f and Chapter 15 by KUMP and LOVELOCK). Does this imply that while rapid and non-linear changes may well occur in the future they are likely to be buffered on long time-scales by strong negative feedbacks? The answer to this question is not known.

The IPCC Executive Summary (HOUGHTON et al., 1990, p xi) states:

> We are certain of the following:
> - there is a natural greenhouse effect which already keeps the Earth warmer than it would otherwise be.
> - emissions resulting from human activities are substantially increasing the atmospheric concentrations of the greenhouse gases: carbon dioxide, methane, chlorofluorocarbons (CFCs) and nitrous oxide. These increases will enhance the greenhouse effect, resulting on average in an additional warming of the Earth's surface. The main greenhouse gas, water vapour, will increase in response to global warming and further enhance it.

and

> Based on current model results, we predict:
> - under the IPCC Business-as-Usual (Scenario A) emissions of greenhouse gases, a rate of increase of global mean temperature during the next century of about 0.3°C per decade (with an uncertainty range of 0.2°C to 0.5°C per decade); this is greater than that seen over the past 10,000 years. This will result in a likely increase in global mean temperature of about 1°C above the present value by 2025 and 3°C before the end of the next century. The rise will not be steady because of the influence of other factors.

Policy development requires information about future climates. For the immediate future 50–100 years, the continuing IPCC assessments provide the clearest available assessment. This book places those predictions in a wider context (Chapters 10 and 11 by ANDREAE and BRASSEUR et al.); provides a broader view of the uncertainties associated with prediction of climate even as far as the next century (Chapter 6 by DIAZ and KILADIS and Chapter 14 by PENG); and places these predictions in a human context by reviewing future urban climates (Chapter 13 by CLEUGH) and the possible impacts of human-induced land-use change on climate (Chapter 12 by HENDERSON-SELLERS).

Evidence from traditional meteorological observations and palæo-indices of climate cannot confirm, with certainty, the existence of an enhanced greenhouse warming signal caused by human activities. The observed temperature rise this century is less than that predicted by numerical climate models (Chapter 5 by JONES) and is commensurate with multi-decadal oscillations in the climate system which are almost certainly due to natural variability (Chapter 6 by DIAZ and KILADIS). However, the Cretaceous climate cannot be simulated by GCMs which have low or even mid-range IPCC-reported sensitivities to doubled CO_2 and the Eocene climate, which is characterized by warm poles but weaker atmospheric dynamics, can only be explained if ocean transport of energy was greatly increased during that period compared with the present day (Chapter 3 by BARRON).

Palæoclimatic evidence indicates that the climate of the North Atlantic region, which is closely coupled to the formation of North Atlantic Deep Water and the deep ocean conveyor belt circulation (Chapter 14 by PENG), has remained remarkably constant since the very rapid transition from cold to warm conditions about 10,000 years ago (the Younger Dryas (3), noted in Fig. 1). The predicted enhanced greenhouse warming is likely not only to

Chapter 2

Modelling the response of the climate system to astronomical forcing

A. BERGER

Introduction

The Earth's climate has always been changing (SHACKLETON and IMBRIE, 1990) and will no doubt continue to change. The future climate is unknown, but increasing attention is being given to its prediction. Over the decades, the reconstruction of past climates has largely improved. Key ingredients to this progress include the large variety of paleoclimatic indicators (geochemical, botanical, biological), the accuracy of dating, the expanding geographic coverage of measurements, and the application of statistical techniques for extracting the climatic signal from other signals – and noise – contained in geologic records.

The aim of studying past climates is (i) to reconstruct the whole range of natural climatic variability, which includes abrupt changes (i.e. occurring after periods shorter than a human lifespan) and which did not occur during the period covered by instrumental observations; (ii) to test models using past conditions in order to understand better how the climate system works; and so (iii) to address the future with more confidence.

The instrumental period database contains primarily details of the climate typical of present-day conditions. If we expect significant changes to occur within the next centuries, this record must be enlarged to cover other climatic types like those which characterize the extreme glacial and interglacial states and other gradual and abrupt climatic changes. Paleoclimatology has clearly demonstrated its ability to reconstruct the gross characteristics of such climatic situations of the past and has so helped the community to discover the natural variability of the climate system at time-scales from years (in ice cores) to hundreds of millennia for the broad geographical regions of the world.

Simultaneously, a hierarchy of climate models of increasing accuracy, dimensionality, and versatility are being developed and applied. These models are an invaluable tool for the quantitative testing of theories and for predicting or suggesting phenomena that can subsequently be studied with observations. The reliability of models which are calibrated and validated for the present day and their power to "predict" significant climatic change in the future, can therefore be tested in many different climatic situations using paleoclimatic data, in particular over a range of known changes in orbitally induced radiation. These tests may reveal weaknesses of the models; they will also suggest further process studies which will help to improve the integrated understanding of the climate system (HECHT, 1985). In addition the reconstruction of climatic time series allows transient experiments to be conducted relating causes and effects of climatic changes, experiments which are expected to aid understanding of the progressive changes of our climate under the forcing of human activities.

Climates of the past

During most of the Earth's history the climate has been warm, even warmer than today. This warm climate has been interrupted by cold times, the so-called Ice Ages, which have, on the geological time-scale, been relatively short, covering only perhaps 5–10% of the Earth's whole history. The most well known are the Pre-Cambrian Ice Ages, the late Ordovician Ice Age, the Permo-Carboniferous Ice Age and our Present Ice Age (CROWLEY and NORTH, 1991).

The current Ice Age, which the Earth entered 2–3 Ma ago (Ma, million years) is called the Quaternary Ice Age and is composed of the Pleistocene and the Holocene, which are usually preferred terminology for this period. It was preceded by the Antarctic ice-sheet formation 25–10 Ma BP (BP, before present) and a long cooling trend through all the Tertiary; in western Europe, for example, the mean temperature went down from 25°C, 50 Ma ago, to 15°C at the end of the Pliocene (LLOYD, 1984). Warm pre-Quaternary climates are discussed in Chapter 3 by BARRON.

The current Ice Age is characterized by multiple switches of the global climate between glacials (with extensive ice sheets) and interglacials (with a climate similar or warmer than today by a few degrees Celsius). These past climates are reconstructed from analysis of proxy data taken from deep-sea cores (e.g. IMBRIE et al., 1992), ice cores (e.g. LORIUS et al., 1990) and land records (e.g. GUIOT et al., 1989). Throughout the past million years (Fig. 1), these data demonstrate that successive glaciation–deglaciation cycles have occurred with a dominant quasi-periodicity of 100 ka (ka, thousand years). The last cycle (Fig. 2) goes from the Eemian interglacial centered roughly 125 ka BP to the present day Holocene interglacial which peaked 6 ka BP, and includes the last glacial maximum (LGM) which occurred 20 ka BP.

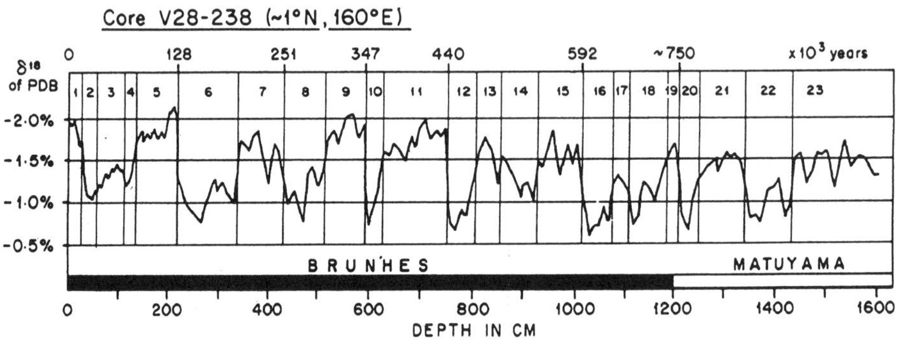

Fig. 1. Oxygen isotope and paleomagnetic record of the last 1.6 million years in core V28–238 from the equatorial Pacific (~1°N, 160°E) as a proxy for long-term climatic changes. Isotope stages are shown in the upper part of the diagram. Isotopic values are from measurements on *Globigerinoides sacculifer* (after SHACKLETON and OPDYKE, 1976). These isotopic stages were defined by EMILIANI (1955). With the recognition of consistent stable isotope signals in the sedimentary records of many different areas, warm periods (interglacials and interstadials) were assigned odd numbers (the present interglacial being number 1) and cold (glacials and stadials) were assigned even numbers (see e.g. BRADLEY, 1985).The date of the Matuyama-Brunhes boundary has been more recently discussed by SHACKLETON et al. (1990) and proposed to be 760 ka BP, it means 5–7% older than the currently adopted radiometric dates. Oxygen isotope is here a proxy for total ice volume over the Earth; this ice volume increases downward in this figure.

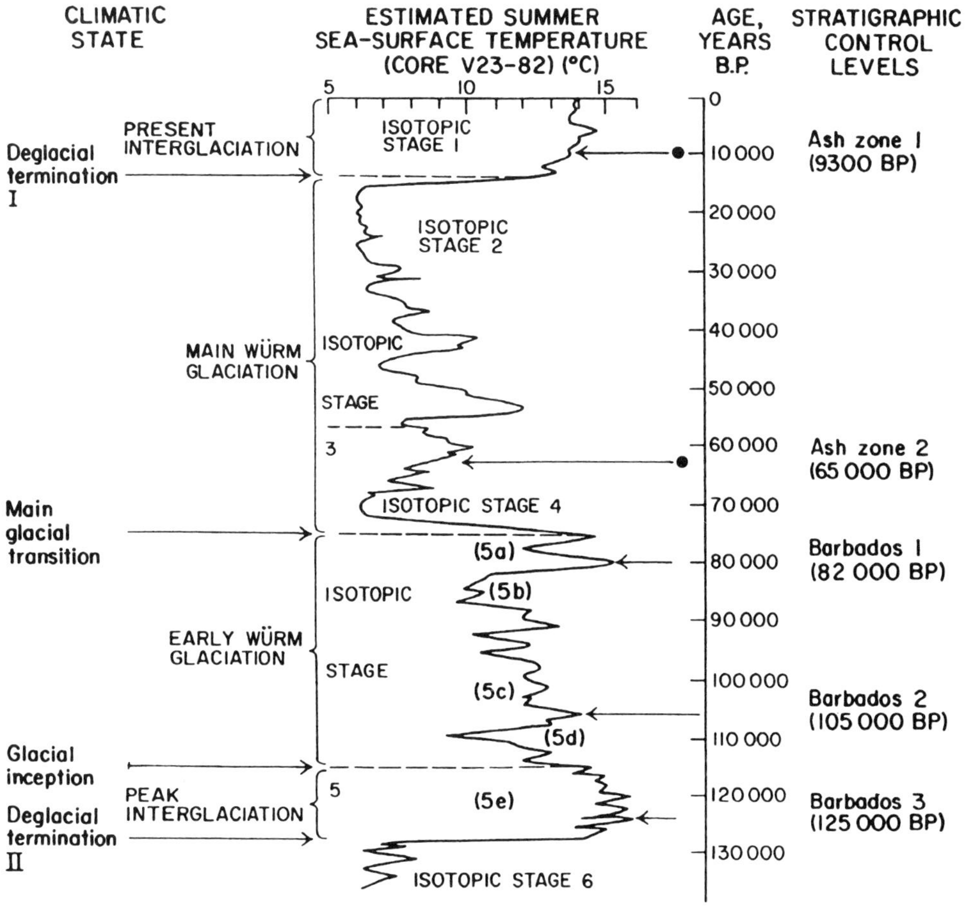

Fig. 2. Summer sea-surface temperature reconstructions for the North Atlantic Ocean based on foraminiferal assemblage paleotemperature estimates, using core V23–82 from ~53°N, 22°W (SANCETTA et al., 1973). Chronological controls used in other cores are shown on the right (tephra layers and Barbados sea-level stands). Generalized climatic conditions and major changes are shown on the left. Isotopic stages are after SHACKLETON and OPDYKE (1973). Isotopic stage 5 was subdivided into several sub-stages, bearing letter designations (5a to 5e). Stages 5a, 5c and 5e were periods of reduced terrestrial ice volume and/or higher temperatures, with sub-stage 5e being the peak of the interglacial called Eemian in NW Europe; stages 5b and 5d were periods of cooler temperature and/or terrestrial ice growth, but on a smaller scale than occurred in stage 4.

These LGM cold conditions were documented in great detail by the Climate Long-Range Investigation Mapping and Prediction group (CLIMAP, 1976, 1981). In the northern hemisphere, the LGM world differed strikingly from the present in the huge land-based ice sheets, reaching approximately 2–3 km in thickness and amounting to about (40–45) × 10^6 km^3 of ice. There was also a dramatic increase in the extent of pack-ice and marine-based ice sheets. In the southern hemisphere, the most striking contrast was the greater extent of sea ice. On land, grasslands, steppes and deserts spread at the expense of forests. This change in vegetation, together with extensive areas of permanent ice and sandy outwash plains, caused an increase in global surface albedo compared to modern values (cf. the possible impact of human-induced land-use change discussed in Chapter 12 by A. HENDERSON-SELLERS). Sea-level was lower by at least 115 m (CHAPPELL and SHACKLE-

TON, 1986; LABEYRIE et al., 1987; TUSHINGHAM and PELTIER, 1991) which represents roughly 50 × 10^6 km^3 of ice more than now, and the global average surface air temperature was 5°C below present. CO_2 levels were less than two-thirds their present value (BARNOLA et al., 1987; JOUZEL et al., 1993) and aerosol loading was higher than present (DE ANGELIS et al., 1987; JOUSSAUME, 1993).

According to CLIMAP (1976), the LGM oceans in August were characterized by: (i) marked steepening of thermal gradients along polar frontal systems, particularly in the North Atlantic and Antarctic; (ii) an equator-ward displacement of polar frontal systems; (iii) general cooling of most surface waters, with a global average of –2.3°C; (iv) increased cooling and upwelling along equatorial divergences in the Pacific and Atlantic; (v) low temperatures extending equator-ward along the western coast of Africa, Australia and South America, indicating increased upwelling and advection of cool waters; (vi) nearly stable positions and temperatures of the central gyres in the sub-tropical Atlantic, Pacific and Indian oceans; and (vii) little temperature change (–2°C) in the tropical oceans. There are, however, geological data suggesting that the sea-surface temperatures in the tropics were lower than reconstructed by CLIMAP, about 5°C lower than today. The evidence came from corals around Barbados (ANDERSON and WEBB, 1994; GUILDERSON et al., 1994) and matches the cooling estimated from terrestrial evidence of lower snowlines and changes in ancient patterns of vegetation at mid to high elevations.

According to the results by Co-operative Holocene Mapping Project Members (COHMAP, 1988), beginning around 16 ka BP, the climate system experienced major change during which the ice sheets retreated, ocean fronts shifted poleward, and the areas covered by sea ice contracted (RUDDIMAN and MCINTYRE, 1981). In the mid-latitudes, vegetation zones shifted poleward and after 14 ka BP, lake levels fell (for example, in the southwestern United States). The time interval 16–13 ka BP is sometimes known as the Late Glacial and is the time of maximum desiccation for African lakes (STREET and GROVE, 1979). Although residual remnants of the Laurentide Ice Sheet lasted until 6–7 ka BP most of the ice disappeared within an interval of about 5 ka, from 14 to 9 ka BP. Deglaciation actually occurred in two main steps: an abrupt warming around 13 ka BP, followed by a climate reversal at about 12 ka BP (termed the Younger Dryas in Europe) and then another abrupt warming at 10 ka BP. The two warming trends are also manifested as rapid changes in sea-level dated 13.5 and 11 ka BP (FAIRBANKS, 1989). This Younger Dryas may have been caused by an interruption of thermohaline circulation due to meltwater-induced changes in the surface waters of the North Atlantic (BROECKER et al., 1989), but the sea-level curve and other evidence from tracers of ocean circulation do not favour this mechanism. The finding that meltwater discharge was minimal during the Younger Dryas led DUPLESSY et al. (1992) to suggest that the surface-water salinity drop might have been caused by a reduction in poleward advection of sub-tropical saline water.

The waxing and waning of ice sheets occurred in a more or less regular way. In Fig. 1, one can recognize a sawtooth shape with a 100-ka quasi-cycle over which shorter quasi-cycles of roughly 41 and 21 ka are superimposed. These kinds of broad climatic features characterizing the last few million years (and probably other climatic variations of non-Ice Ages, like those which occurred during the late Triassic (OLSEN, 1986)) are those explained by the astronomical theory of paleoclimates (BERGER, 1988). Proponents of this theory claim that the changes in the Earth's orbital and rotational parameters have been sufficiently large as to in-

duce significant changes in the seasonal and latitudinal distributions of irradiation received from the Sun and so, to force glacials and interglacials to recur in the manner deduced from geological records. This topic is the main theme of this chapter.
In addition to these climatic changes at the millennia time-scale, climate shifts over periods as short as a few decades to a few centuries have also been detected. Sediments in the North Atlantic ocean contain a series of layers (named after their discoverer, Heinrich) deposited between 14 ka BP and 70 ka BP (BOND et al., 1992). These layers are rich in ice-rafted debris, which may reflect repeated rapid advances of the Laurentide ice sheet, perhaps associated with reductions in air temperature, although temperatures from Greenland ice cores appear to exhibit only a weak corresponding signal. Similar events are seen in land records; for example, major vegetation shifts seem to have occurred during the last 50 ka in Florida (GRIMM et al., 1993). Results from a deep-ice core drilled at the summit of the Greenland ice sheet, also reveal irregular but well-defined episodes of relatively mild climatic conditions during the mid and late parts of the last glaciation; these interstadials (the Oeschger-Dansgaard events) lasted from 500 to 2000 years (JOHNSEN et al., 1992). Climate instability is not confined to the last glaciation: isotope and chemical analyses of the GRIP ice core (GRIP, GREENLAND ICE-CORE PROJECT MEMBERS, 1993) reveal that climate in Greenland during the last interglacial period was characterized by a series of several cold periods, which began extremely rapidly and lasted from decades to centuries (although the intensity of these cold events needs to be confirmed, in particular by the analysis of the results obtained from the US Greenland Ice Sheet Project, GISP2; BOULTON, 1993; GROOTES et al., 1993; TAYLOR et al., 1993). In contrast, the past 8 ka has been strangely stable. Therefore, apart from the Holocene, instability seems to have dominated the North Atlantic climate over the past 230 ka (DANSGAARD et al., 1993). This recognition highlights the question of whether the Holocene will remain stable in spite of the growing human disturbance of the atmosphere. This question is tackled in other parts of this book, particularly Chapter 8 by MCAVANEY and HOLLAND and Chapter 9 by WANG et al.

Astronomical theory of paleoclimates

Perturbations of Earth's orbital and rotational parameters

The incoming solar radiation received over the Earth has an annual periodic variation due to the Earth's elliptic translation motion around the Sun. In addition, the seasonal and latitudinal distributions of this solar radiation have a long-period variation due to the so-called long-term variations in the orbital elements. The total solar energy received by the whole Earth over 1 year also varies, but by a very small amount, in relation to one of these orbital parameters, the eccentricity of the Earth's orbit. Other long-period variations affecting the solar constant and attributable to changes in the physical conditions of the Sun and/or in the opacity or transparency of the interplanetary medium have been postulated from time to time (e.g. LEAN, 1991; LEAN et al., 1992), but their effects remain difficult to prove.
Analysis of the incoming solar radiation and of its latitudinal distribution shows that there exist three main orbital parameters whose long-term variations must be calculated (BERGER, 1993): the eccentricity, e, a measure of the shape of the Earth's orbit around the Sun, the

obliquity, ε, the tilt of the equator with respect to the plane of the Earth's orbit, and the climatic precession, $e \sin \tilde{\omega}$, a measure of the Earth–Sun distance at the summer solstice ($\tilde{\omega}$ being the longitude of the perihelion measured from the moving equinox; seasons referred to in this chapter are always specific to the northern hemisphere).

As determined from celestial mechanics, the secular variations of these elements of the Earth's orbit and rotation are due to the gravitational perturbations which the Sun, the other planets and the Moon exert on the Earth's orbit and on its axis of rotation (BERGER, 1976a, 1977a).

For the long-term variations over the last few million years, the equations for the eccentricity, for the precession and for the obliquity cannot be integrated analytically, but their numerical solutions can be expressed in trigonometrical form as quasi-periodic functions of time:

$$e \sin \tilde{\omega} = \sum P_i \sin(\alpha_i t + \eta_i) \tag{1}$$

$$\varepsilon = \varepsilon^* + \sum A_i \cos(\gamma_i t + \zeta_i) \tag{2}$$

$$e = e^* + \sum E_i \cos(\lambda_i t + \phi_i) \tag{3}$$

where the amplitudes P_i, A_i, E_i, the frequencies α_i, γ_i, λ_i and phases η_i, ζ_i, ϕ_i, are given in BERGER (1978) for calculations over one million years. BERGER and LOUTRE (1991) have calculated a new solution which can be used for paleoclimatic studies over 3×10^6 years in the time domain and over 10×10^6 years, at least, in the frequency domain. Its accuracy has been demonstrated from a comparison with two other similar solutions (BERGER and

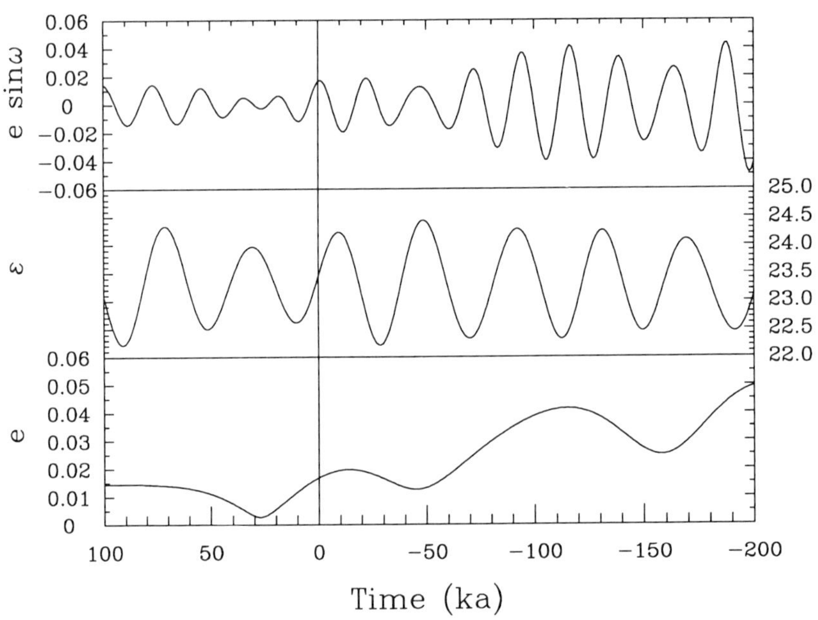

Fig. 3. Long-term variations of precession, obliquity and eccentricity from 200 ka BP to 100 ka AP according to the astronomical solution given in BERGER (1978).

TABLE I

AMPLITUDES, MEAN RATES, PHASES AND PERIODS OF THE 6 LARGEST AMPLITUDE TERMS IN THE TRIGONOMETRICAL EXPANSIONS OF CLIMATIC PRECESSION (A), OBLIQUITY (B) AND ECCENTRICITY (C)

Amplitude	Mean rate ("/year)	Phase (°)	Period (years)
(a) *Climatic precession* $e \sin \tilde{\omega}$			
0.018970	54.66624	32.2	23708
0.016318	57.87275	201.3	22394
0.012989	68.33975	153.4	18964
0.008136	67.79501	311.4	19116
0.003870	55.98574	78.6	23149
0.002557	67.52198	38.1	19194
(b) *Obliquity* ε			
−1969.00	31.54068	247.14	41090
−903.50	32.62947	288.79	39719
−631.67	32.08588	265.33	40392
−602.81	24.06077	129.70	53864
−352.88	30.99683	43.20	41811
−266.00	44.78645	18.77	28937
(c) *Eccentricity* e			
0.011268	3.20651	169.2	404178
0.008819	13.67352	121.2	94782
0.007419	10.46700	312.0	123818
0.005600	13.12877	279.2	98715
0.004759	9.92226	110.1	130615
0.003861	0.54475	202	2379077

It must be stressed that this very limited number of terms do not allow the exact numerical values of these astronomical parameters to be computed; more than 50 terms in equations (1), (2) and (3) are necessary to reach an acceptable accuracy (BERGER and LOUTRE, 1991)

LOUTRE, 1992). The six largest amplitude terms of this new solution are listed in Table I. Figure 3 shows the long-term variations of these three astronomical parameters over the past 200,000 years and into the future for the next 100,000 years.

Over the past 3×10^6 years, the eccentricity of the orbit varies between near circularity ($e = 0$) and slight ellipticity ($e = 0.07$) at a period whose mean is about 100 ka. The most important terms in the series expansion occur, however, at 404, 95, 124, 99, 131 and 2380 ka (in decreasing order of the amplitudes). The tilt of the Earth's axis varies between about 22° and 25° at a period of nearly 41 ka. Although this period corresponds in equation (2) to the amplitude which is by far the largest, there are two other important terms with periods of 54 and 29 ka. As far as precession is concerned, two components must be considered. The first is the axial precession in which the torque of the Sun, the Moon and the planets on the Earth's equatorial bulge causes the axis of rotation to wobble like that of a spinning top. The net effect is that the North Pole describes clockwise a circle in space (provided the nutations, i.e. terms with small amplitude and much smaller periods, are neglected) with a period of ~25,800 years corresponding to the period of the vernal equinox against a fixed reference point. The second is related to the fact that the elliptical figure of

solation for each latitude is only a function of precession. At the solstices, both precession and obliquity influence insolation, although precession is dominating for most of the latitudes (Fig. 5).

Changes in incoming solar radiation due to changes in tilt are the same in both hemispheres during the same local season: an increase of ε leads to an increase of insolation in the summer hemisphere and a decrease in the winter hemisphere. As the strength of the effect is small in the tropics and maximum at the poles, obliquity increases tend to amplify the seasonal cycle in the high latitudes of both hemispheres simultaneously.

The precession effect can cause warm winters and cool summers in one hemisphere while causing the opposite effects in the other hemisphere. For example, portions of the northern hemisphere receive during present winter as much as 10% more insolation than 11 ka BP when the perihelion occurred in the northern hemisphere summer.

It must be stressed that the pattern of solar irradiation at the top of the atmosphere differs significantly from the pattern of radiation absorbed by the surface of the Earth, particularly in high latitudes where the surface albedo is large (TRICOT and BERGER, 1988).

Milankovitch theory

The orbital hypothesis of climatic change was first quantitatively formulated by the astronomer Milutin Milankovitch in the 1920s and 1930s. His early calculations provided information on the variations in incident solar radiation, as a function of latitude, for the last million years in winter and summer (MILANKOVITCH, 1920, 1941).

Unlike CROLL (1864), one of the very first, with ADHÉMAR (1842), to suggest that the prime mover of the glacial-interglacial cycles might be variations in the way the Earth moves around the Sun (IMBRIE and IMBRIE, 1979), Milankovitch argued that insolation changes in the high northern latitudes during the summer season were critical to the formation of continental ice sheets. During periods when insolation in the summer was reduced, the snow of the previous winter would tend to be preserved – a tendency that would be enhanced by the high albedo of the snow and ice areas. Eventually, the effect of this positive feedback would lead to the formation of persistent ice sheets.

According to the mathematics of insolation, a minimum in the northern hemisphere caloric summer[1] insolation at high latitudes requires a northern hemisphere summer occurring at the aphelion, a maximum eccentricity which leads to a large distance between the Earth and the Sun at the aphelion, and a minimum obliquity implying a weak seasonal contrast and an increased latitudinal energy gradient between the equator and the poles. Given these conditions, it is suggested that, not only would the northern high latitudes remain cool enough in summer for preventing snow and ice from melting, but also that mild winter would allow a substantial evaporation in the intertropical zone and, thus, abundant snowfalls in temperate

[1] Milankovitch defined the caloric summer to avoid the variations of the length of the astronomical summer, as the half-year which comprises all the days of stronger radiation and consequently experiences the greatest possible irradiation; the other half-year is the caloric winter. This raises two problems: first, the beginning of these caloric seasons changes with time; second, it is difficult to define it in the equatorial regions where the seasonal march of insolation has two maxima and two minima.

and polar latitudes, the humidity being supplied there by an intensified general circulation due to a maximum latitudinal energy gradient.

A simple linear version of the Milankovitch model would therefore predict that the total ice volume and climate over the Earth would vary with the same regular pattern as the insolation; this means that the proxy record of climate variations would contain the frequencies of the astronomical parameters that are responsible for changing the seasonal and latitudinal distributions of the incoming solar radiation. It happens that investigations during the last 20 years have indeed demonstrated that the 19 ka, 23 ka and 41 ka periodicities actually occur in long records of the Quaternary climate (HAYS et al., 1976), that the climatic variations observed in these frequency bands are linearly related to the orbital forcing functions and that there is a fairly consistent phase relationship between insolation, sea-surface temperature and ice volume (IMBRIE et al., 1984, 1989). The geological observation of the bipartition of the precessional peak, confirmed in astronomical computations by BERGER (1977b), was one of the first delicate and impressive tests of the Milankovitch theory. Another impressive confirmation is the discovery in environmental spectra of other astronomical peaks predicted in 1977: the 54 ka and 30 ka periods of the obliquity were found by RUDDIMAN et al. (1986) in the North East Atlantic winter sea-surface temperature record; the 412 ka eccentricity period was found in particular by BRISKIN and HARRELL (1980) and by MOORE et al. (1982) using 2 Ma sediment cores from the Atlantic and the Pacific.

Despite these discoveries in support of the Milankovitch hypothesis, the same investigations identified the largest climatic cycle as being 100 ka. This eccentricity cycle is very weak in the insolation. So it cannot be related to the orbital forcing by any simple linear mechanism. Actually, the variance components centred near this 100-ka cycle seem to be in phase with the eccentricity cycle, but its exceptional strength in the climatic record demands non-linear amplification (i) by the glacial ice sheets themselves involving mechanisms such as the ice-albedo feedback and the isostatic rebound of the lithosphere (BIRCHFIELD and WEERTMAN, 1978); (ii) by the carbon dioxide atmospheric concentration (LORIUS et al., 1990); and/or (iii) by the oceanic circulation (BROECKER and DENTON, 1990).

Moreover, the determination of climatic periodicities from the geological records is not always as clear. The 100-ka cycle, the most dominant feature in the ice volume record for the late Pleistocene, does not exhibit a constant amplitude over the past (2–3) $\times 10^6$ years; it disappears before 10^6 years ago, at a time when the ice sheets were much less extensive across the Earth (PRELL, 1982; SHACKLETON et al., 1988). The SPECMAP group (Spectral Mapping Project) has also shown that the shape of the spectra and the phase lags in the climate response to orbital forcing depend upon the location of the deep-sea core and the nature of the climatic parameter determined (IMBRIE et al., 1989, 1992, 1993).

Since the publication of the papers by HAYS et al. (1976) and IMBRIE and IMBRIE (1980), a number of modelling studies have attempted to explain the relationship between astronomical forcing and climatic change. Both general circulation models, which provide snapshot views of the climate in equilibrium with given boundary conditions, and statistical-dynamical models, which can take into account long-period forcing of the climate system, have been used.

"Snapshot" simulations of past climate

Equilibrium models

Climate models can be divided into three very broad categories: (1) general circulation models (GCMs); (2) statistical-dynamical models, which include 2-D zonally or sectorally averaged dynamics; and (3) thermodynamic and energy balance (EBMs) models (KUTZBACH, 1985). GCMs, statistical-dynamical models and EBMs have been used for the modelling of paleoclimates (KUTZBACH, 1985, 1992). GCMs have been used for simulating geographic features of paleoclimates and for including, in the most explicit form, processes such as precipitation that depend on details of the atmospheric flow. However, because of the large computational costs that are involved, GCMs have been used to provide "snapshot" views of the climate in equilibrium with specific boundary conditions. These conditions include the seasonal and latitudinal changes of solar radiation at the top of the atmosphere (resulting from modifications in the Earth's orbit about the Sun), as well as variations in sea-surface temperatures, major ice sheets, atmospheric composition and opacity (see Chapter 4 by RAMPINO and 10 by ANDREAE), and the locations of continents (see chapter 3 by BARRON) for simulations distant enough in the past. On the other hand, statistical-dynamical models have great potential for long-period climate simulations. As they are less complex than GCMs, they consume less computer time, thus allowing account to be taken of additional parts of the climate system in an interactive and time-dependent way. Therefore, these models can be used to simulate long-period transient responses of the climate system to any "external" forcing or internal perturbation. For example, some of them have been coupled to slowly responding portions of the climate system (e.g. ice sheets) in order to investigate the temporal evolution of climate on millenia time-scales.

Energy balance models, which are relatively simple one-dimensional models of the Earth's radiation balance, were also used for snapshot simulations of paleoclimates (e.g. DONN and SHAW, 1977; SCHNEIDER and THOMPSON, 1979; SUAREZ and HELD, 1979; NORTH et al., 1983; SELLERS, 1984; ADEM et al., 1984; ADEM, 1988), as well as 2-D zonally averaged dynamical models (e.g. OERLEMANS and VERNEKAR, 1981). For example, NORTH et al. (1983) used such an energy budget climate model with a realistic distribution of the continents and the oceans to show that for large eccentricity, January perihelion and minimum axial tilt, summer temperatures could be several degrees Celsius lower than present in northern continental interiors where present-day summer temperatures are only a few degrees Celsius above freezing. When an ice-albedo feedback was included, the cooling was enhanced. While this result lent some support to the hypothesis that certain orbital configurations favour initiation of glaciation, the role of the ice-albedo feedback mechanism remained uncertain because this model had no explicit hydrological cycle.

Even where EBMs are used to model time dependent behaviour (HARVEY and SCHNEIDER, 1985) many important components such as snow and ice feedbacks, clouds and the deep oceans, may be grossly simplified or ignored. They can thus only be used to investigate isolated elements of the climate response to orbital forcing, such as the differential response of land and sea (SHORT et al., 1991).

Over the last 25 years, the development of detailed three-dimensional global climate models (GCMs) has occurred in concert with high performance computing. These models, based on physical laws of radiative transfer and thermodynamics (e.g. WASHINGTON and PARKINSON, 1986; SCHLESINGER, 1988; TRENBERTH, 1992), have been evaluated extensively against present climate (GATES et al., 1992). They show some skill in reproducing the main features of the general circulation of the atmosphere, and are being used to understand and predict the climatic effects of increases in greenhouse gases (MITCHELL, 1989; HOUGHTON et al., 1990, 1992; and Chapter 9 by WANG et al.), ozone depletion (Chapter 11 by BRASSEUR et al.) and human disturbance (e.g. Chapter 12 by HENDERSON-SELLERS).

In atmospheric general circulation models (AGCMs), prognostic equations governing the conservation of momentum, heat, mass and water are solved numerically at a set of discrete grid-points over the globe and at multiple levels in the vertical. The largest source of uncertainty is associated with the parameterization of the sub-gridscale processes, such as cloud, cumulus convection, surface evaporation and surface drag. Ocean general circulation models are constructed in a similar way to AGCMs, except that salinity rather than specific humidity is used and the parameterization of physical processes is generally simpler. When atmospheric and oceanic models are coupled together, errors in both components can lead to the accumulation of substantial errors in the simulation of sea-surface temperature and sea-ice extents and, often to "climatic drifts" away from present-day conditions. In order to produce a realistic simulation of present climate, a technique known as "flux correction" was introduced. This means that the coupled model is continually restored to observations to counteract the rate at which it tends to drift (MANABE and STOUFFER, 1988; SAUSEN et al., 1988; HADLEY CENTER, 1992; MURPHY, 1992). Recent improvements in coupled A-OGCMs so that these corrections will not be required have been fairly successful.

Although there is a broad agreement among different models on the nature of the simulated changes at the largest scales (e.g. all GCMs produce a warmer climate when the amount of CO_2 in the atmosphere is doubled, e.g. HOUGHTON et al., 1990, 1992), the regional and local response of these models is often dependent on the parameterizations used. Two models can actually simulate similar present climates, but substantially different responses to increases in greenhouse gases, for example (SCHLESINGER and MITCHELL, 1987; HOUGHTON et al., 1990, 1992; SENIOR and MITCHELL, 1993); consequently, validation against the present climate is not sufficient to discriminate between reliable and unreliable models. These models must, therefore, be tested against a larger set of different climatic situations which can be provided by the reconstruction of the climatic history of the Earth (CROWLEY and NORTH, 1991; BRADLEY, 1989; and Chapter 6 by DIAZ and KILADIS).

The ability of a model to simulate a past climate does not necessarily guarantee that it will correctly predict future climates as the mechanisms of change may be different. Nevertheless, as stressed by MITCHELL (1993), failure to simulate a past climate successfully may help to identify shortcomings in model formulation. On the other hand, there have been numerous attempts to look for geologic analogs in the climatic history of the Earth in order to better understand future enhanced greenhouse warming, for example. However, as demonstrated by CROWLEY (1990) and MITCHELL (1990), there may be no warm time period that is a satisfactory past analog for future climate because the mechanisms producing such warm periods in the past may have been totally different from those that would occur following an increase in the concentration of greenhouse gases. Paleoclimates must therefore be primarily studied to provide important insights into processes operating in the climate system and for model testing rather than to provide analogs for future climates.

vegetation. Reducing CO_2 had the largest effect, producing a cooling in both hemispheres, but particularly in the southern hemisphere where there is a strong sea ice-temperature feedback.

The cooling of the low latitudes ocean simulated by BROCCOLI and MANABE (1987) is larger than that reconstructed from paleoclimatic data (CLIMAP, 1981), particularly in the sub-tropics, a model result which is in agreement with data from mountains at these latitudes. RIND and PETEET (1985) found that a simulation with CLIMAP (1981) sea temperatures reduced everywhere by a further 2°C gives a better fit to the observed temperature reconstruction from tropical mountains. They also obtained a marked decrease in the strength of the hydrological cycle over the tropical continents, in agreement with paleoclimatic reconstructions and with the simulation of MANABE and HAHN (1977) who used the CLIMAP (1976) estimate of sea-surface temperatures; the latter indicating a greater cooling at low latitudes. On the other hand, accepting that the tropical ocean was no more than about 1°C cooler at 18 ka BP than at present and the snow lines on high peaks had descended 1 km, SUN and LINDZEN (1993) conclude that the lapse rate in the lower half of the tropical troposphere during the last glaciation would have been about 20% greater than at present.

According to these simulations, it therefore became possible that CLIMAP reconstructions of the sea-surface temperature of the tropical ocean may have to be revised (as done by GUILDERSON et al., 1994), a recommendation which was made by the Paleoclimate Modelling Intercomparison Project in May 1991 at the NATO Advanced Research Workshop on Paleoclimate Modelling in Paris. This project, chaired by S. Joussaume, was initiated in 1991, and has as its goal to make model–model and model–data comparisons in the field of paleoclimate in order to understand better the mechanisms of climate change.

Tracer studies are another Earth system linkage now being explored as a tool for improving our understanding of past climates and hence improving the prediction skills of models applied to future climates. The spatial and temporal distribution of isotope species, such as $\delta^{18}O$ in precipitation and aerosols are important paleoclimatic indicators and are generally termed "tracers". Various modelling studies have used AGCMs to investigate the link between the ice age climate and the atmospheric cycle of windblown dust material from desert areas (e.g. JOUSSAUME, 1993) and water isotope cycles (e.g. JOUSSAUME and JOUZEL, 1993). These tracer studies can provide a test for evaluating models of present-day conditions and are also useful for checking paleoclimatic inferences made from the distribution of isotopes and dust in geologic records. For example, the reasons for which the LMD (Laboratoire de Météorologie Dynamique, Paris) model used in the experiment by JOUSSAUME (1993) simulates only a weak increase of the global atmospheric dust content at the LGM, contrary to the observations from ice cores, might include inaccuracies of the circulation patterns, of the dust model or more likely, of the actual sources of dust. On the other hand, the overall patterns of the simulated $\delta^{18}O$-temperature relationship for the LGM climate are practically the same as for present day, which tends to support the use of water isotopes in paleoclimatology. This is not, however, true for deuterium excess.

Climate since the LGM

The period from the LGM to present has also been the subject of many studies. There is a wide range of data for this time interval (COHMAP, 1988). Moreover, the errors in dating are relatively small (BARD et al., 1990) and the known changes in boundary conditions are simple. Since about 20 ka BP, the time of perihelion shifted from January to July (about 10 ka BP) and back to January (present day). These changes, along with changes in the tilt of the rotational axis, led after 18 ka BP to increasingly extreme solar radiation at the solstices in the northern hemisphere which reached a maximum around 10 ka BP, followed by decreasing extremes from 10 ka BP to present. Combining the changes in solar radiation due to orbital forcing with the changes in prescribed boundary conditions (the ice-sheet configuration, sea-ice extent, ocean temperature, atmospheric CO_2 concentration and aerosols), paleoclimate "snapshots" have been obtained from 18 ka BP to present at 3000-year intervals for both January and July conditions (KUTZBACH and STREET-PERROTT, 1985; KUTZBACH and GUETTER, 1986). Some comparisons between the early model results and the data revealed areas of disagreements. For example, the KUTZBACH and STREET-PERROTT (1985) model shows moisture increases in the northern tropics at 15 and 12 ka BP when the data indicate relatively dry conditions. Work continues to study these areas and times of disagreements and to use them to suggest re-interpretation of the data; improvements in the model and in the boundary conditions used; and better methods for comparing the data with the model results (WEBB et al., 1987).

There have been attempts to use model simulations to also explain the Younger Dryas, a temporary reversion to colder conditions at around 11–10 ka BP. The GISS (Goddard Institute for Space Studies) general circulation model (RIND et al., 1986) shows that colder North Atlantic sea temperatures produce cooling over western and central Europe in good agreement with paleoclimatic evidence. On the other hand, MANABE and STOUFFER (1988) produced two stable equilibria in a coupled ocean-atmosphere GCM using present-day boundary conditions but different initial conditions. One is similar to present, with deep-water formation in a relatively warm, salty North Atlantic. In the other equilibrium, the northern North Atlantic becomes fresher and colder and meridional overturning almost ceases. This circulation is more or less similar to that believed to have prevailed during the Younger Dryas (BROECKER et al., 1988) and was also simulated by STOCKER and WRIGHT (1991) using a more simple atmosphere-ocean coupled model (cf. discussions in Chapter 14 by PENG). BIRCHFIELD (1989) has examined similar kinds of bimodality in coupled atmosphere-ocean box models. These oceanic changes seem likely to be linked to changes in biogeochemical cycling that may, ultimately, explain the glacial-interglacial difference of about 80 ppmv in the atmospheric concentration of CO_2. Some of these indicators of ocean climate show significant amplitude variability and phase relationships consistent with orbital cycles (IMBRIE et al., 1992).

Around 10 ka BP, changes in the Earth's orbital parameters – increased obliquity and a change in perihelion from northern winter to northern summer – led to an increase of up to 8% in northern hemisphere summer insolation and a decrease in winter, giving an enchancement of the seasonal cycle. For that reason, early simulations of the climate at the peak of the Holocene have focussed on 9 ka BP when the insolation changes are large but there was still a substantial Laurentide ice-sheet, instead of 6 ka BP when the ice sheets were similar to present and the insolation changes smaller. More recent model simulations show that results for 6 ka BP are similar to those 9 ka BP, but the effects are less pronounced (KUTZBACH and GUETTER, 1986; MITCHELL, 1993).

Experiments with both low- and high-resolution GCMs show that the increased seasonality at 9 ka BP resulted in continental interiors that were warmer in summer and cooler in winter than they are today. This intensified the simulated summer monsoon circulation of Africa and southern Asia, because ocean temperatures during these seasons were much the same as today (KUTZBACH, 1981; KUTZBACH and OTTO-BLIESNER, 1982; KUTZBACH and GUETTER, 1984b). This stronger monsoon flow over the northern Indian Ocean in summer generates an increased and more extensive upwelling in the Arabian sea (LUTHER et al., 1990), consistent with paleoclimatic data.

In northern winter, insolation is reduced, which has a number of consequences, according to the modelling experiment of MITCHELL et al. (1988): cooling of the continents; increasing the surface pressure in the tropics; and reducing precipitation over tropical land areas. In summer, the model simulates increases in soil moisture in the northern sub-tropics because of increased precipitation, but in midlatitudes, the increase in evaporation is greater than the increase in precipitation so that soil moisture decreases (MITCHELL, 1990).

The simulated climate at 9 ka BP is also sensitive to changes in other boundary conditions. MITCHELL et al. (1988) found that including an idealised Laurentide ice-sheet reduced the warming over land in July by 20%, and STREET-PERROTT et al. (1990) found that reducing surface albedo over north Africa and Arabia, as suggested by PETIT-MAIRE et al. (1988), enhanced the intensification of monsoon precipitation. It must be said, however, that both numerical experiments and observations indicate that tropical precipitation is sensitive to small changes in sea–surface temperatures (FOLLAND et al., 1986) stressing even further the importance of an accurate paleoclimatic reconstruction of these sea–surface temperatures.

The Eemian interglacial

Data from the previous (Eemian) interglacial (~125 ka BP) indicate a climate significantly warmer than the peak of the Holocene (6 ka BP). The Vostok ice core records show relative maxima in atmospheric concentrations of CO_2 and CH_4 during this period (JOUZEL et al., 1993). In the tropics, there is evidence of stronger monsoons (PETIT-MAIRE, 1989). The seasonal and latitudinal changes in solar radiation caused by orbital changes are approximately twice as large at 125 ka BP as at 6–9 ka BP. This is because the eccentricity of the Earth's orbit was significantly larger then (0.04), so that the Earth–Sun distance was significantly smaller (by 3%) at perihelion and because perihelion passage occurred in northern summer around 125 ka BP, whereas it occurs now in northern winter. In the northern hemisphere high latitudes, summer radiation was increased by more than 50 W m^{-2} (12–13%) compared to present, winter radiation was slightly decreased and the annual average insolation was increased by 3–4 W m^{-2}. According to ROYER et al. (1984) and PRELL and KUTZBACH (1987), warmer temperature in high latitudes, decreased sea-ice extent, increased mid-continent aridity and strengthened tropical monsoons are among the simulated responses in sensitivity experiments with AGCMs using 125 ka orbital parameters.

Finally, PRELL and KUTZBACH (1987) have used a GCM to investigate monsoon variability over the past 150,000 years. The agreement between simulated and observed paleoclimatic time series suggests that both orbitally produced solar radiation changes and glacial age boundary condition changes are necessary to explain the major regional features of monsoon climates at millennial or longer time-scales. Moreover, some non-linearity of response is required to produce annual mean warmings and coolings as the changes in the Earth's orbital parameters alter the seasonal distribution of insolation substantially, but modify the total amount of radiation reaching the top of the atmosphere over the year very little.

Initiation of the last glaciation

The same snapshot approach has been used to determine whether or not the simulated climate is in long-term equilibrium with the prescribed boundary conditions. For example, it has been used to see if there is a net annual increase in snow depth in high latitudes over land at the initiation of the last glaciation at isotopic stage 5d which is characterized by a large volume of ice peaking at 110 ka BP (MARTINSON et al., 1987).This was a time with particularly large changes in summer insolation in high northern latitudes, with a maximum of insolation occurring around 125 ka BP, a minimum near 115 ka BP and a further maximum near 105 ka BP; at 65°N in June, the amplitude of that variation was ~100 W m^{-2}, the insolation going from 540 W m^{-2} to 440 W m^{-2} in about 10,000 years (BERGER, 1979). It is therefore a particularly well-suited period to test if these changes in the insolation may have led to an extension of perennial snowfields over northern North America and northwestern Europe, with the help of the temperature-albedo feedback, the orographic and continental effects of the growing ice sheets and changes in the concentration of atmospheric CO_2.
ROYER et al. (1984) simulated the annual cycle of climate at 125 and 115 ka BP using a GCM with the appropriate seasonal distribution of insolation and boundary conditions similar to the present day. The change of insolation from 125 ka BP to 115 ka BP produced a cooling and an increase of soil water content over eastern Canada in the model. These are favourable conditions for the accumulation of an ice cover, provoking an abortive glaciation. OGLESBY (1990) was able to obtain an extended permanent snow cover over part of the region occupied by the major ice sheets by reducing CO_2 to 200 ppmv or by prescribing an initial snow depth of 1 m over the areas of interest although his results are very sensitive to the fraction of the grid square assumed to be covered by snow. However, in a similar experiment for the same isotopic stage 5d and using the GISS GCM, RIND et al. (1989) failed to maintain snow cover through the summer at locations of suspected initiation of the major ice sheets, despite the reduced summer and fall insolation. This failure to initiate ice sheets in response to applied orbital insolation variations seems to be related to serious flaws in the sensitivity of the hydrologic cycles in the GISS A-OGCM (SANTER et al., 1993).

Conclusions

In all these snapshot experiments, many of the characteristics of the paleoclimate data are reproduced, but all these results must be regarded as tentative given the uncertainty associated with the parameterization of sub-grid scale processes and the treatment of sea-ice and of the cloud feedbacks (of which even the sign may vary from model to model in global warming experiments) (CESS et al., 1989).

Simulations of temporal evolution of climate

As seen in the previous section, numerical simulations in powerful computers have allowed the first climatic simulations to be made using atmosphere, ocean and coupled atmosphere-ocean general circulation models. These resolve processes whose scales extend in time from days to decades and in space from hundreds of kilometres to the circumference of the Earth. They are typically used for transient CO_2 (STOUFFER et al., 1989; WASHINGTON and MEEHL, 1989; MURPHY, 1992; MANABE and STOUFFER, 1993) or equilibrium Milankovitch (MANABE and BROCCOLI, 1985b; KUTZBACH and GALLIMORE, 1988) experiments. However, these paleoclimate studies carried out with GCMs and EBMs, being equilibrium studies, do not model the time-dependent behaviour of the climate system, which is crucial if we are to understand the processes of glacial initiation and decay. Because model boundary conditions, such as ice-sheet location and size and sea-surface temperature, are prescribed from the paleoclimate data, they are therefore limited in what they can reveal about the mechanisms linking orbital forcing and climate.

Models which consider the physics of processes with larger characteristic time-scales of thousands to hundreds of thousands of years can be developed by relying on the results obtained by these atmosphere-ocean general circulation models for the parameterization of processes that they cannot resolve explicitly, such as the transport by the atmospheric eddies, for example. These models which simulate the time-dependent evolution of the slow-response climatic variables over time-scales longer than the damping times of the fast-response variables can be constructed by an essentially inductive process (SALTZMAN, 1983, 1990) or by deriving statistical-dynamical equations directly from the conservation of mass, momentum and energy (SALTZMAN, 1978). Such models can be used to simulate the transient response of the climate system to any external forcing or internal perturbation.

Since the publication of the historic HAYS et al. (1976) paper, a number of modelling efforts have attempted to explain the relation between astronomical forcing and climate change. Most of these modelling studies have focused on the origin of the 100-ka cycle in ice volume. These studies have been motivated by the recognition that the amount of insolation perturbation at 100 ka is not enough to cause a climate change of ice-age magnitude. Specific explanations for the large 100-ka variance generally fall into two classes. In the first case, orbital variations are essential to ice volume fluctuations, but non-linear interactions in the air–sea–ice system modify the signal. In the second case, ice volume fluctuations result from inherently non-linear interactions in the air–sea–ice system, with orbital variations serving only to phase-lock the variations at the preferred time-scales.

For models where the ice volume fluctuations are primarily driven by orbital forcing, IMBRIE et al. (1984) showed that at the 23-ka and 41-ka periods, ice volume responds linearly to orbital forcing, but at 100 ka the effect is non-linear. In particular, the 100-ka power can be generated by transmission of 19-ka and 23-ka frequencies through a non-linear system (WIGLEY, 1976), producing substantial power in both harmonics and 100-ka subharmonics. Specific details of this non-linear interaction vary considerably, but many mechanisms focus on the abrupt terminations of ice ages, which occur over an interval of about 10,000 years (RUDDIMAN and MCINTYRE, 1981; HUGHES, 1987). Other approaches to the 100-ka cycle focus on non-linear interactions between accumulation/ablation, ice-sheet flow, elastic lithosphere, and viscoelastic mantle (OERLEMANS, 1982; BIRCHFIELD et al., 1982; POLLARD, 1982; HYDE and PELTIER, 1985). In particular, POLLARD (1983) coupled an axisymmetric ice-sheet model to a one-level energy balance climate model containing land-ocean contrasts and both north-south and east-west heat exchange. The ice-sheet model had provisions for bedrock response and for rapid melting at the southern tip of the ice sheet due to proglacial lakes or marine incursions. Solutions were obtained by using the solar radiation variations associated with orbital forcing for the past 700,000 years; the cumulative addition of bedrock lag and ice calving improved the fit to the oxygen isotope record of ice-volume fluctuations. In a similar experiment, WATTS and HAYDER (1984) found that a strong 100-ka cycle is present in the simulated ice volume, but only if a mechanism that enhances ice-sheet collapse is introduced.

A more recent model incorporates an improved scheme for glacial isostasy which takes into account mantle viscosity (PELTIER, 1987). The latest version is able to reproduce a reasonable 100-ka cycle over the last 500 ka, but is less successful in reproducing the record of the earlier period between 800 ka and 500 ka BP. The 100-ka model cycle has a sawtooth shape reflecting the typical slow growth and rapid melting of ice sheets (PELTIER, 1987; HYDE and PELTIER, 1987). The realistic representation of the rapid termination of glacial episodes is attributed to a two-stage mechanism, which reflects the greater sensitivity of ice sheets to changes in ablation as opposed to accumulation. First, during the glacial episode itself, the stability of the ice sheet decreases as it increases in area and height. As the model climate warms at the end of the glacial period, the ablation zone becomes progressively and disproportionately larger. In the second stage, a critical level of warming is reached and a large ice-free area develops on land equator-ward of the former ice-sheet margin, in a region which is still depressed due to glacial loading. Ice from the poleward area of accumulation then begins to slide and surge into this depression and rapidly melts. The 100-ka cycle of ice-sheet build-up and decay simulated by the model is not caused directly by the eccentricity forcing (HYDE and PELTIER, 1987). Since the strongest eccentricity forcing occurs at a point in each climate cycle when the model ice sheets respond purely to precession, and because ice-sheet decay is triggered once orbitally induced warming reaches a critical level, it is considered that eccentricity phase-locks the climate response to precession. Because of the presence of additional orbital periodicities (obliquity and the lower-frequency eccentricity cycles) the phase-locking is not perfect. This model is, therefore, an example of an internally driven system, the oscillations of which are phase-locked by external forcing (cf. discussion of oscillations in the deep-ocean circulation in Chapter 14 by PENG).

In contrast, results from a model incorporating ice sheets, but with different representations of the physical processes, suggest that the external orbital forcing, together with internal forcing, drives the observed 100-ka cycles. Ghil and colleagues (GHIL and LE TREUT, 1981; LE TREUT et al., 1988) modelled such internally generated glacial-interglacial fluctuations which, when orbitally forced, produce characteristic Milankovitch periods, including harmonics (e.g. 10 ka) and sub-harmonics (e.g. 100 ka). It is particularly significant that PESTIAUX et al. (1988) detected in deep-sea cores records 10.3-ka, 4.7-ka and 2.5-ka periods which were shown to be combination tones of the classical 41, 23 and 19 ka.

We see therefore that the most recent ice-sheet models are capable of reproducing both the spectrum and the broad pattern of sawtooth glacial-interglacial cycles found in the geological record of the last few hundred thousand years. They indicate that the climate response to orbital forcing is non-linear and that both external and internal forcing is involved. Despite the improvements in the performance of ice-sheet models, a number of problems remain (BROECKER and DENTON, 1990): the models are unrealistic, lacking marine ice sheets for example; they show obvious errors and they leave out many of the proposed mechanisms

and the ice slope. The lateral ice flow is represented by a diffusivity term calculated for each latitude by assuming that the east-west profile of the ice sheet is double parabolic. This parabolic shape has been calibrated against the 18-ka BP ice sheet. The net mass balance is the difference between the calculated local snow precipitation and the local ablation computed from the balance of the heat fluxes at the snow or ice surfaces. The isostatic rebound (i.e. the bedrock-surface elevation responding to the changing ice load) is calculated using a time-dependent diffusive equation of the asthenosphere along latitude.

Over the ice sheets, the model includes the elevation-desert feedback which reduces the accumulation to low values as the ice sheet grows high. The snow albedo is dependent on the surface temperature and the snow age as in DANARD et al. (1984). This snow aging-albedo feedback is based on the fact that during daytime, the heat applied to the snow cover by radiative processes is used for recrystallization processes and subsequently for a decrease in the many facets of the minute crystals of fresh fallen snow. Thus the scattering processes in the uppermost layer of the snow are diminished, and radiation can penetrate deeper into the snow where increased absorption can take place. This results in a general decrease of the reflected portion of the shortwave solar and scattered radiation. Since metamorphism is an irreversible process, a snow cover which is once undergoing a change by metamorphism cannot recover during the night. The decrease in albedo due to metamorphism thus results in an overall decrease in the albedo of the snow cover from day to day (DIRMHIRN and EATON, 1975). It must be noted, however, that despite a very low accumulation rate, current snow albedo over central Antarctica is very high throughout the year because of low temperatures. This observation would possibly be in conflict with the model assumption if this present-day observation is confirmed for longer time-scale. This is why, in order to take into account high snow albedo at low surface temperatures, the snow albedo in our model is maintained at its maximum (0.85) for daily mean surface temperatures lower than −10°C.

Simulation of the last 200 ka

The model was first run with ice-sheet feedback by forcing it only with the astronomical insolation over the past 122 ka (BERGER, 1978) keeping the CO_2 concentration at the Eem level (~270 ppmv). The atmosphere–land–ocean seasonal model is run every day during 20 years for reaching quasi-equilibrium; the output is then used to force the ice-sheets model which is run with a time step of 1 year for 1,000 years. Sensitivity tests have confirmed the adequacy of such choices. Large variations of ice volume are simulated between 122 and 55 ka BP, with a rapid latitudinal extension of the North American and Eurasian ice sheets starting at 120 ka BP. The model simulates a maximum ice volume of 40×10^6 km^3 at roughly 20 ka BP and a total deglaciation as well, this simulation of both the glaciation and deglaciation being a crucial test of the efficiency of the model. The simulated evolution of the three northern ice sheets is generally in phase with geological reconstructions. The major discrepancy between the simulation and observations lies in the temperature variations and in the ice-sheet extent, a discrepancy linked to an overestimation of the CO_2 concentration which was assigned an interglacial value throughout the run.

A set of experiments therefore addresses the effect of CO_2 on the climate of the last glacial-interglacial cycle. This experiment made by forcing the model with both insolation and CO_2 variations over the last 122 ka (BARNOLA et al., 1987) significantly improved the temperature results and the ice-volume reconstructions obtained in the insolation-only experiment (GALLÉE et al., 1992). The simulated total ice-volume deviation from present-day value (assumed to be 30.5 × 10^6 km^3, that is, 27.9 × 10^6 km^3 for Antarctica and 2.6 × 10^6 km^3 for Greenland; HUGHES et al., 1981) is displayed in Fig. 6 as a function of time compared to the variation of the global seawater oxygen isotopic ratio given by LABEYRIE et al. (1987) and DUPLESSY et al. (1988). The latter values are scaled and plotted such that their maximum corresponds to the deviation from present-day of the LGM total ice volume assumed to be 48.6 × 10^6 km^3, as reconstructed by MARSIAT and BERGER, 1990). The simulated deviations are obtained by adding the northern hemisphere ice-volume changes simulated by the model for each of the individual ice sheets (Eurasia, Greenland, North America) and the changes in Antarctica reconstructed as follows. We assumed (1) that the Antarctic ice volume at 122 ka BP and from 6 ka BP to the present was the same as now (J. OERLEMANS and D. RAYNAUD, personal communication, 1987); (2) that 18 ka ago, it was 9.8 × 10^6 km^3 larger than today (HUGHES et al., 1981); and (3) that intermediate values are given by linear interpolation. Assumption (2) seems to be in agreement with the results obtained by HUYBRECHTS (1990) from a three-dimensional ice-sheet model.

The main characteristic of Fig. 6 is the similarity between the reconstructed and the simulated global ice volumes, except for variations with time-scales shorter than 5 ka and for an overestimation of the calculated ice volume during isotopic stage 3 when compared to LABEYRIE et al. (1987). Large ice-volume oscillations are simulated between 122 ka BP and 55 ka BP; in reasonable agreement with deep-sea records, two stadials are culminating at 110 ka BP and 90 ka and isotopic stage 5a is followed by a dramatic expansion to mid-Würm ice-volume values between 75 ka and 60 ka BP. The deglaciation is abrupt both in reconstruction and in simulation. The ice volumes for each ice sheet simulated by the LLN model (BERGER et al., 1990a; GALLÉE et al., 1992) are also in good agreement with independent reconstructions made from proxy data by MANGERUD (1991) and BOULTON et al. (1985). At 19 ka BP, the simulated temperature of the northern hemisphere is 3.4°C colder than the present day simulated value, and the simulated total ice-volume deviation is 53 × 10^6 km^3.

An experiment (called the 200 ka experiment) with the LLN model was undertaken in which the model was forced by astronomically driven insolations and by prescribed CO_2 concentrations over the last 200 ka (GALLÉE et al., 1993), in order to assess the ability of the model to sustain glacial-interglacial cycles. At the time this experiment was done, the Vostok ice core (BARNOLA et al., 1987; RAYNAUD et al., 1993) did not provide a reconstruction of CO_2 concentration variations before 150 ka BP (data are now extending to 220 ka BP; JOUZEL et al., 1993), so that the model was forced by CO_2 concentrations adapted from SHACKLETON et al. (1992). The simulation starts by assuming that the ice sheets are melted at 200 ka BP, because this time corresponds to an interglacial in marine $\delta^{18}O$ records. Two glacial maxima comparable in amplitude are simulated at 134 and 15 ka BP as found in the marine $\delta^{18}O$ records (e.g. IMBRIE et al., 1984; see Fig. 7). The phase lags between the insolation minima at 65°N and ice volume maxima are 6, 5, 6, 11 and 8 ka, respectively, for the ice volume maxima at 180, 134, 109, 59 and 15 ka BP.

A further comparison of the results of the 200 ka experiment with the SPECMAP record is made in Fig. 7. The variations correlate well and a spectral analysis of both time series gives spectral peaks of similar amplitude for the 100-, 41-, 23- and 19-ka periods. The correct

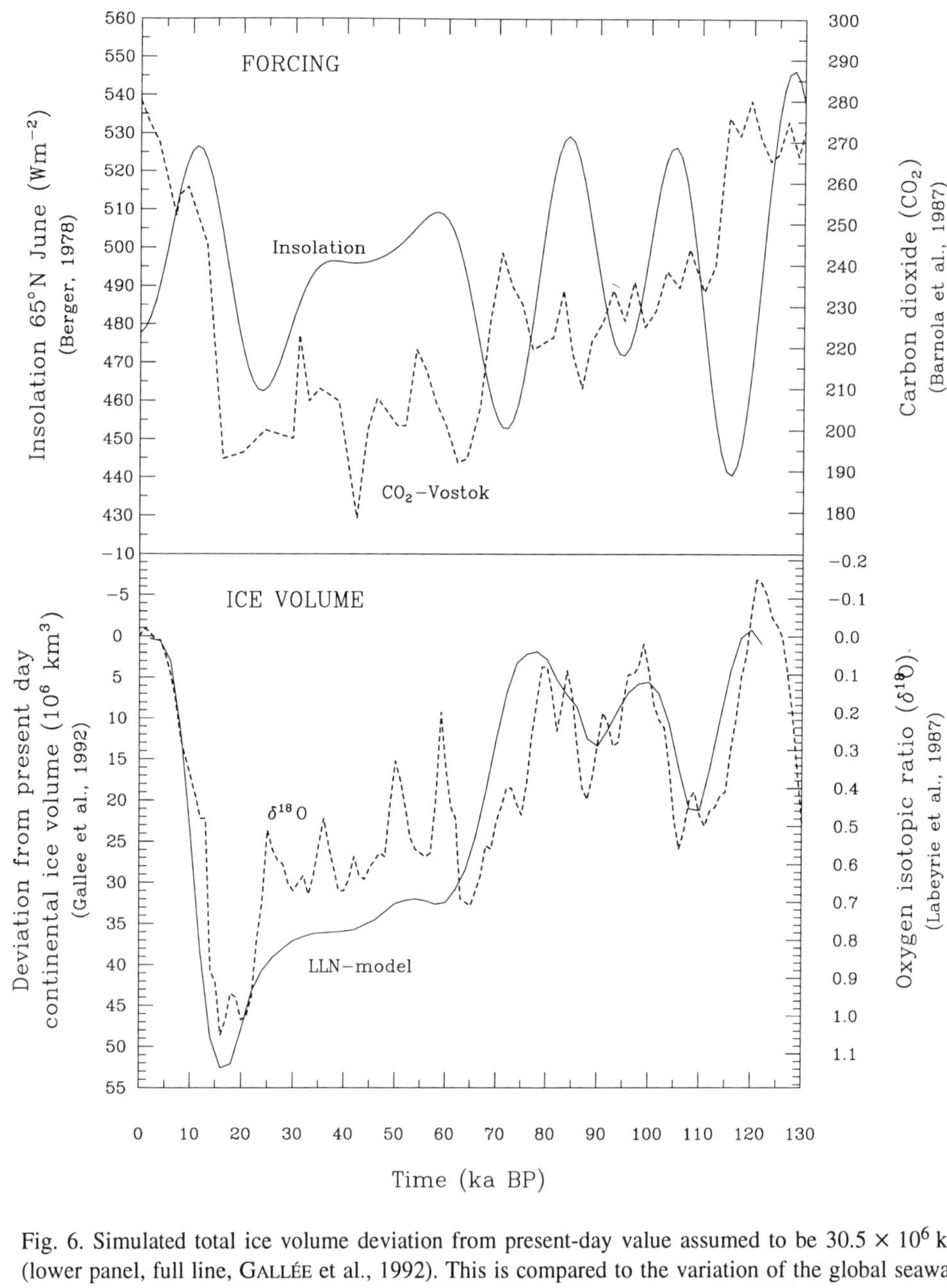

Fig. 6. Simulated total ice volume deviation from present-day value assumed to be 30.5 × 10^6 km^3 (lower panel, full line, GALLÉE et al., 1992). This is compared to the variation of the global seawater oxygen isotopic ratio (lower panel, dashed line) given by LABEYRIE et al. (1987). These values are scaled and plotted such that their maximum corresponds to the deviation from present-day of the last glacial maximum total ice volume taken as 48.6 × 10^6 km^3 (MARSIAT and BERGER, 1990). The upper panel provides the long-term variations of the June insolation at 65°N (full line, BERGER, 1978) and of the Vostok CO_2 concentration (dashed line, BARNOLA et al., 1987).

relative intensity of the spectral peaks for the simulated ice volume is probably the most important result of this 200 ka experiment as it demonstrates the ability of the LLN model to reproduce the 100-ka cycle. Moreover, the LLN model simulates a deglaciation centred on

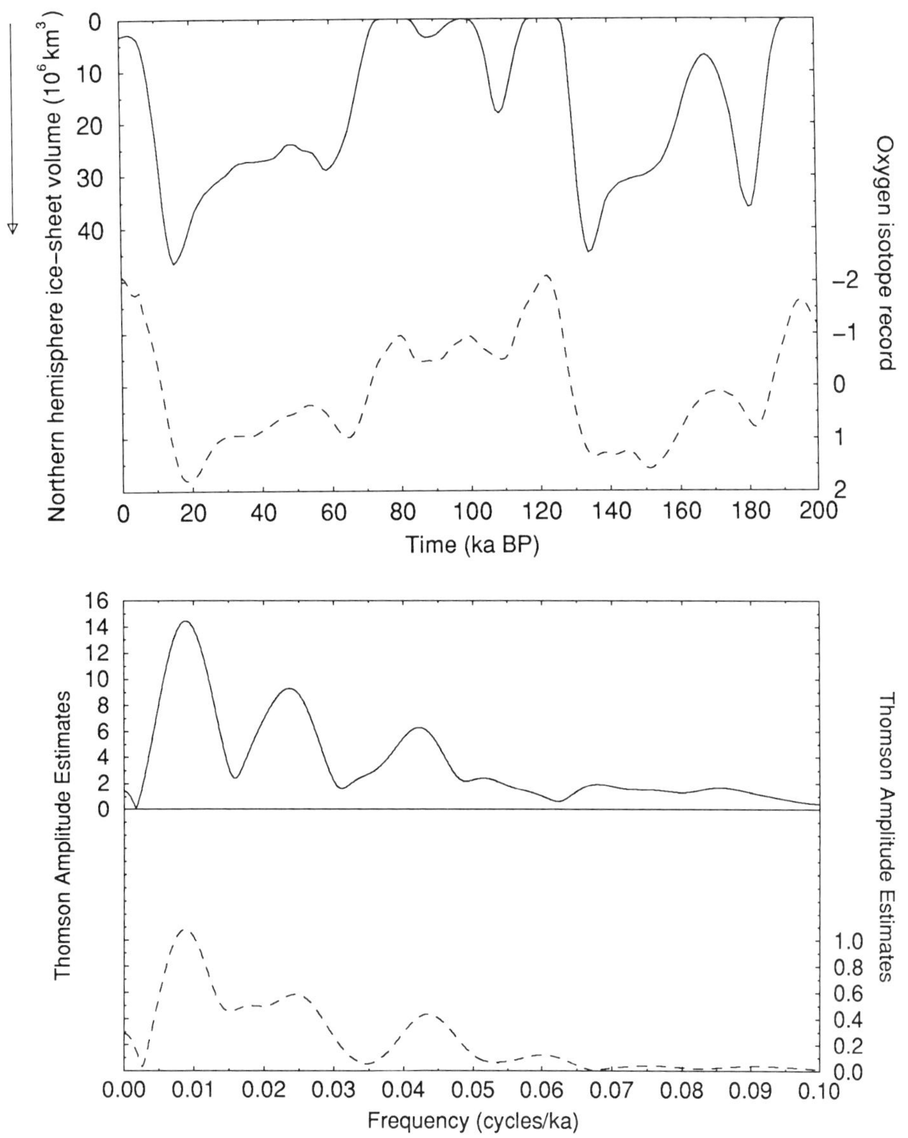

Fig. 7. (a) Upper panel: variations over the last 200 ka of (i) the simulated ice volume of the northern hemisphere (solid curve) (GALLÉE et al., 1993), and (ii) $\delta^{18}O$ variation as reconstructed in the SPECMAP time series (short-dashed curve) (IMBRIE et al., 1984; MARTINSON et al., 1987). The forcing used in the simulation is the insolation variation at the top of the atmosphere and the CO_2 variation adapted from SHACKLETON et al. (1992). (b) Lower panel: spectral amplitude in the Thomson multitaper harmonic analysis of (i) the simulated ice volume of the northern hemisphere for the 200 ka experiment (solid curve), and (ii) $\delta^{18}O$ variation over the last 200 ka as reconstructed in the SPECMAP time series (short-dashed curve).

130 ka BP and a stable interglacial with high sea-level lasting from about 127 ka BP to 115 ka BP. This duration agrees well with geological evidence for the last interglacial (MÜLLER, 1974), whereas a major weakness both of the SPECMAP record and of previous model simulations (e.g. IMBRIE and IMBRIE, 1980; BERGER et al., 1981) is precisely their inability to depict the several thousand-year long stable interglacials.

melting is faster than ice flow and isostatic rebound is not fast enough to level up the ice sheet which could prevent future net ablation. North, the ice sheets remain high allowing the North to South flux of ice to continue to be efficient. As a consequence, the ice-sheet volumes decrease, the height of the ice sheets decreases, temperature at their surfaces increases which increases the global ablation. This leads to a decrease of the zonal surface albedo and a further replacement of tundra by taiga which feedbacks positively on the albedo decrease, directly and by reducing the albedo of the continental surfaces covered by snow. This leads finally to a minimum of ice volume reached about 5 ka later than the insolation maximum.

The CO_2–water vapour–albedo feedback mechanisms at the LGM

Although uncertainty still surrounds the mechanisms by which CO_2 concentrations have varied over the last interglacial-glacial cycle, there is little doubt that such changes did occur. Modelling evidence supports the role of CO_2 as a positive feedback mechanism. Using an energy balance model, HARVEY (1989) and HOFFERT and COVEY (1992), for example, find that the decrease in atmospheric CO_2 at the LGM enhanced global cooling by up to 1.5°C. GENTHON et al. (1987) estimate that the direct radiative effect of the CO_2 changes over the last glacial-interglacial cycle, as recorded in the Vostok core, is 0.6°C. The direct effect is amplified by various feedback mechanisms such as the water vapour feedback. Results from a multivariate analysis of the Vostok CO_2 and temperature records, using a range of amplification factors, imply that the CO_2 changes could account for 27–85% of the observed temperature variance (GENTHON et al., 1987). But analyses of air bubbles in the Vostok and Byrd cores, also reveal changes in the atmospheric concentration of methane (CH_4). As a greenhouse gas, CH_4 has a direct radiative effect, but also has indirect photochemical effects involving ozone and water vapour. Allowing for both direct and indirect effects, CHAPPELLAZ et al. (1990) estimate that CO_2 and CH_4 changes may have accounted for about 2.3°C of a total global warming during the last glacial-interglacial transition of 4.5 ± 1.0°C. Another study, based on multivariate analysis of the Vostok records, suggests that the total contribution of greenhouse gas fluctuations to the regional warming seen in the Vostok temperature record is 3°C, or 50 ± 10% (LORIUS et al., 1990). From these two studies, it follows that CO_2 and CH_4 may be responsible for 50% of the observed glacial-to-interglacial warming, methane alone accounting for 10%. Finally, new results on the change of N_2O levels during the last deglaciation indicate that N_2O has also contributed significantly to the change in radiative forcing (up to 15% of that of CO_2, i.e. as much as CH_4, LEUENBERGER and SIEGENTHALER, 1992).

A similar analysis of the relative importance of the feedback mechanisms involving CO_2, water vapour and albedo in the LLN experiments was therefore relevant. The comparison of the results obtained from the LLN transient experiments using the Milankovitch and the Milankovitch + CO_2 forcings shows that, at the Last Glacial Maximum, the long-term CO_2 changes are responsible for roughly 50% of the temperature change and 30% of the ice-volume change. In an attempt to understand the role of the temperature–albedo feedback and the importance of the greenhouse gases (water vapour and CO_2), several sensitivity experiments have been made (GALLÉE et al., 1992; TRICOT, 1992; BERGER et al., 1993b).

In particular, the physical processes responsible for the cooling simulated at the LGM under the forcing of both insolation and CO_2 were investigated with a 1-D radiative convective

(RCM) climate model (TRICOT, 1992). A full description of this RCM, its validation and sensitivity tests are given in TRICOT (1992) and BERGER et al. (1993b). In particular, when the CO_2 concentration is doubled (from 330 ppmv to 660 ppmv), the instantaneous perturbations of the net radiative budget (before the action of feedbacks) are estimated to be 3.72 W m^{-2} at the tropopause and 1 W m^{-2} at the surface. At equilibrium, the temperature increases by 1.8 K at the surface and in the troposphere and decreases in the stratosphere above 20 km. The sensitivity of the RCM to CO_2 change is thus very similar to that calculated in other sensitivity studies using RCMs with the same basic assumptions (e.g. RAMANATHAN, 1981). Without allowing the water vapour feedback to operate, the warming is reduced to 1.2 K.

As the purpose here is to quantify the relative contribution of each of the main climatic feedbacks, some improvements in the LLN model were first made by TRICOT (1992) in such a way that the sensitivity of this modified version, designated as LLN1, to changes in trace gas concentrations will be closer to the sensitivity of the RCM itself. These improvements concern mainly the radiative part of the model, involving new infrared and solar radiation schemes, and an explicit computation of the stratospheric temperature. Climatological data about cloud and aerosols also have been revised. LLN1 has been validated for the present-day climate. Its sensitivity to a doubling of CO_2 concentration ($\Delta T = 1.97$ K) is similar to the sensitivity of the original version of the LLN model ($\Delta T = 2$ K).

The cooling at the LGM simulated by LLN1 using the surface boundary conditions calculated by the LLN transient simulation, the seasonal cycle of insolation at 18 ka BP and a CO_2 concentration of 194 ppmv amounts to 4.5 K. This cooling is associated with an increase of the planetary albedo from 31 to 32.6%, an increase of the surface albedo from 16.2 to 19.3% and a decrease in the annually (averaged vertically) integrated water vapour concentration from 2.44 to 1.99 g cm^{-2}.

At the LGM, in addition to the insolation and CO_2-concentration variations, the hemispherically averaged changes in the surface albedo and in the vertical distribution of water vapour are available as part of the LLN1 model diagnostics. These changes were inserted one by one or in combination into the RCM.

All these sensitivity tests lead to the following conclusions for the northern hemisphere:

(1) The cooling using only the astronomical forcing but, allowing for the water vapour feedback, amounts to 3°C, 67% of the 4.5°C cooling resulting from both the astronomical and CO_2 forcings. It must be stressed that what we call astronomical forcing includes both the change in insolation and the consequent changes in surface and planetary albedo.

(2) If the CO_2 concentration is kept fixed to its interglacial level (330 ppmv), the astronomical-albedo forcing (AA) without the water vapour feedback (WVF) explains 60% of the 3°C cooling, the WVF being therefore responsible for 40% of the total cooling in this Milankovitch alone experiment.

(3) In the astronomical + CO_2 experiment, the direct effect of these forcings (i.e. without the WVF) explains 60% (2.7°C) of the 4.5°C cooling. As in (2), the WVF is therefore responsible for 40% of the cooling. If we discriminate between the AA and the CO_2 contribution (CO_2), we can see that without the WVF, AA explains 40 of the 60% of the total cooling. The contributions to the WVF (40% of the total cooling) are 27% for AA and 13% for CO_2.

(4) In this astronomical + CO_2 cooling, it may also be interesting to analyse the proportion

which is due to water vapour for each sub-experiment. Considering only the AA, the WVF is responsible for 27 of the 67% of the global cooling, the remaining 40% is the direct effect of AA. In the CO_2 alone experiment, the WVF explains 13 of the 33% of the global cooling, the direct effect of CO_2 accounting for the remaining 20%.

Other feedbacks and future work

A number of improvements to the Louvain-la-Neuve model are under way, such as coupling the climate and ice-sheet models with deep-ocean and global carbon-cycle models and including a better chemical composition of the atmosphere. The importance of aerosols in particular is confirmed by the Vostok records which provide information about changes in the aerosol content of past atmospheres (LEGRAND et al., 1988). The atmospheric loading of aerosols from marine and terrestrial sources is generally greater during colder periods, such as 100–15 ka BP, than during warmer periods, such as 130 ka and 10 ka BP. The increased loading is related to increases in both the sources and transport of aerosols during glacial episodes and has a positive feedback effect. The particles reflect radiation back to space, thus enhancing surface cooling. Using an energy balance model, HARVEY (1989) estimates that the increased loading associated with the Last Glacial Maximum may have caused an additional cooling of 2.2°C.

LEGRAND et al. (1988) have examined the potential causes and effects of changes in the concentration of marine aerosols. Dimethylsulphide (DMS) is produced by planktonic algae and oxidizes in the atmosphere to form a sulphase aerosol called marine non-sea-salt sulphate (CHARLSON et al., 1987). Initial analyses of the data suggested that the concentration of marine non-sea-salt sulphate in the Antarctic atmosphere increased by 20–46% during the Last Glacial Maximum (LEGRAND et al., 1991).

The peak concentrations of non-sea-salt sulphate in the Vostok core are believed to be due not only to more efficient poleward atmospheric transport, but also to increased productivity. The record shows a strong spectral peak at 21 ka and a weaker 42 ka peak. The existence of these peaks at the orbital frequencies, and the correspondence with the Vostok CO_2 record, suggests a change in the surface ocean circulation which would lead to increased marine productivity and, hence, to increased DMS production and a reduction in atmospheric CO_2 concentrations.

It has been suggested (CHARLSON et al., 1987) that increased DMS production by marine microorganisms leads to increased atmospheric concentrations of cloud condensation nuclei (CCN). The increased CCN concentrations in turn lead to an increase in cloud albedo, thus more radiation is reflected back to space. Since increased marine productivity leads to surface cooling, the proposed feedback effect would be positive during glacial episodes. Assuming that the proposed feedback mechanism is plausible, and that concentrations of CCN increase in direct proportion to marine non-sea-salt sulphate, LEGRAND et al. (1988) have estimated that a 46% increase in CCN would increase cloud albedo by about 0.03 and that the direct radiative effect of this increased albedo would cause an additional surface cooling of 1°C.

A number of potential feedback effects involving changes in atmospheric composition over glacial-interglacial cycles have thus been identified. These range from highly plausible (CO_2) to highly speculative (ocean productivity, cloud albedo) mechanisms. In terms of helping to explain the links between orbital forcing and climate response, all the proposed feedbacks operate in the required direction (they are positive feedbacks amplifying the initial forcing) and on the required scale (global or hemispheric). On the basis of multivariate analyses, LORIUS et al. (1990) conclude that three of these mechanisms (changes in the concentration of greenhouse gases, aerosol loading and marine non-sea-salt sulphate), together with solar insolation and ice-volume changes, are sufficient to explain over 90% of the variance in the Vostok temperature record.

Although the results presented here may be considered only preliminary, they confirm the results of many of the simpler and physically less realistic models discussed in the section on simulations of temporal evolution of climate. Orbital forcing is confirmed as a highly plausible cause of the Quaternary glacial-interglacial cycles. The modelling evidence also implies that internal processes and feedbacks, such as ice-sheet dynamics and changes in atmospheric composition, play an important part in the climate response.

Future climates under astronomical forcing

Simulation of the waxing and waning of the ice sheets over the past 1 million years seems to indicate that the climate is sensitive to the orbital parameter variations, or at least that these are able to trigger feedbacks which then induce significant climatic variations at that timescale. No general circulation model has been able to generate such long-term climatic variations, because the complexity of the different parts of the climate system themselves and of their mutual interactions requires considerable computer time and space. Moreover, the nonlinearities in the climate system can easily introduce spurious response (which can be masked sometimes by compensating errors) if one process is given too great a weight compared to the others (including those which are missing).

Simplified 0-D and 1-D models have thus largely been used to simulate these long-term climatic variations. The problem with such a basic approach is that it is possible to obtain almost any desired sensitivity by tuning procedures which involve parameterizations not tied firmly to first principles (although their general formulation derives sometimes directly from such first principles).

The LLN 2-D climate model has however reproduced fairly well the seasonal and latitudinal distributions of surface air temperature and of ice volume for both present day situation and for the last 200 ka. In particular, a good agreement was found between the low frequency parts of the simulated long-term climatic variations and of the most recent reconstructions of sea-level and ice-volume changes from geological observations; the 100-ka cycle was well reproduced by the model as well as the timing of the glacial maxima and the length of the interglacials; and its sensitivity to greenhouse gases (water vapour and CO_2) is well within the range of GCMs' results. Such a validation allows this model to be used for predicting future climates.

Although climate models have been widely used to explore the reality of orbital-climate relationships and the potential mechanisms linking cause and effect, comparatively few model studies have attempted to predict future climate on the basis of orbital variations. In a paper for studying the stability of repository sites for nuclear waste disposal, BERGER et al. (1991) have identified seven statistical/dynamical models and time-dependent ice-sheet models which do attempt such a prediction. The principal characteristics of the seven models are also presented in GOODESS et al. (1992).

Climate reconstructions for the last 120 ka and projections for the next tens to hundreds of thousands of years have been computed and, despite some shortcomings, the models provide a reasonably coherent guide to the range of future climate change caused by periodicities in the orbital parameters. Although there are some discrepancies between the future sequences of climate predicted by the models, the major changes are common to all seven studies. These models suggest that, setting aside the potential for anthropogenic change, the world's climate, under the astronomical forcing alone, should soon begin a slow deterioration towards glacial conditions. Oscillatory cooling is expected to continue with progressively colder episodes at about 5 ka, 23 ka and 60 ka after the present-day. The final extreme glaciation around 60 ka from now, is expected to rival the intensity of the LGM and should be followed by a gradual shift towards warmer conditions. All the models indicate that climates as warm as the present day are relatively rare; the world is not expected to return to conditions matching the Holocene thermal optimum before 120 ka in the future.

Preliminary results from the LLN climate model, forced by insolation variations only (i.e. that the CO_2 concentration is kept fixed at its pre-industrial level), indicate also that in the absence of enhanced greenhouse warming, the next glacial maximum in the northern hemisphere will happen about 55 ka AP (after present). The principal features of the simulation for the next 80 ka AP are shown in Fig. 8 (BERGER et al., 1991); the long-term cooling trend which began 6,000 years ago will continue for the next 5,000 years; this first slight minimum will be followed by a stabilisation at around 15 ka AP, by a cold interval centred at

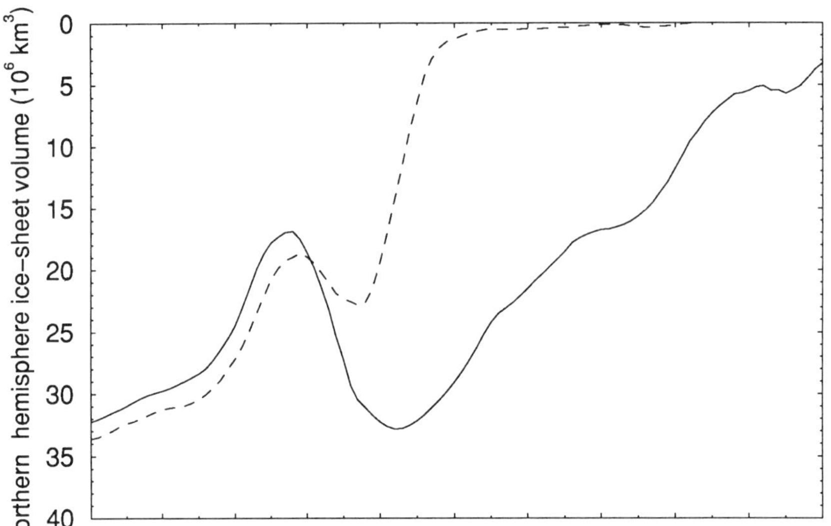

Fig. 8. The future ice volume in the northern hemisphere resulting from the astronomical forcing alone without anthropogenic disturbances (full line) and assuming no Greenland ice sheet today (dashed line) (GALLÉE, 1989; BERGER et al., 1991). The range between these two curves may be assumed to provide a range of uncertainties related to the approximations made in the model, assuming that the greenhouse gas concentrations are constant and equal to their pre-industrial values. Ice volume increases downward in this figure.

25 ka AP and finally by a major glaciation at 55 ka AP. This tendency for our natural climate to be steered astronomically towards the next glacial maximum leads to an average cooling of 0.01°C per century. At the human time-scale, this is negligible when compared to the 1–5°C warming expected from the increase of greenhouse gas concentrations in the course of the 21st century (HOUGHTON et al., 1990, 1992). This is not the case for the related sea-level change. The build up of 27 × 10^6 km^3 of ice in the northern hemisphere at 55 ka AP would indeed be responsible for a eustatic sea-level drop of about 70 m which represents a falling of the sea-level at an average rate of 10 cm per century. This is the same order of magnitude as the 15 cm rise over the past century, but in the reverse direction. If we assume that this long-term trend in equivalent sea-level fall is already underway since the Holocene climatic peak 6,000 years ago, the absolute rise over the past century, whatever its cause may be, must have been larger. This may have implications on the prediction for the next century, as it is expected that sea-level may rise between 20 and 60 cm at the end of the 21st century under the influence of human activities alone (WARRICK et al., 1993).
These experiments consider only the response of the climate system to orbital forcing, the influence of shorter-term forcing factors, including anthropogenic effects, being ignored. However, recently, the LLN model has been used in a series of simulations using different scenarios of CO_2 concentration and of possible impacts of large CO_2 concentration expected to result from human activities.
In a first experiment for the next 5,000 years (LOUTRE, 1993), it has been assumed that the pre-industrial concentration of CO_2 will rise from 280 to 710 ppmv within the next 500 years and then decreases progressively to reach 450 ppmv and 350 ppmv, respectively 1,000 years and 1,500 years from now. A new equilibrium might then maintain this concentration roughly constant for the future. The temperature response of the LLN model (Fig. 9)

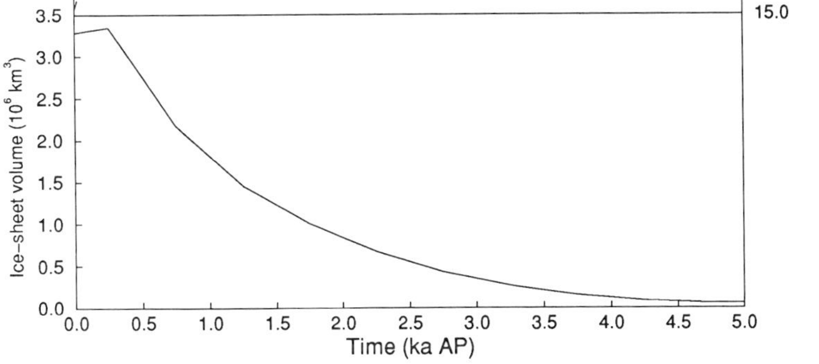

Fig. 9. Simulation of the northern hemisphere temperature (dashed line, right-hand scale) and of the Greenland ice sheet volume (full line, left-hand scale) over the next 5000 years using the LLN model (LOUTRE, 1993).

is what we could expect from its sensitivity to CO_2 changes; surface-air temperature in the northern hemisphere increases from 15°C to 18°C over the next 500 years. The changes are particularly important in the high latitudinal zones (north of 65°N) during spring and fall. These changes are related to a parallel reduction in the extent of the areas covered by snow, a phenomena which acts as a positive feedback in the system. This result is similar to the findings of GROISMAN et al. (1994) who claim that a 10% decline in northern hemisphere snow cover over the past 20 years caused by a general warming of spring temperatures may itself have forced spring temperatures to rise even higher. Between 500 and 1,500 years from now, the air temperature decreases to reach a minimum of 16.4°C. Temperature will afterwards increase slightly but steadily up to 5000 years into the future. This is related to the melting of the Greenland ice sheet which disappears at the end of the integration period, a disappearance which, through the ice-albedo-temperature feedback, explains the warming. This is in line with previous sensitivity experiments for which the CO_2 concentration was kept constant (HUYBRECHTS et al., 1991; GALLÉE et al., 1992, 1993) and which show that a concentration larger than the pre-industrial value leads to a progressive decay of the Greenland ice sheet, every other external forcing being kept constant. It is also worth mentioning that the sensitivity of the LLN model is in agreement with a 500-year integration made by MANABE and STOUFFER (1993) using a coupled ocean-atmosphere model. They assume that the CO_2 concentration increases by 1% year^{-1} (compounded) until it reaches two times (after 70 years) and four times (after 140 years) the initial value and remains unchanged thereafter. During the 500-year period of the doubling and quadrupling experiments, the global mean surface air temperature increases by about 3.5°C and 7°C, respectively. The rise of sea-level due to the thermal expansion of seawater is about 1 and 1.8 m, respectively, but they conclude that "it could be much larger if the contribution of meltwater from continental ice sheets were included".

As an extreme case to test the sensitivity to the possible impact of human activities, it was therefore assumed that the greenhouse warming of the 21st century would melt the Greenland ice sheet totally. The main results of an integration of the LLN model for the next 80 ka with this initial condition (Fig. 8 and GALLÉE, 1989) are that the ice sheets do not reappear in the northern hemisphere before 15 ka AP and that the next glaciation occur 60 ka after present (5,000 years later than in the simulation without the enhanced greenhouse assumption) and is less extensive (19 instead of 27×10^6 km^3). The main difference between the "normal" and enhanced greenhouse simulations, as far as the behaviour of the individual ice sheets is concerned, arises mainly for the Fennoscandian ice sheet which appears only 50 ka after present (35,000 years later than in the "normal" case) and reaches only half its "natural" size (4 instead of 7×10^6 km^3); the Greenland and Laurentide ice sheets reappear 15 ka after present in the no-initial-Greenland ice-sheet run, the Laurentide ice sheet reaches a maximum size of 11 instead of 14×10^6 km^3 and the Greenland ice sheet recovers its 5×10^6 km^3 quite rapidly. This is in agreement with the GRIP results (DANSGAARD et al., 1993) which show that the Greenland ice sheet did not disappear during the Eemian Interglacial and therefore is a resilient part of the natural climate system under conditions similar to present-day.

However, the representation of enhanced greenhouse warming in this experiment is very crude and the model does not yet have a carbon cycle. This simulation must therefore be considered only as a guide to the possible sensitivity of orbital forcing effects to potential changes in cryospheric boundary conditions due to the human-induced enhanced greenhouse.

There remain clearly great uncertainties surrounding the relationship between greenhouse and orbital forcing; three possible patterns describing the relationship between enhanced greenhouse warming and orbital forcing have been identified by GOODESS et al. (1992). First, the simplest assumption is that a relatively short (say 1 ka) period of greenhouse gas-induced warming will be followed by a switch back into the "natural pattern" of glacial-interglacial cycles. The second possibility is that, following a period of warming, the next glaciation will be delayed and will be less severe. The third possibility is that enhanced greenhouse warming will so weaken the positive feedback mechanisms which transform the relatively weak orbital forcing into global interglacial-glacial cycles, that the initiation of future glaciations will be prevented. This is the so-called "irreversible greenhouse effect".

On the basis of the limited evidence available, GOODESS et al. (1992) and BERGER et al. (1991) considered that the second possibility, delayed and reduced glaciation, had the highest probability. However, new experiments based upon a more recent version of the LLN model (GALLÉE et al., 1992) show that the future climate behaviour is very sensitive to the greenhouse gas concentrations and to their hypothetical changes in time. According to these new results, the third possibility of an "irreversible greenhouse effect" could become the most likely future climate.

Modelling results presented here confirm that, potential anthropogenic forcing apart, the pattern and range of climatic conditions likely to be experienced over the next tens to hundreds of thousands of years will be close to those experienced over the last million years. This implies that it is reasonable to use the reconstructed record of Quaternary climate to better understand the behaviour of models of the climate system and as a guide to future conditions. These studies also provide some indication of the ways in which future climate sequences may diverge from past sequences because of enhanced greenhouse warming.

References

ADEM, J., 1988. Possible causes and numerical simulations of the Northern Hemisphere climate during the last deglaciation. *Atmosfera*, 1: 17–38.

ADEM, J., BERGER, A., GASPAR, PH., PESTIAUX, P. and VAN YPERSELE, J. P., 1984. Preliminary results on the simulation of climate during the last deglaciation with a thermodynamic model. In: A. BERGER, J. IMBRIE, J. HAYS, G. KUKLA and B. SALTZMAN (Editors), *Milankovitch and Climate, Part 2.* Reidel, Dordrecht, pp. 801–820.

ADHÉMAR, J. A., 1842. *Révolution des mers, déluges périodiques.* Publication privée, Paris.

ALYEA, F. N., 1972. Numerical simulation of an Ice Age paleoclimate. *Atmos. Sci. Pap.*, No. 193, Colorado State University, Fort Collins, CO.

ANDERSON, D. M. and WEBB, R. S., 1994. Ice-age tropics revisited. *Nature*, 367: 23–24.

BARD, E., HAMELIN, B., FAIRBANKS, R. G. and ZINDLER, A., 1990. Calibration of the ^{14}C timescale over the past 30,000 years using mass spectrometric U-Th ages from Barbados corals. *Nature*, 345: 405–410.

BARNOLA, J.-M., RAYNAUD, Y., KOROTKEVICH, Y. S. and LORIUS, CL., 1987. Vostok ice core provides 160,000-yr record of atmospheric CO_2. *Nature*, 329: 408–414.

BERGER, A., 1976a. Obliquity and general precession for the last 5,000,000 years. *Astron. Astrophys.*, 51: 127–135.

BERGER, A., 1976b. Long-term variations of daily and monthly insolation during the last Ice Age. *EOS*, 57: 254.

BERGER, A., 1977a. Long-term variations of the Earth's orbital elements. *Celestial Mech.*, 15: 53–74.

BERGER, A., 1977b. Support for the astronomical theory of climatic change. *Nature*, 268: 44–45.

BERGER, A., 1978. Long-term variations of daily insolation and Quaternary climatic changes. *J. Atmos. Sci.*, 35: 2362–2367.

BERGER, A., 1979. Insolation signatures of Quaternary climatic changes. *Il Nuovo Cimento*, 2C(1): 63–87.

BERGER, A., 1988. Milankovitch theory and climate. *Rev. Geophys.*, 26: 624–657.

BERGER, A., 1993. Orbital parameters, 1993. In: S. SCHNEIDER (Editor), *Encyclopedia of Climate and Weather*. R. Ubell Associates, New York.

BERGER, A. and LOUTRE, M. F., 1990. Origine des fréquences des éléments astronomiques intervenant dans le calcul de l'insolation. *Bull. Cl. Sci., Acad. R. Belg.*, 6e série, Tome I, 1–3: 45–106.

BERGER, A. and LOUTRE, M. F., 1991. Insolation values for the climate of the last 10 million years. *Quat. Sci. Rev.*, 10: 297–317.

BERGER, A. and LOUTRE, M. F., 1992. Astronomical solutions for paleoclimate studies over the last 3 million years. *Earth Planet. Sci. Lett.*, 111: 369–382.

BERGER, A., GUIOT, J., KUKLA, G. and PESTIAUX, P., 1981. Long term variations of monthly insolation as related to climate change. *Geol. Rundsch.*, 70: 748–758.

BERGER, A., GALLÉE H., FICHEFET, TH., MARSIAT, I. and TRICOT, C., 1990a. Testing the astronomical theory with a coupled climate-ice sheet model. In: L. D. LABEYRIE and C. JEANDEL (Editors), *Geochemical Variability in the Oceans, Ice and Sediments. Palaeogr., Palaeoclimatol., Palaeoecol.*, 89(1/2), *Global Planet. Change Section*, 3: 125–141.

BERGER, A., FICHEFET, TH., GALLÉE, H., MARSIAT, I., TRICOT, C. and VAN YPERSELE, J. P., 1990b. Physical interactions within a coupled climate model over the last glacial-interglacial cycle. *Trans. R. Soc. Edinburgh, Earth Sci.*, 82: 357–369.

BERGER, A., GALLÉE, H. and MÉLICE, J. L., 1991. The Earth's future climate at the astronomical time scale. In: CL. GOODESS and J. PALUTIKOF (Editors), *Future Climate Change and radioactive Waste Disposal*. NIREX Safety Series, NSS/R257, pp. 148–165.

BERGER, A., FICHEFET, TH., GALLÉE H., TRICOT, C. and VAN YPERSELE, J. P., 1992. Entering the glaciation with a 2-D coupled climate model. *Quat. Sci. Rev.*, 11: 481–493.

BERGER, A., LOUTRE, M. F. and TRICOT, C., 1993a. Insolation and Earth's orbital periods. *J. Geophys. Res.*, 98(D6): 10341–10362.

BERGER, A., TRICOT, C., GALLÉE, H. and LOUTRE, M. F., 1993b. Water vapour, CO_2 and insolation over the last glacial-interglacial cycles. *Philos. Trans. R. Soc., London, Ser. B*, 341: 253–261.

BERGER, A., GALLÉE, H. and TRICOT, C., 1993c. Glaciation and deglaciation mechanisms in a coupled 2-D climate-ice sheet model. *J. Glaciol.*, 39(131): 45–49.

BIRCHFIELD, G. E., 1989. A coupled ocean-atmosphere climate model: temperature versus salinity effects on the thermohaline circulation. *Clim. Dynamics*, 4: 57–71.

BIRCHFIELD, G. E. and WEERTMAN, J., 1978. A note on the spectral response of a model continental ice sheet. *J. Geophys. Res.*, 83 (C8): 4123–4125.

BIRCHFIELD, G. E., WEERTMANN, J. and LUNDE, A. T., 1982. A model study of high-latitude topography in the climatic response to orbital insolation anomalies. *J. Atmos. Sci.*, 39: 71–87.

BOND, G., HEINRICH, H., BROECKER, W., LABEYRIE, L., MCMANUS, J., ANDREWS, J., HUON, S., JANTSCHIK, R., CLASEN, S., SIMET, C., TEDESCO, K., KLAS, M., BONANI, G. and IVY, S., 1992. Evidence for massive discharges of icebergs into the North Atlantic ocean during the last glacial period. *Nature*, 360: 245–249.

BOULTON, G. S., 1993. Two cores are better than one. *Nature*, 366: 507–508.

BOULTON, G. S., SMITH, G. D., JONES, A. S. and NEWSOME, J., 1985. Glacial geology and glaciology of the last mid-latitude ice sheets. *J. Geol. Soc., London*, 142: 447–474.

BRADLEY, R. S., 1985. *Quaternary Paleoclimatology. Methods of paleoclimatic reconstruction.* Allen and Unwin, Boston, MA, 472 pp.

BRADLEY, R. S. (Editor), 1989. *Global Changes of the Past.* UCAR/Office for Interdisciplinary Earth Studies, Boulder, CO.

BRISKIN, M. and HARRELL, J., 1980. Time series analysis of the Pleistocene deep-sea paleoclimatic record using periodic regression. *Mar. Geol.*, 36: 1–22.

BROCCOLI, A. J. and MANABE, S., 1987. The influence of continental ice, atmospheric CO_2, and land albedo on the climate of the last glacial maximum. *Clim. Dynamics*, 1: 87–89.

BROECKER, W. S. and DENTON, G. H., 1990. The role of ocean-atmosphere reorganizations in glacial cycles. *Quat. Sci. Rev.*, 9: 305–341.

BROECKER, W. S., PETEET, D. M. and RIND, D., 1985. Does the ocean-atmosphere system have more than one stable mode of operation ? *Nature*, 315: 21–26.

BROECKER, W. S., ANDREE, M., WOLFI, W., OESCHGER, H., BONANI, C., KENNETT, J. and PETEET, D., 1988. The chronology of the last deglaciation: implications to the cause of the Younger Dryas event. *Paleoceanography*, 3: 1–19.

BROECKER, W. S., FLOWER, B. P., TRUMBORE, S., WOLFLI, W., KENNETT, J. P., TELLER, J. T. and BONANI, G., 1989. Routing of meltwater from the Laurentide ice sheet during the Younger Dryas cold episode. *Nature*, 341: 318–321.

CESS, R. D., POTTER, G. L., BLANCHET, J. P., BOER, G. J., GHAN, S. J., KIEHL, J. T., LE TREUT, H., LI, Z. X., LIANG, X. Z., MITCHELL, J. F. B., MORCRETTE, J. J., RANDALL, D. A., RICHES, M. R., ROECKNER, E., SCHLESE, U., SLINGO, A., TAYLOR K. E., WASHINGTON, W. M., WETHERALD, R. T. and YAGAI, I., 1989. Interpretation of cloud-climate feedback as produced by 14 atmospheric general circulation models. *Science*, 245: 513–516.

CHAPPELL, J. and SHACKLETON, N. J., 1986. Oxygen isotopes and sea level. *Nature*, 324: 137–140.

CHAPPELLAZ, J., BARNOLA, J. M., RAYNAUD, D., KOROTKEVITCH, Y. S. and LORIUS, C., 1990. Ice-core record of atmospheric methane over the past 160,000 years. *Nature*, 345: 127–131.

CHARLSON, R. J., LOVELOCK, J. E., ANDREAE, M. D. and WARREN, S. G., 1987. Oceanic phytoplankton, atmospheric sulphur, cloud albedo and climate. *Nature*, 326: 655–669.

CLIMAP PROJECT MEMBERS, 1976. The surface of the Ice-Age Earth. *Science*, 191: 1131–1136.

CLIMAP PROJECT MEMBERS, 1981. Seasonal reconstructions of the Earth's surface at the last glacial maximum. *Geol. Soc. Am., Map Chart Ser.,* MC-36. Geological Society of America, Boulder, CO.

COHMAP MEMBERS, 1988. Climatic changes of the last 18,000 years: observations and model simulations. *Science*, 241: 1043–1052.

CROLL, J., 1864. On the physical cause of the changes of climate during geological epochs. *Philos. Mag.*, 28: 121–137.

CROWLEY, T. J., 1989. Paleoclimate perspectives on a greenhouse warming. In: A. BERGER, S. SCHNEIDER and J. CL. DUPLESSY (Editors), *Climate and Geo-Sciences.* Kluwer, Netherlands, pp. 179–207.

CROWLEY, T. J., 1990. Are there any satisfactory geologic analogs for a future greenhouse warming? *J. Climate*, 3: 1282–1292.

CROWLEY, J. and NORTH, G. R., 1991. *Paleoclimatology.* Oxford University Press, Oxford, 352 pp.

CROWLEY, T. J., KIM, K. Y., MENGEL, J. G. and SHORT, D. A., 1992. Modeling 100,000-year climate fluctuations in Pre-Pleistocene time series. *Science*, 255: 705–707.

DANARD, M., GRAY, M. and LYV, G., 1984. A model for prediction ice accretion and ablation in water bodies. *Mon. Weather Rev.*, 112: 1160–1169.

DANSGAARD, W., JOHNSEN, S. J., CLAUSEN, H. B., DAHL-JENSEN, D., GUNDESTRUP, N. S., HAMMER C. U., HVLDBORG, C. S., STEFFENSEN, J. P., SVELNBJÖRNSDOTTIR, A. E., JOUZEL, J. and BEND, G., 1993. Evidence for general instability of past climate from a 250-ka ice-core record. *Nature*, 364: 218–220.

DE ANGELIS, M., BARKOV, N. I. and PETROV, V. N., 1987. Aerosol concentrations over the last climatic cycle (160 ka) from an Antarctic ice core. *Nature*, 325: 318–321.

DEBLONDE, G. and PELTIER, W. R., 1990. A model of late Pleistocene ice sheet growth with realistic geography and simplified cryodynamics and geodynamics. *Clim. Dynamics*, 5: 103–110.

DEBLONDE, G. and PELTIER, W. R., 1991a. Simulations of continental ice sheet growth over the last glacial-interglacial cycle: experiments with a one level seasonal energy balance model including realistic geography. *J. Geophys. Res.*, 96: 9189–9215.

DEBLONDE, G. and PELTIER, W. R., 1991b. A one-dimensional model of continental ice volume fluctuations through the Pleistocene: implications for the origin of the Mid-Pleistocene climate transition. *J. Climate*, 4: 318–344.

DENTON, G. H. and HUGHES, T. J. (Editors), 1981. *The Last Great Ice Sheets.* Wiley, New York.

DIRMHIRN, I. and EATON, F. D., 1975. Some characteristics of the albedo of snow. *J. Appl. Meteorol.*, 14: 375–379.

DONN, W. L. and SHAW, D. M., 1977. Model of climate evolution based on continental drift and polar wandering. *Geol. Soc. Am. Bull.*, 88: 390–396.

DUPLESSY, J. CL., LABEYRIE, L. and BLANC, P. L., 1988. Norwegian sea deep water variations over the last climatic cycle: paleo-oceanographical implications. In: H. WANNER and U. SIEGENTHALER (Editors), *Long and Short Term Variability of Climate.* Earth Science Series, Springer-Verlag, Berlin, pp. 83–116.

DUPLESSY, J. CL., LABEYRIE, L., ARNOLD, M., PATERNE, M., DUPRAT, J. and VAN WEERING, T. C. E., 1992. Changes in surface salinity of the North Atlantic Ocean during the last deglaciation. *Nature*, 358: 485–488.

EMILIANI, C., 1955. Pleistocene temperatures. *J. Geol.*, 63: 538–578.

FAIRBANKS, R. G., 1989. A 17,000-year glacio-eustatic sea level record: influence of glacial melting rates on Younger Dryas event and deep-ocean circulation. *Nature*, 343: 637–642.

FOLLAND, C. K., PALMER, T. N. and PARKER, D. E., 1986. Sahel rainfall and worldwide sea temperature, 1901–85. *Nature*, 320: 602–607.

FRANKIGNOUL, C. and HASSELMANN, K., 1977. Stochastic climate models, II. Application to sea-surface temperature anomalies and thermocline variability. *Tellus*, 29: 289–305.

GALLÉE, H., 1989. Conséquences pour la prochaine glaciation d'une disparition éventuelle de la calotte glaciaire recouvrant le Groenland. *Scientific Report, 1989/7.* Institut d'Astronomie et de Géophysique G. Lemaître, Université Catholique de Louvain, Louvain-la-Neuve, Belgium.

GALLÉE, H., VAN YPERSELE, J. P., FICHEFET, TH., TRICOT, C. and BERGER, A., 1991. Simulation of the last glacial cycle by a coupled sectorially averaged climate-ice-sheet model. I. The climate model. *J. Geophys. Res.*, 96: 13139–13161.

GALLÉE, H., VAN YPERSELE, J. P., FICHEFET, TH., MARSIAT, I., TRICOT, C. and BERGER, A., 1992. Simulation of the last glacial cycle by a coupled, sectorially averaged climate-ice-sheet model. II. Response to insolation and CO_2 variation. *J. Geophys. Res.*, 97(D14): 15713–15740.

GALLÉE, H., BERGER, A. and SHACKLETON, N. J., 1993. Simulation of the climate of the last 200 ka with the LLN 2D-model. In: R. PELTIER (Editor), *Ice in the Climate System.* NATO ASI Series I, Global Environmental Change, Vol. 12, Springer-Verlag, Berlin, pp. 321–341.

GATES, W. L., 1976a. Modeling the Ice Age climate. *Science*, 191: 1138–1144.

GATES, W. L., 1976b. The numerical simulation of Ice-Age climate with a global general circulation model. *J. Atmos. Sci.*, 33: 1844–1873.

GATES, W. L., MITCHELL, J. F. B., BOER, G. J., CUBASCH, U. and MELESHKO, V. P., 1992. Climate modelling, climate prediction and model validation. In: J. T. HOUGHTON, B. A. CALLENDER and S. K. VARNEY (Editors), *Climate Change 1992. The Supplement to the IPCC Scientific Assessment.* Cambridge University Press, Cambridge, UK, pp. 97–134.

GENTHON, C., BARNOLA, J. M., RAYNAUD, D., LORIUS, CL., JOUZEL, J., BARKOV, N. I., KOROTKEVITCH, Y. S. and KOTLYAKOV, V. M., 1987. Vostok ice core: climatic response to CO_2 and orbital forcing changes over the last climatic cycle. *Nature*, 239: 414–418.

GHIL, M. and LE TREUT, H., 1981. A climate model with cryodynamics and geodynamics. *J. Geophys. Res.*, 86: 5262–5270.

GOODESS, C. M., PALUTIKOF, J. P. and DAVIES, T. D. (Editors), 1992. *The Nature and Causes of Climate Change.* Studies in Climatology Series. Lewis, Belhaven Press, London.

GRIMM, E. C., JACOBSON, G. L., WATTS, W. A., HANSEN, B. C. and MAASCH, K. A., 1993. A 50,000-year record of climate oscillations from Florida and its temporal correlation with the Heinrich events. *Science*, 261: 198–200.

GRIP, GREENLAND ICE-CORE PROJECT MEMBERS, 1993. Climate instability during the last interglacial period recorded in the GRIP ice core. *Nature*, 364: 203–207.

GROISMAN, P. Y., KARL, T. R. and KNIGHT, R. W., 1994. Observed impact of snow cover on the heat balance and the rise of continental spring temperatures. *Science*, 263: 198–200.

GROOTES, P. M., STUIVER, M., WHITE, J. W. C., JOHNSEN, S. and JOUZEL, J., 1993. Comparison of oxygen isotope records from the GISP2 and GRIP Greenland ice cores. *Nature*, 366: 552–554.

GUILDERSON, TH. P., FAIRBANKS, R. G. and RUBENSTONE, J. L., 1994. Tropical temperature variations since 20,000 years ago: simulating interhemispheric climate change. *Science*, 263: 663–665.

GUIOT, J., PONS, A., DE BEAULIEU, J. L. and REILLE, M., 1989. A 140,000-year continental climate reconstruction from two European pollen records. *Nature*, 338: 309–313.

HADLEY CENTRE, 1992. *The Hadley Centre Transient Climate Change Experiment.* Hadley Centre for Climate Prediction and Research, Bracknell, UK.

HANSEN, J., LACIS, A., RIND, D., RUSSEL, G., STONE, P., FUNG, I., RUEDY, R. and LERNER, J., 1984. Climate sensitivity: analysis of feedback mechanisms. In: J. E. HANSEN and T. TAKAHASHI (Editors), *Climate Processes and Climate Sensitivity.* Maurice Ewing Ser. No 5. Am. Geophys. Union, Washington, DC, pp. 130–163.

HARVEY, L. D. D., 1989. An energy balance climate model study of radiative forcing and temperature response at 18 ka. *J. Geophys. Res.*, 94(D10): 12873–12884.

HARVEY, L. D. D. and SCHNEIDER, S. H., 1985. Transient climate response to external forcing on 10^0–10^4 year time scales, 2, Sensitivity experiments with a seasonal, hemispherically averaged, coupled atmosphere, land, and ocean energy balance model. *J. Geophys. Res.*, 90(D1): 2207–2222.

HASSELMANN, K., 1976. Stochastic climate models, I. Theory. *Tellus*, 28: 473–484.

HAYS, J. D., IMBRIE, J. and SHACKLETON, N. J., 1976. Variations in the earth's orbit: pacemaker of the ice ages. *Sciences*, 194: 1121–1132.

HECHT, A., 1985. Paleoclimatology. A retrospective of the past 20 years. In: A. HECHT (Editor), *Paleoclimate Analysis and Modelling.* Wiley, New York, pp. 1–26.

HERTERICH, K. and HASSELMANN, K., 1987. Extraction of mixed layer advection velocities, diffusion coefficients, feedback factors and atmospheric forcing parameters from the statistical analysis of north Pacific SST anomaly fields. *J. Phys. Oceanogr.*, 17: 2145–2156.

HOFFERT, M. I. and COVEY, C., 1992. Deriving global climate sensitivity from palaeoclimate reconstructions. *Nature*, 360: 573–576.

HOUGHTON, J. T., JENKINS, G. J. and. EPHRAUMS, J. J., (Editors), 1990. *Climatic Change: The IPCC Scientific Assessment.* Cambridge University Press, Cambridge, UK, 365 pp.

HOUGHTON, J. T., CALLENDER, B. A. and VARNEY, S. K., 1992. *Climate Change 1992. The Supplement to the IPCC Scientific Assessment.* Cambridge University Press, Cambridge, UK, 200 pp..

HOVINE, S., 1993. Variabilité à long terme de la circulation océanique mondiale : une étude à l'aide d'un modèle aux équations primitives à deux dimensions. Dissertation doctorale, Université Catholique de Louvain, Louvain-la-Neuve. Unpublished.

HUGHES, T. J., 1987. Ice dynamics and deglaciation models when ice sheets collapsed. In: W. F. RUDDIMAN and H. E. WRIGHT JR. (Editors), *North America and Adjacent Oceans During the Last Deglaciation. The Geology of North America.* v.K-3 Geol. Soc. Am., Boulder, CO, pp. 183–220.

HUGHES, T. J., DENTON, G. H., ANDERSON, B. G., SCHILLING, D. H., FASTHOOK, J. L. and LINGLE, C. S., 1981. The last great ice sheets: a global view. In: G. H. DENTON and T. J. HUGHES (Editors), *The Last Great Ice Sheets.* Wiley Interscience, New York, pp. 275–317.

HUYBRECHTS, PH., 1990. The Antarctic ice sheet during the last glacial-interglacial cycle: a three-dimensional experiment. *Ann. Glaciol.*, 14: 115–119.

HUYBRECHTS, P., LETREGUILLY, A. and REEH, N., 1991. The Greenland ice sheet and greenhouse warming. *Palaeogeogr., Palaeoclimatol., Palaeoecol.*, 89(4), *Global Planet. Change*, 3(4): 399–412.

HYDE, W. T. and PELTIER, W. R., 1985. Sensitivity experiments with a model of the Ice Age: the response to harmonic forcing. *J. Atmos. Sci.*, 42: 2170–2188.

HYDE, W. T. and PELTIER, W. R., 1987. Sensitivity experiments with a model of the Ice Age cycle: the response to Milankovitch forcing. *J. Atmos. Sci.*, 44: 1351–1374.

IMBRIE, J. and IMBRIE, K. P., 1979. *Ice Ages, Solving the Mystery.* Enslow, Short Hills, NJ.

IMBRIE, J. and IMBRIE, J. Z., 1980. Modeling the climatic response to orbital variations. *Science*, 207: 943–953.

IMBRIE, J., HAYS, J. D., MARTINSON, D. G., MCINTYRE, A., MIX, A. C., MORLEY, J. J., PISIAS, N. G., PRELL, W. L. and SHACKLETON, N. J., 1984. The orbital theory of Pleistocene climate: support from a revised chronology of the marine $\delta^{18}O$ record. In: A. BERGER, J. IMBRIE, J. HAYS, G. KUKLA and B. SALTZMAN (Editors), *Milankovitch and Climate.* D. Reidel, Dordrecht, pp. 269–305.

IMBRIE, J., MCINTYRE, A. and MIX, A., 1989. Oceanic response to orbital forcing in the Late Quaternary: observational and experimental strategies. In: A. BERGER, S. SCHNEIDER and J. CL. DUPLESSY (Editors), *Climate and Geo-Sciences.* Kluwer, Dordrecht, pp. 121–164.

IMBRIE, J., BOYLE, E. A., CLEMENS, S. C., DUFFY, A., HOWARD, W. R., KUKLA, G., KUTZBACH, J., MARTINSON, D. G., MCINTYRE, A., MIX, A. C., MOLFINO, B., MORLEY, J. J., PETERSON, L. C.,

MITCHELL, J. F. B., GRAHAME, N. S. and NEEDHAM, K. J., 1988. Climate simulations for 9000 years before present: seasonal variations and effect of the Laurentide ice-sheet. *J. Geophys. Res.*, 93: 8283–8303.

MOORE, T. C., PISIAS, N. G. and DUNN, D.A., 1982. Carbonate time series of the Quaternary and late Miocene sediments in the Pacific Ocean: a spectral comparison. *Mar. Geol.*, 46: 217–233.

MÜLLER, H., 1974. Polleanalytische Untersuchungen und Jahresschitenzählungen an der Eem-zeitlichen Kieselgure von Bispingen-Luhe. *Geol. Jahrb.*, A21: 149–169.

MURPHY, J. M., 1992. A prediction of the transient response of climate. *Climate Research Technical Note, 32.* September, 27 pp.

NICOLIS, C., 1984. Self-oscillations, external forcings, and climate predictability. In: A. BERGER, J. IMBRIE, J. HAYS, G. KUKLA and B. SALTZMAN (Editors), *Milankovitch and Climate.* D. Reidel, Dordrecht, pp. 637–652.

NORTH, G. R. and CROWLEY, T. J., 1985. Application of a seasonal climate model to Cenozoic glaciation. *J. Geol. Soc., London*, 142: 475–482.

NORTH, G. R., MENGEL, J. G. and SHORT, D. A., 1983. Simple energy balance model resolving the seasons and the continents: application to the astronomical theory of Ice Ages. *J. Geophys. Res.*, 88: 6576–6586.

OERLEMANS, J., 1982. Glacial cycles and ice-sheet modelling. *Clim. Change*, 4: 353–374.

OERLEMANS, J. and VERNEKAR, A. D., 1981. A model study of the relation between Northern Hemisphere glaciation and precipitation. *Contrib. Atmos. Phys.*, 54: 352.

OGLESBY, R., 1990. Sensitivity of glaciation to initial snow cover, CO_2, snow albedo, and oceanic roughness in the NCAR CCM. *Clim. Dynamics*, 4: 219–236.

OLSEN, P. E., 1986. A 40-million-year lake record of Early Mesozoic orbital climatic forcing. *Science*, 234: 842–848.

PELTIER, W. R., 1987. A relaxation oscillator model of the ice age cycle. In: C. NICOLIS and G. NICOLIS (Editors), *Irreversible Phenomena and Dynamical Systems Analysis in Geosciences. NATO ASI Series C, Mathematical and Physical Sciences*, Vol. 192. D. Reidel, Dordrecht, pp. 399–416.

PESTIAUX, P., DUPLESSY, J. C., VAN DER MERSCH, I. and BERGER, A., 1988. Paleoclimatic variability at frequencies ranging from 1 cycle per 10,000 years to 1 cycle per 1,000 years: evidence for nonlinear behaviour of the climate system. *Clim. Change*, 12: 9–37.

PETIT-MAIRE, N., 1989. Interglacial environments in presently hyperarid Sahara: paleoclimatic implications. In: M. LEINEN and M. SARNTHEIN (Editors), *Paleoclimatology and Paleometeorology: Modern and Past Patterns of Global Atmospheric Transport.* Kluwer, Dordrecht, pp. 637–661.

PETIT-MAIRE, N., RISER, J., FABRE, J. and COMMELIN, D., 1988. *Le Sahara à l'Holocène: Mali 1: 1,000,000.* CCGM, Institut Géographique National, Paris.

PISIAS, N. G. and SHACKLETON, N. J., 1984. Modelling the global climate response to orbital forcing and atmospheric carbon dioxide changes. *Nature*, 310: 757–759.

POLLARD, D., 1982. A simple ice sheet model yields realistic 100 ka glacial cycles. *Nature*, 296: 334–338.

POLLARD, D., 1983. A coupled climate-ice sheet model applied to the Quaternary Ice Ages. *J. Geophys. Res.*, 88: 7705–7718.

PRELL, W. L., 1982. Oxygen and carbon isotope stratigraphy for the Quaternary of Hole 502B: evidence for two modes of isotopic variability. In: W. L. PRELL, J. V. GARDNER et al. (Editors), *Initial Reports of the Deep Sea Drilling Project*, Vol. LXVIII, Washington DC, pp. 455–464.

PRELL, W. L. and KUTZBACH, J. E., 1987. Monsoon variability over the past 150,000 years. *J. Geophys. Res.*, 92: 8411–8425.

RAMANATHAN, V., 1981. The role of ocean-atmosphere interactions in the CO_2 climate problem. *J. Atmos. Sci.*, 38: 918–930.

RAYNAUD, D., JOUZEL, J., BARNOLA, J. M., CHAPPELLAZ, J., DELMAS, R. J. and LORIUS, CL., 1993. The ice record of greenhouse gases. *Science*, 259: 926–934.

RIND, D., 1987. Components of the Ice Age circulation. *J. Geophys. Res.*, 92(D4): 4241–4280.

RIND, D. and PETEET, D., 1985. Terrestrial conditions at the last lgacial maximum and CLIMAP sea-surface temperature estimates: are they consistent ? *Quat. Res.*, 24: 1–22.

RIND, D., PETEET, D., BROECKER, W., MACINTYRE A. and RUDDIMAN, W., 1986. The impact of cold North Atlantic sea temperatures on climate: implications for the Younger Dryas cooling (11–10K). *Clim. Dynamics*, 1: 3–33.

RIND, D., PETEET, D. and KUKLA, G., 1989. Can Milankovitch orbital variations initiate the growth of ice sheets in a general circulation model ? *J. Geophys. Res.*, 94, D10: 12851–12871.

ROOTH, C., 1982. Hydrology and ocean circulation. *Prog. Oceanogr.*, 11: 131–149.

ROYER, J. F., DEQUE, M. and PESTIAUX, P., 1984. A sensitivity experiment to astronomical forcing with a spectral GCM: simulation of the annual cycle at 125,000 BP and 115,000 BP. In: A. BERGER, J. IMBRIE, J. HAYS, G. KUKLA and B. SALTZMAN (Editors), *Milankovitch and Climate. Part 2*. Reidel, Dordrecht, pp. 733–763.

RUDDIMAN, W. F. and KUTZBACH, J. E., 1989. Forcing of Late Cenozoic Northern Hemisphere climate by plateau uplift in Southern Asia and the American West. *J. Geophys. Res.*, 94(D15): 18409–18428.

RUDDIMAN, W. F. and MCINTYRE, A., 1981. Oceanic mechanisms for amplification of the 23,000-year ice-volume cycle. *Science*, 212: 617–627.

RUDDIMAN, W. F., KIDD, R. B., THOMAS, E. et al., 1986. Site 609. In: W. F. RUDDIMAN, R. B. KIDD, E. THOMAS et al. (Editors), *Initial Reports of the Deep Sea Drilling Project*, Vol. XCIV. Washington, DC.

SALTZMAN, B., 1978. A survey of statistical dynamical models of the terrestrial climate. *Adv. Geophys.*, 20: 183–304.

SALTZMAN, B., 1983. Climatic systems analysis. *Adv. Geophys.*, 25: 173–233.

SALTZMAN, B., 1985. Paleoclimate modeling. In: A. D. HECHT (Editor), *Paleoclimate Analysis and Modeling*. Wiley-Interscience, New York, pp. 341–396.

SALTZMAN, B., 1990. Three basic problems of paleoclimatic modeling: a review. *Clim. Dynamics*, 5: 67–78.

SALTZMAN, B. and MAASCH, K. A., 1988. Orbital forcing and the Vostok ice core. *Nature*, 333: 123–124.

SALTZMAN, B. and SUTERA, A., 1987. The mid-Quaternary climatic transition as the free response of a three-variable dynamical model. *J. Atmos. Sci.*, 44: 236–241.

SALTZMAN, B., HANSEN, A. R. and MAASCH, K. A., 1984. The late quaternary glaciations as the response of a three-component feedback system to earth-orbital forcing. *J. Atmos. Sci.*, 41: 3380–3389.

SANCETTA, C., IMBRIE, J. and KIPP, N. G., 1973. Climatic record of the past 130,000 years in the North Atlantic deep-sea core V23–83: correlation with the terrestrial record. *Quat. Res.*, 3: 110–116.

SANTER, B., BERGER, A., EDDY, J. A., FLOHN, H., IMBRIE, J., LITT, T., SCHNEIDER, S. H., SCHWEINGRUBER, F. H. and STUIVER, M., 1993. How can paleodata be used in evaluating the forcing mechanisms responsible for past climate changes ? In: J. A. EDDY and H. OESCHGER (Editors), *Global Changes in the Perspective of the Past*. Dahlem Workshop, Report ES/2. Wiley, Chichester, UK.

SAUSEN, R., BARTHEL, K. and HASSELMANN, K., 1988. Coupled ocean-atmosphere models with flux corrections. *Clim. Dynamics*, 2: 154–163.

SCHLESINGER, M. (Editor), 1988. Proceedings of the NATO Advanced Study Institute on physically-based modelling and simulation of climate and climate change. Erice, Italy, 11–23 May 1986. *Series C: Mathematical and Physical Sciences*, 243. Kluwer, Dordrecht.

SCHLESINGER, M. E. and MITCHELL, J. F. B., 1987. Climate model simulations of the equilibrium climatic response to increased carbon dioxide. *Rev. Geophys.*, 25: 760–798.

SCHNEIDER, S. H. and THOMPSON, S. L., 1979. Ice ages and orbital variations: some simple theory and modelling. *Quat. Res.*, 12: 188–203.

SELLERS, W. D., 1984. The response of a climate model to orbital variations. In: A. BERGER, J. IMBRIE, J. HAYS, G. KUKLA and B. SALTZMAN (Editors), *Milankovitch and Climate, Part 2*. Reidel, Dordrecht, pp. 765–788.

SENIOR, C. A. and MITCHELL, J. F. B., 1993. CO_2 and climate: the impact of cloud parameterizations. *J. Climate*, 6: 393–418.

SHACKLETON, N. J. and IMBRIE, J., 1990. The $\delta^{18}O$ spectrum of oceanic deep water over a five-decade band. *Clim. Change*, 16: 217–230.

SHACKLETON, N. J. and OPDYKE, N. D., 1973. Oxygen isotope and paleomagnetic stratigraphy of equatorial Pacific core V28–238: oxygen isotope temperatures and ice volumes on a 10^5 year and 10^6 year scale. *Quat. Res.*, 3: 39–55.

SHACKLETON, N. J. and OPDYKE, N. D., 1976. Oxygen-isotope and paleomagnetic stratigraphy of Pacific core V28–239. Late Pliocene to latest Pleistocene. In: R. M. CLINE and J. D. HAYS (Editors),

Investigation of Late Quaternary Paleooceanography and Paleoclimatology. Geological Society of America Memoir 145, Boulder, CO, pp. 449–464.

SHACKLETON, N. J., IMBRIE, J. and PISIAS, N. G., 1988. The evolution of oceanic-isotope variability in the North Atlantic over the past three million years. *Philos. Trans. R. Soc. London Ser. B*, 318: 679–688.

SHACKLETON, N. J., BERGER, A. and PELTIER, W. R., 1990. An alternative astronomical calibration of the lower Pleistocene timescale based on ODP site 677. *Trans. R. Soc. Edinburgh, Earth Sciences*, 81: 251–261.

SHACKLETON, N. J., LE, J., MIX, A. and HALL, M. A., 1992., Carbon isotope records from Pacific surface waters and atmospheric carbon dioxide. *Quat. Sci. Rev.*, 11: 387–400.

SHORT, D. A., MENGEL, J. G., CROWLEY, T. J., HYDE, W. T. and NORTH, G. R., 1991. Filtering of Milankovitch cycles by Earth's geography. *Quat. Res.*, 35: 157–173.

STOCKER, T. W. and WRIGHT, D. G., 1991. Rapid transitions of the ocean's deep circulation induced by changes in surface water fluxes. *Nature*, 351: 729–732.

STOCKER, T. W., WRIGHT, D. G. and MYSAK, L. A., 1992. A zonally averaged coupled ocean-atmosphere model for paleoclimate studies. *J. Climate*, 5: 773–797.

STOUFFER, R. J., MANABE, S. and BRYAN, K., 1989. Interhemispheric asymmetry in climate response to a gradual increase of atmospheric CO_2. *Nature*, 342: 660–662.

STOUFFER, R. J., MANABE, S. and VINNIKOV, K. Y., 1994. Model assessment of the role of natural variability in recent global warming. *Nature*, 367: 634–636.

STREET, F. A. and GROVE, A. T., 1979. Global maps of lake-level fluctuations since 30,000 yr BP. *Quat. Res.*, 12: 83–118.

STREET-PERROTT, F. A., 1991. General circulation (GCM) modelling of paleoclimates: a critique. *The Holocene*, 1: 74–80.

STREET-PERROTT, F. A. and HARRISON, S. P., 1984. Temporal variations in lake levels since 30,000 Yr BP – an index of the global hydrological cycle. In: J. E. HANSEN and T. TAKASHI (Editors), *Climate Processes and Climate Sensitivity*. AGU, Geophysical Monogr. 29, pp. 118–129.

STREET-PERROTT, F. A., MITCHELL, J. F. B., MARCHAND, D. S. and BRUNNER, J. S., 1990. Milankovitch and albedo forcing of the tropical monsoons: a comparison of geological evidence and numerical simulations for 9000 y BP. *Trans. R. Soc. Edinburgh*, 81: 407–427.

SUAREZ, M. J. and HELD, I. M., 1979. The sensitivity of an energy balance climate model to variations in the orbital parameters. *J. Geophys. Res.*, 84: 4825–4836.

SUN, D. Z. and LINDZEN, R. S., 1993. Water vapor feedback and the ice age snowline record. *Ann. Geophys.*, 11: 204–215.

TAYLOR, K. C., HAMMER, C. U., ALLEY, R. B., CLAUSEN, H. B., DAHL-JENSEN, D., GOW, A. J., GUNDESTRUP, N. S., KIPSTUHL, J., MOORE, J. C. and WADDINGTON, E. D., 1993. Electrical conductivity measurements from the GISP2 and GRIP Greenland ice cores. *Nature*, 366: 549–552.

TRENBERTH, K. E. (Editor), 1992. *Climate System Modelling*. Cambridge University Press, Cambridge, UK, pp. 788.

TRICOT, C., 1992. Sur la contribution des gaz à effet de serre dans les changements climatiques à long terme. Thèse de doctorat, Faculté des Sciences, Université Catholique de Louvain, Louvain-la-Neuve. Unpublished.

TRICOT, C. and BERGER, A., 1988. Sensitivity of present-day climate to astronomical forcing. In: H. WANNER and U. SIEGENTHALER (Editors), *Long and Short-Term Variability of Climate*. Earth Sciences Series, Springer-Verlag, Berlin, pp. 132–152.

TUSHINGHAM, A. M. and PELTIER, W. R., 1991. A new global model of late Pleistocene deglaciation based upon geophysical predictions of post-glacial relative sea level change. *J. Geophys. Res.*, 96(B3): 4497–4523.

WARRICK, R. A., BARROW, E. M. and WIGLEY, T. M. L. (Editors), 1993. *Climate and Sea Level Change: Observations, Projections and Implications*. Cambridge University Press, Cambridge, UK, 424 pp.

WASHINGTON, W. M. and MEEHL, G. A.,1989. Climate sensitivity due to increased CO_2: experiments with a coupled atmosphere and ocean general circulation model. *Clim. Dynamics*, 4: 1–38.

WASHINGTON, W. M. and PARKINSON, C. L., 1986. *An Introduction to Three-dimensional Climate Modeling*. University Science Books, Mill Valley, CA and Oxford University Press, New York, 422 pp.

WATTS, R. G. and HAYDER, M. E., 1984. A two-dimensional, seasonal, energy balance climate model with continents and ice sheets: testing the Milankovitch theory. *Tellus*, 36A: 120–131.

WEBB, T., STREET-PERROTT, F. A. and KUTZBACH, J. E., 1987. Late-Quaternary paleoclimatic data and climate models. *Episodes*, 10: 4–6.

WIGLEY, T. M. L., 1976. Spectral analysis and the astronomical theory of climatic change. *Nature*, 264: 629–631.

WILLIAMS, J., BARRY, R. G. and WASHINGTON, W. M., 1974. Simulation of the atmospheric circulation using the NCAR global circulation model with Ice Age boundary conditions. *J. Appl. Meteorol.*, 13: 305–317.

Chapter 3

Warmer worlds: global change lessons from earth history

ERIC J. BARRON

Introduction

The climate over the last 600 million years of Earth history ranges from periods of extensive polar glaciation to episodes of extreme warmth, lacking any evidence of permanent ice. The most recent occurrences of these extremes have the best documented climates. The mid-Cretaceous, about 100 million years ago (the Cretaceous extends from approximately 144 million years ago to 65 million years ago), is the warmest climate which can be well-documented from both terrestrial and marine observations. BARRON (1983) estimates that the globally averaged surface temperature was 6–12°C higher than the present day and that the planet was warmer at every latitude. Some authors (e.g. FRAKES et al., 1992) suggest that the entire record of climate for the last 600 million years can be divided into warm and cold climate modes of which the Cretaceous is a primary example of a warm climate (Fig. 1). However, closer examination of past climates reveals major differences between warm climates. For example, the early Eocene (approximately 50 million years ago) appears to be characterised by very warm polar regions yet cooler tropics. The globally averaged surface temperature has been estimated at only 2°C higher than at present (BARRON, 1987) yet regions above the Arctic circle are characterised by abundant floras and temperature sensitive cold-blooded animals. The large differences between the Cretaceous and the early Eocene are suggestive of very different explanations of global warmth.

The climatic changes recorded over the last 600 million years may have been influenced by a wide variety of climatic forcing factors. These forcing factors can be divided into three major categories: (1) changes in solar input due to factors such as solar luminosity variations and modifications to the Earth's orbital elements; (2) changes in the opacity of the Earth's atmosphere due to variations in the concentrations of greenhouse gases or aerosol and particulate loading; and (3) changes to the surface of the Earth including different positions of the continents (area of land at the poles, barriers or gateways to ocean circulation), changes in total continental area due to sea-level variations and differences in continental topography due to mountain building events.

The Cretaceous and the early Eocene may be viewed as two "case studies" of global warmth. How might these two cases contribute to the understanding of future climates? There are two possibilities. First, extensive analysis of the geologic observations would yield a reconstruction of global temperatures which could be utilised as an analogue for future warm climates. Unfortunately, so many conditions are different, including the distribu-

whether both CO_2 and changes in oceanic heat transport are required to explain the early Eocene record.

Comparison with observations

The explanation of the early Eocene climate provides two potential opportunities for validation. First, the decreased intensity of the atmospheric circulation is supported by the dramatic decrease in size of aeolian material transported to the deep sea (JANACEK and REA, 1983; JANACEK, 1985; REA et al., 1985). Second, the transition from the Paleocene to the early Eocene is characterised by a major extinction of benthic foraminifera (MILLER et al., 1987; THOMAS, 1989). Benthic foraminifera distributions closely reflect the characteristics (temperature and salinity) of water masses, and water mass changes are a logical result of a major change in the site of deep water formation from a cold source to a subtropical warm saline source. Both these data are indicative of abrupt change, matching the proposed mechanism of a change in the thermohaline circulation.
The sea surface temperature record of the Eocene does not represent an independent validation of model efforts since the entire focus of the RIND and CHANDLER (1991) and BARRON et al. (1993b) experiments was to increase oceanic heat transport or change specified sea surface temperatures in order to explain the SST distribution. The only additional alternative for model validation is the continental record.
SLOAN and BARRON (1992) have synthesised the quantitative observations from the early Eocene for North America. Based on floral assemblages of both macrofossils and palynomorphs and on fossil leaf physiognomic characteristics, minimum temperatures, mean an-

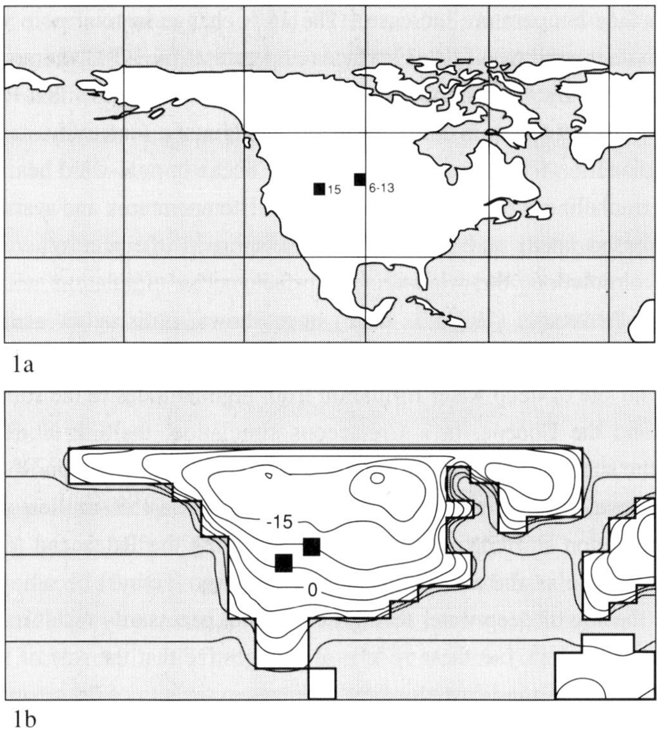

1a

1b

Fig. 10, 1a,b.

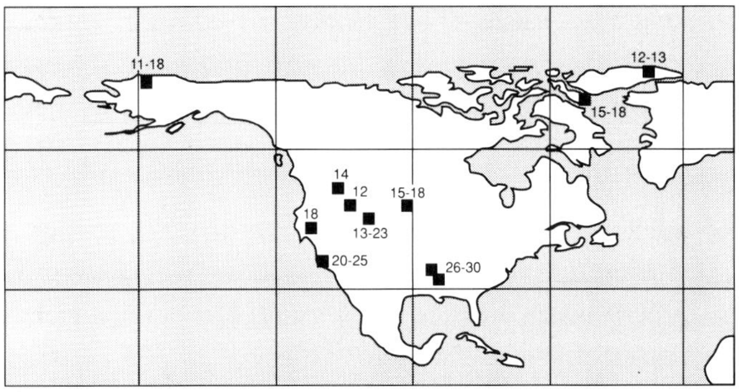

2a

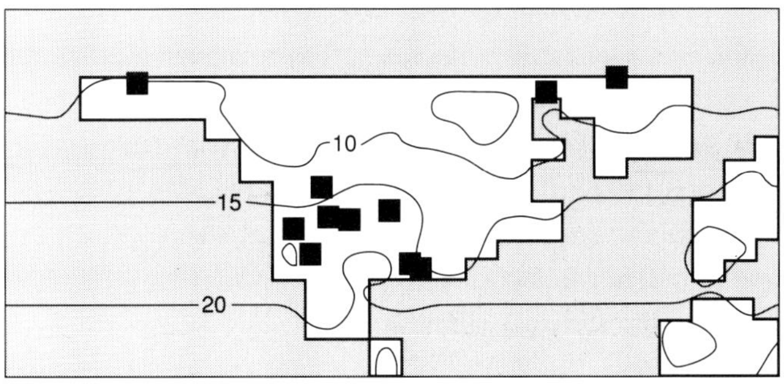

2b

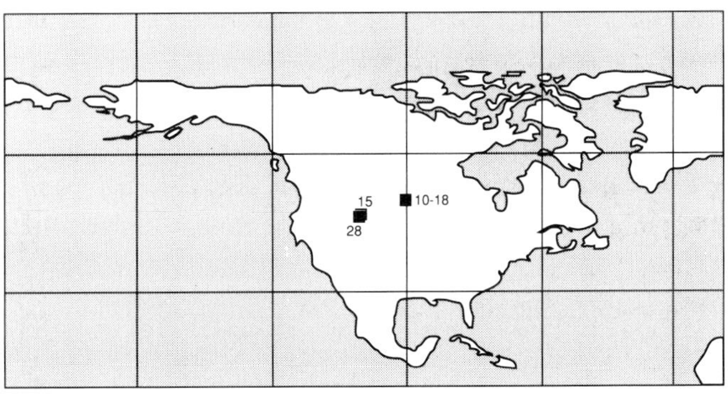

3a

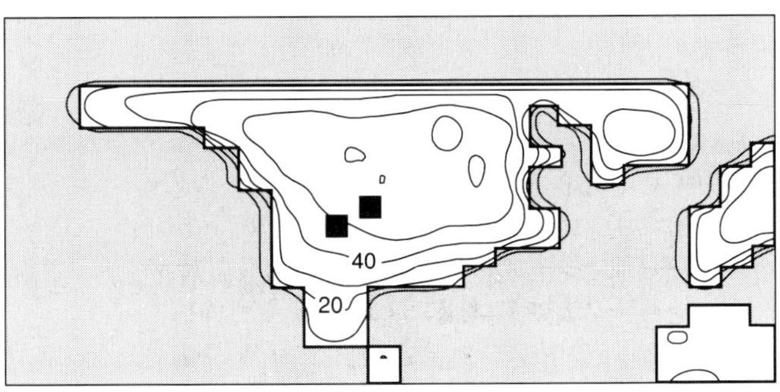

3b

Fig. 10. Comparisons of observations (a) with an Eocene simulation (b) with specified sea surface temperatures: (1) minimum surface temperatures in °C, (2) mean annual surface temperatures, and (3) annual surface temperature range for North America (SLOAN and BARRON, 1992).

nual temperatures and maximum ranges of temperatures have been reconstructed for the early Eocene. These data can be compared (Fig. 10) with Eocene GCM simulations with specified sea surface temperatures for both modern equator-to-pole gradients and for a gradient similar to SHACKLETON and BOERSMA (1981). In both cases the model-predicted January surface temperatures are too low to explain the observations. This may partly reflect the nature of the experiment performed by SLOAN and BARRON (1992) which was a perpetual January simulation, but the differences are as much as 20°C. The predicted mean annual temperatures are a reasonable match with the observations, but the high latitude sites are only explained in the low gradient case with warm polar oceans. The seasonal ranges of temperatures derived from the observations are less than half those of the model simulation. The low gradient case appears to be slightly closer to the observations.

In summary, the observations qualitatively support an interpretation that the early Eocene climate may have been characterised by a different role of the oceans in pole-ward heat transport. However, the data from continental interiors have not been adequately explained. Continental climates are not predicted to be warm enough to explain the observations and the amplitude of the seasonal cycle is predicted to be too large.

Implications for the prediction of future climates

The early Eocene case study provides two "lessons" for future climate prediction. First, the early Eocene introduces the possibility that changes in the role of the ocean, either through changes in geography or in response to increases in carbon dioxide, may yield substantially different equilibrium climates than currently simulated by GCMs with mixed layer oceans. The most likely mechanism for a change in the role of the ocean in pole-ward heat transport involves the intensity and location of deep water formation. The potential importance of the thermohaline circulation in climatic change is also evident from a variety of other studies. Modern observations from tracer studies in the North Atlantic (OSTLUND et al., 1976; TALLEY and MCCARTNEY, 1982; BROECKER et al., 1985; AAGAARD and CARMACK, 1987; DICKSON et al., 1988) also indicate that the production of North Atlantic Deep Water is sensitive to variations in North Atlantic moisture fluxes. Geochemical studies of glacial-interglacial cycles support the concept of significant changes in deep water production (e.g. BOYLE and KEIGWIN, 1987; DUPLESSY et al., 1988). Climate model studies (BRYAN, 1986; MANABE and STOUFFER, 1988) demonstrate that changes in the hydrologic balance can result in dramatic changes in the strength and character of the thermohaline circulation, suggestive of two modes of ocean circulation. The geologic record and these additional studies suggest that explicit incorporation of the role of the ocean in future climate model predictions is essential. A different role for the ocean could result in a substantially different future climate.

Second, the early Eocene study highlights substantial discrepancies between model predictions and the observations from continental interiors. There are many possible explanations of these differences. The model simulations for continental interiors may have improper sensitivities due to poor land-vegetation-atmosphere parameterisations (see also Chapter 11 by Brasseur et al.). Additional forcing factors may be involved (e.g. higher CO_2). The fossil data may be misinterpreted if floras, such as palm trees, had different environmental limits. The early Eocene case study suggests that each possibility warrants further examination.

Discussion of implications for future climate and conclusions

The geologic record preserves the integrated response of the Earth system to a large number of perturbations. Two case studies, the mid-Cretaceous and the early Eocene, have been utilised to illustrate global change lessons from the Earth's history. In each case study, the climate is characterised by extreme warmth in comparison with the present day. Although the older the record, the less certain are the interpretations, the observations are sufficiently dramatic to provide challenging problems. The cases of extreme warmth also represent large system responses. A large climatic "signal" with identifiable forcing factors provides an excellent opportunity for climate model study. With the exception of the last ice age, the mid-Cretaceous and the early Eocene are probably the two most studied intervals of geologic time involving the application of climate models. Furthermore, in each case, increased atmospheric carbon dioxide concentration has been suggested as a primary or partial explanation of global warmth.

The two case studies provide interesting insights into climatic change. If our understanding of the direct effect of major changes in paleogeography on the atmosphere is correct, then the explanation of Cretaceous warmth rests largely with the climatic response to increased levels of atmospheric carbon dioxide. The evidence for a mid-Cretaceous atmosphere with higher levels of carbon dioxide is strong. However, even a GCM with a sensitivity in the mid-range of IPCC estimates (see also Chapter 8 by McAvaney and Holland) cannot achieve the degree of Cretaceous warmth without resorting to CO_2 levels which exceed the estimates from geochemical models and observations. Sensitivity at the lower range of IPCC estimates (e.g. 1.5°C warming for a CO_2 doubling) would make explanations of warm time periods like the Cretaceous much more difficult.

Of course there is substantial room for error. All the estimates of CO_2 concentration for the mid-Cretaceous may be underestimates. Additional factors, such as an increased role of oceanic heat transport, may mask the true role of CO_2 during the Cretaceous unless the change in ocean circulation (e.g. changes in sites of deep water formation) are in response to CO_2 increases. The geologic observations of the environment may also be in error. However, the climate models cannot readily achieve even the most conservative estimates of Cretaceous warmth for plausible CO_2 concentrations. The evidence that time periods like the Cretaceous demand a climatic sensitivity which is in the upper half of the IPCC estimates is further supported by a survey of all paleoclimatic simulations. In no case to date has a climate model over-predicted geologic observations for warming or cooling. It becomes increasingly difficult to assume that in each of these cases the geologic data have been over-interpreted or that in each case additional forcing factors are required which, in total, yield a colder or warmer climate. The fact that in all previous paleoclimate experiments the observations have been under-simulated provides growing evidence that climate model sensitivity for doubled CO_2 is far from excessive.

The early Eocene provides a very different lesson. The vast majority of projections of future climate depend on atmospheric GCMs with simple mixed layer oceans. The nature of the deficiency has been clearly recognised and the incorporation of an explicit dynamic ocean is a major goal of climate research (as yet, these models do not produce very accurate simulations of the present day). The early Eocene is suggestive of how significant the lack of a realistic ocean may be for future climate predictions. One plausible explanation of early Eo-

that of a 10 km asteroid. In such an impact, >10^{16} g of excavated rock and vaporized asteroid, crust (and possibly ocean water) can be launched ballistically to ≥100 km and ~10–20% of these ejecta might be distributed globally (O'KEEFE and AHRENS, 1982a,b; MCLAREN and GOODFELLOW, 1990).

The large amount of fine ejecta that was apparently distributed worldwide by the K/T boundary impact could have created a dense dust cloud while in the atmosphere and ALVAREZ et al. (1980) suggested that the darkness and short-term surface cooling beneath such a cloud led to cessation of photosynthesis and frigid conditions (impact winter), resulting in the observed mass extinctions. Calculations suggest that the solar transmission reduction resulting from a dust injection of ≥10^{16} g (only 0.01 times the mass of a 10 km diameter object) uniformly over the surface of the Earth would reduce photosynthesis by a factor of 10^5 (GERSTL and ZARDECKI, 1982) and could lead to a collapse of marine and terrestrial food chains.

Initially, the Alvarez group estimated that the duration of the global dust cloud would have been several years, based on atmospheric opacity increases after historical volcanic eruptions. However, such relatively long-lived volcanic effects are the result of volcanogenic sulphuric acid aerosols in the stratosphere (RAMPINO and SELF, 1984) – volcanic dust settles out largely within 3–6 months – and thus the K/T boundary dust cloud would most likely have dissipated in a few months (POLLACK et al., 1982; TOON et al., 1982). Calculations (MILNE and MACKAY, 1982) and experiments (GRIFFIS and CHAPMAN, 1988) suggest that even a few months of such darkness would have been sufficient to produce the degree of extinction seen among pelagic phytoplankton at the K/T boundary, breaking the food chain. Analysis of carbon and nitrogen in the K/T boundary clay suggests that the amounts of incorporated marine and terrestrial biomass are approximately equivalent to the entire global standing biomass that would have been swept out of the ocean and buried by the ejecta and/or burned in global wildfires (GILMOUR et al., 1990).

Predicted climatic effects from impact produced dust clouds are always severe, but to some extent model dependent. TOON et al. (1982) used a 1-D model of aerosol physics and a one-dimensional radiative/convective climate model to calculate the evolution and climatic effects of such a dense dust cloud with an atmospheric lifetime of 3–6 months. Model results suggested that light levels would have remained too low for visibility for up to 6 months and too low for photosynthesis for up to 1 year after the impact. Calculations of surface cooling showed reductions in continental temperatures to below freezing for up to 2 years, whereas ocean-surface temperatures fell only a few degrees (Fig. 3). Increased reflectivity from snow cover in continental interiors might provide a positive feedback process that could prolong cooling once the dust settled out of the atmosphere (TOON et al., 1982).

Possible changes in atmospheric thermal structure were also investigated in a similar model study by POLLACK et al. (1982). In normal times, solar energy is absorbed by the ground and the infrared emission back to space comes largely from the upper troposphere. At times of heavy dust loading, however, both the solar energy deposition level and the infrared emission level of the atmosphere are controlled by the dust and both lie within the stratosphere. The surface no longer receives any solar radiation and has a net energy deficit. Under these conditions, drastic surface cooling is inevitable and POLLACK et al. (1982) predicted land-surface temperature decreases of ~30°C in the first 1–2 months after impact of a 10 km diameter body. Furthermore, these modelling studies did not include the possible ef-

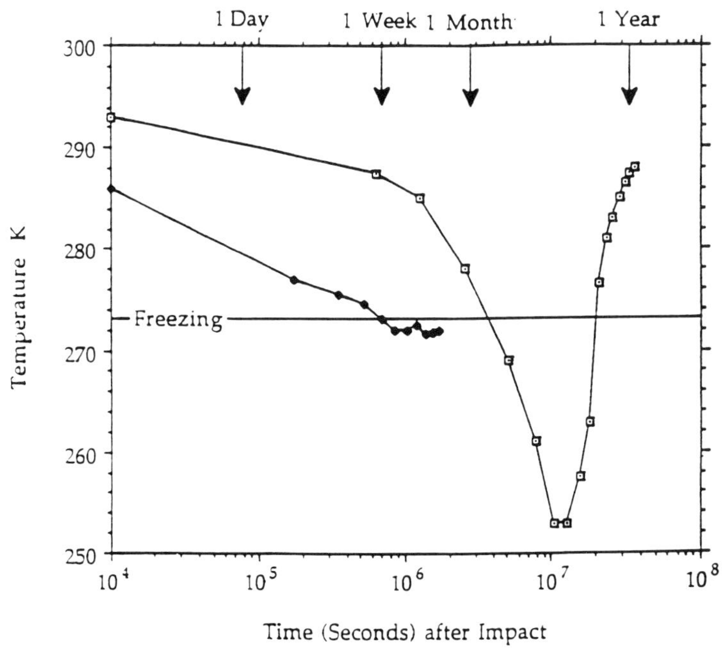

Fig. 3. Estimates of the reduction in global temperatures as a result of a global cloud of fine dust produced by a 10 km diameter asteroid collision, using a one-dimensional radiative-convective climate model (open squares) (after TOON et al., 1982), and a 3-D Global Climate Model (diamonds) (after COVEY et al., 1990).

fects of soot produced by global fires, which should exacerbate the situation (CROWLEY and NORTH, 1991).

More recently, COVEY et al. (1990) utilized a more sophisticated three-dimensional global climate model (GCM) to study the effects of an impact-induced global dust cloud on climate. For a dust loading of 5×10^{15} g after a 10 km diameter impact, their results were qualitatively similar to the earlier one-dimensional model results, although in the GCM land-surface cooling was somewhat mitigated by the heat capacity of the oceans and by atmospheric circulation patterns set up by the dust cloud that transported additional heat from the oceans to land areas. In this simulation, however, fixed sea-surface temperatures exaggerated the oceanic heat reservoir, providing an unrealistic situation. Land-surface cooling was less than that of POLLACK et al. (1982) by more than a factor of two, but the land-surface temperatures fell below freezing in less than a week (decreases of about 15°C), compared to nearly 2 months in the one-dimensional simulation (Fig. 3). GCM studies also suggest an almost complete collapse of the hydrologic cycle after the impact. The GCM model results indicated that impacts of smaller objects (~1 km diameter) could cause less severe, but still quite significant decreases in temperature (–5°C) over the continents within 20 days of impact. Such rapid and severe decreases in temperatures could have devastated global vegetation (TINUS and RODDY, 1990).

Water vapour in the atmosphere

In the case of an oceanic impact, water would probably comprise a significant fraction of the ejecta lofted to high altitudes. For a 10 km impactor hitting the ocean, it is estimated that the expansion of a ~500 km^3 steam bubble would drive a plume of mixed impactor vapour, debris and water vapour more than 100 kilometres above the surface (CROFT, 1982). Condensation of the cloud of water vapour and meteoroid vapour could produce a combination of dust and ice particles in the upper atmosphere (MELOSH, 1982). MCKAY and THOMAS (1982) considered the consequences of enhanced water vapour concentrations on the middle atmosphere (50–100 km) chemistry and heat budget. They estimated that the increased mixing ratio of hydrogen would have caused a decrease of ozone concentration above 60 km. The ozone reduction would lead to a lowering of the average height of the mesopause and lowering of average temperatures. These conditions were predicted to produce a long-lived (possibly 10^5–10^6 years) global layer of mesospheric ice clouds.

An increase in global albedo of up to several percent is predicted from such clouds, but the net climatic effect (cooling or warming) depends upon how much infrared radiation from below is also trapped by the clouds, which is a function of the mean size of the ice crystals in the clouds. KYTE et al. (1988) suggested that similar ice clouds in the upper stratosphere following a relatively small (0.5 km diameter) ocean impact in the Late Pliocene might have triggered Northern Hemisphere Pleistocene glaciation. The size of the dust and ice grains produced in the atmosphere is also critical to the mean lifetime of the dust and ice clouds – if the grains are too large, rapid fallout would occur and long-term darkness and climatic effects would be less likely. A measure of the initial grain-size distribution of material in the K/T boundary fallout layer would help in improving the understanding of the likely outcomes.

The water vapour should locally supersaturate the stratosphere and thus much of the vapour might rapidly recondense and precipitate out of the atmosphere (CROFT, 1982). If the coagulation time of the water vapour is similar to that of the dust as calculated by TOON et al. (1982), then most of the water would leave the atmosphere within weeks to months. However, during this time, an enhanced greenhouse effect and reduction of stratospheric ozone could result (O'KEEFE and AHRENS, 1982b). EMILIANI (1980) and EMILIANI et al. (1982) suggested that such a water vapour greenhouse effect would cause sudden global heating and that an average global rise of only a few degrees in surface temperatures would be intolerable to many species, including reptiles weighing more than 25 kg. On the other hand, some evidence suggests that dinosaurs may have been quite adaptable to changing environmental conditions and hence climatic change alone may not have been responsible for their demise (LEHMAN, 1987).

Formation of NO and acid rain effects

The shock waves from the K/T bolide and the transfer of energy from the fine ejecta would have heated the global atmosphere. The fraction of energy transferred to the atmosphere is estimated to be ~40% (O'KEEFE and AHRENS, 1982a,b). Calculations suggest that, for a 10^{24} J impact, this could result in an immediate heat pulse with a global average temperature increase of ≥15°C and ocean-surface temperature increases of ~5°C.

The high-temperature shock waves produced by the passage of the impactor through the atmosphere (TURCO et al., 1981; LEWIS et al., 1982) and interaction of the high velocity plume of ejecta with the atmosphere (O'KEEFE and AHRENS, 1982) could create large amounts of NO; PRINN and FEGLEY (1987) estimated as much as 3×10^{18} g. They predicted that the global atmosphere could be loaded with 100 ppmv of NO_2, about 1,000 times more than during the heaviest air-pollution episodes today . This amount of NO_2 is enough to poison plants and animals, but it could also produce destructive nitric acid rain with a pH of ~1. NO_x in the stratosphere would also rapidly remove the ozone layer.

The large amount of fixed nitrogen deposited on the earth's surface after an impact might be denitrified to produce N_2O and NO. Smog reactions, involving the oxidation of CH_4 produced by anaerobically decaying organic matter, might then lead to significant amounts of ozone in the troposphere (CRUTZEN, 1987). Production of N_2O (along with CO_2, HNO_3 and CH_4) could result in an enhanced greenhouse effect over the ~150 year lifetime of N_2O in the atmosphere (CRUTZEN, 1987) and the N_2O could also interact with the stratospheric ozone layer, depleting it over about a decade (TURCO et al., 1981).

More recently, ZAHNLE (1990) estimated a much smaller total NO production of ~10^{14} mol from the impact of a 10 km diameter asteroid travelling at 20 km s^{-1}. In his model, most of the global NO production came not from the initial shock wave or plume, but from the dispersed ejecta re-entering the earth's atmosphere. MELOSH et al. (1990) performed calculations on the heat emitted from these ejecta as they re-entered the atmosphere and found values that might reach 50–150 times the solar output for periods up to several hours. According to Zahnle's calculations, larger yields of NO would accompany only relatively rare grazing impacts.

MACDOUGAL'S (1988) report of a global Sr-isotope spike at the K/T boundary that might indicate enhanced continental weathering supports the idea of acid rain at the boundary. The bleached white limestone beds below the boundary in Italy could have had a similar acid-leached origin (LOWRIE et al., 1990).

Oblique impacts were specifically investigated by SCHULTZ and GAULT (1990), who found that a 10 km object with a velocity of 20 km s^{-1}, impacting at 10° to the horizontal, could cause ricocheting fragments of size 0.1–1 km at hypervelocity, producing a global swarm of Tunguska-scale events. Nitrate production from such a swarm could greatly exceed that of a single large impactor. Energy partitioned to the target apparently increases for impact angles between 45° and 15°, increasing the ability of the impact to vaporize potentially volatile targets (e.g. ocean water, CO_2-rich terrains, sulphates). Oblique impacts also provide the possibility for inserting substantial amount of material in orbit, possibly forming a temporary debris ring around the planet. The ring shadow on the Earth's surface could create additional and extended climate changes through seasonal effects on solar insolation.

Global wildfires

The immediate creation of a large, heated mass of low density air with peak temperatures of ~20,000 K at the impact site and the intense heat emitted globally by the re-entering impact ejecta (MELOSH et al., 1990), are calculated to ignite combustible material and to create widespread wildfires. Large amounts of dead vegetation, killed by the lack of sunlight and the abrupt cooling (TINUS and RODDY, 1990), would have provided abundant fuel for the wildfires. The soot produced by the fires could add to the opacity of the atmosphere, exacerbating the darkness and cooling following an impact. Large amounts of such soot were discovered at the K/T boundary (estimated at >10^{17} g worldwide), which supports the burning of a significant fraction of the terrestrial biomass (ANDERS et al., 1986; WOLBACH et al., 1985, 1988; GILMOUR et al., 1989). It has been estimated that such forest fires could produce more than 10^{19} g of CO_2, 10^{18} g of CO, 10^{17} g of CH_4, 10^{16} g of N_2O together with reactive hydrocarbons and oxides of nitrogen ($NO + NO_2$). This might have led to an intense photochemical smog and an increase in the greenhouse effect estimated to produce a heating of ~10°C (CRUTZEN, 1987).

Poisoning by noxious chemicals, such as polynuclear aromatic hydrocarbons (PAH) and CO from the wildfires (VENKATESAN and DAHL, 1989), trace metals released by the impactor, the fires, and/or by acid-enhanced weathering (ERICKSON and DICKSON, 1987; LEARY and RAMPINO, 1990), or possibly cyanide released in a cometary impact (HSU, 1980; HSU et al., 1982) could have contributed to the extinctions on land and in the surface oceans (DAVENPORT et al., 1990).

Impact into carbonates or evaporites

Several recent studies suggest that the target material of the impact might be important in terms of climatic impact. For example, impact of a 10 km diameter asteroid into a carbonate-rich terrain (like that at Chicxulub) might lead to a release of CO_2 that could increase atmospheric carbon dioxide levels by factors of 2–10. This increase in CO_2 has been estimated to cause a possible rise in global temperatures of 2–10°C for 10^4–10^5 years after the impact (O'KEEFE and AHRENS, 1989).

SIGURDSSON et al. (1993) noted that the proposed K/T impact site at Chicxulub in the Yucatan was underlain by thick Cretaceous evaporites including $CaSO_4$ and that some K/T impact glasses were rich in Ca and contained up to 1% SO_3. They predicted that the total sulphur degassing from the evaporites at the Chicxulub impact site could have led to up to 10^{19} g of sulphate aerosols, enough to cause significant cooling after the dissipation of the short-lived dust/soot cloud and to create acid fallout equal to that proposed for the nitric acid produced by bolide–atmosphere interactions.

Other possible effects

Comet showers on the Earth might be triggered by the passage of the Solar System through interstellar clouds, so that effects of the cloud encounter might coincide with comet impacts (RAMPINO and STOTHERS, 1984a, 1986). YABUSHITA and ALLEN (1989) suggested that hydrogen and dust accreted from interstellar clouds would form dust grains coated with ice in the upper atmosphere leading to global cooling and proposed that this may have happened at the end of the Cretaceous. Dust from large comets can also be swept up by the Earth both before and after collisions and ZAHNLE and GRINSPOON (1990) suggested this for the origin of extraterrestrial amino acids reported from the K/T boundary.

It has also been proposed that large impacts might cause changes in the earth's rotation and tilt that could have global climatic effects (DAOHAN et al., 1986), although major changes in orbital parameters from impact of a 10 km object seem unlikely.

Climatic changes from perturbations of biogeochemical cycles in the aftermath of impact-induced extinctions

The Cretaceous/Tertiary boundary is marked by a major perturbation of the global carbon cycle, marked by severe reduction in pelagic carbonate deposition, decrease in biomass and oceanic productivity and changes in organic carbon deposition (HSU and MCKENZIE, 1990). The carbon cycle involves the transfer of carbon between the solid Earth and the ocean/atmosphere system. Carbon dioxide releases associated with metamorphic decarbonation of sediments at subduction zones, mantle outgassing at mid-ocean ridges and the chemical weathering of carbonate rocks, represent the primary sources of carbon dioxide to the oceans and atmosphere. The primary source of alkalinity (excess positive charge) to the oceans is the riverine cation flux (primarily Ca^{2+} and Mg^{2+}) derived from the chemical weathering of carbonate and silicate rocks. The two major sinks for alkalinity and carbon in the oceans and atmosphere are the deposition of shallow-water carbonate sediments and the accumulation of deep-water carbonates derived from the calcareous tests of pelagic plankton, primarily since their global dissemination in the Jurassic Period (~150 Ma) (BERNER et al., 1983).

A major disruption of the pelagic alkalinity and carbon sinks may have occurred in the aftermath of the mass extinction and reduction in populations of calcareous plankton (~97% of planktonic foraminifera and ~88% of calcareous nannoplankton species disappeared) at, and/or very close to, the designated K/T boundary (THIERSTEIN, 1982a,b; KELLER, 1988a,b; BARRERA and KELLER, 1990; SMIT, 1990) and the apparent reduction of ocean primary productivity in the earliest Paleocene, the so-called "Strangelove Ocean" interval (HSU et al., 1982a,b; HSU and MCKENZIE, 1985; ZACHOS and ARTHUR, 1986).

KASTING et al. (1986) performed early model calculations with a 2-box ocean model of the carbon-cycle perturbations that might accompany a large planetesimal impact. They investigated Hsu's suggestion that an impact event would cause ocean surface waters to become undersaturated with respect to calcium carbonate, leading to the extinction of calcareous organisms and also to the delay in recovery of calcareous plankton during the Strangelove Ocean period of up to 500,000 years.

Global darkening could have killed off most of the existing phytoplankton within several weeks to months, while deposition of atmospheric NO_x created in the impact would have lowered the pH of ocean surface waters and released CO_2 into the atmosphere (HSU et al., 1982a,b). However, model results indicated that, because of mixing, ocean surface waters would only remain acidified for ~20 years or less. KASTING et al. (1986) also looked at the direct input of atmospheric CO_2 from a comet or from oxidation of terrestrial and marine organic matter. The net results might be a greenhouse effect raising temperatures by several degrees and perhaps a surface ocean uninhabitable by calcareous organisms for a couple of decades, but the model runs suggested that the surface waters would not remain corrosive for long and thus could not have led to the Strangelove Ocean conditions of decreased carbonate productivity.

However, a longer lived event is predicted if mixing of the deep ocean was involved. For example, if all life in the oceans were to die off suddenly and if sufficient time (>1,000 years) were allowed for the deep waters to mix upward and equilibrate with the atmosphere, then the model predicted that the CO_2 partial pressure of the atmosphere might increase ~2–3 times. Ocean mixing normally takes ~1,000 years (perhaps longer in the latest Cretaceous), but an oceanic impact(s) might temporarily speed up mixing. Model results indicated that these events could increase atmospheric temperatures by perhaps 3–4°, but still would not lead to a surface ocean undersaturated in calcium carbonate for a significant period of time.

HSU and MCKENZIE (1990) thought that blooms of opportunistic species of nannoplankton might have caused repeated rapid fluctuations in CO_2 exchange between the ocean and atmosphere, creating unstable climatic conditions during the early Tertiary Strangelove Ocean period, although the blooms may have been partly in response to unstable climates, suggesting possible feedbacks between biota and climate.

Ocean alkalinity crisis at the K/T boundary?

At the time of the K/T mass extinctions, the rate of pelagic carbonate accumulation apparently fell by about a factor of three (ZACHOS and ARTHUR, 1986). The latest Cretaceous deep-water carbonate accumulation rate was about 20% of the total carbonate accumulation rate in the oceans (OPDYKE and WILKINSON, 1988). This contrasts markedly with modern carbonate sedimentation, in which more than 60% occurs in deep-water environments. The estimated reduction in calcium carbonate accumulation in the earliest Tertiary means that ~6×10^{12} equivalents of alkalinity and 3×10^{12} mol of carbon could have accumulated in the oceans and atmosphere each year.

CALDEIRA et al. (1990) utilized a carbonate-silicate cycle model coupled to a three-box ocean/atmosphere model to investigate the effects of the reduction in pelagic productivity at the K/T boundary. Their model results predicted that curtailed pelagic carbonate deposition could have caused ocean carbon and alkalinity concentrations to increase drastically, leading to an "alkalinity crisis" in the oceans and causing major fluctuations in atmospheric carbon-dioxide levels and climate.

In order to investigate possible processes that might mitigate such perturbations, CALDEIRA and RAMPINO (1993) developed a five-box biogeochemical model of the oceans and atmosphere, incorporating a new parameterization for carbonate deposition in shallow waters, where the flux of calcium carbonate from the mixed layer to shallow-water sediments depends on the carbonate-ion concentration and temperature of the mixed layer.

The model was run in a time-dependent mode, in which all pelagic carbonate and organic carbon production was instantaneously set to zero. Because the carbonate flux from the ocean-surface layer to the deep ocean was eliminated and because this flux carries more alkalinity than carbon, the short-term (less than ~300 years) effect of this productivity crash was to cause a slight reduction in deep-ocean carbonate-ion concentration (Fig. 4a), which caused a reduction in the area of carbonate accumulation (Fig. 4b). This is consistent with evidence of increased dissolution of pelagic carbonate sediments just after the K/T boundary event (ZACHOS and ARTHUR, 1986; STOTT and KENNETT, 1990). In shallow continental shelf sections, carbonate dissolution at the K/T boundary seems to have ranged from negligible (KELLER, 1989a,b) to considerable, most likely depending on local conditions.

In the model, as the pelagic biological carbon pump is turned off, CO_2 begins to outgas from the surface oceans, increasing atmospheric CO_2 levels (Fig. 4c). The elevated atmospheric CO_2 causes an increase in the rate of chemical weathering of silicate and carbonate rocks. The elimination of the biological carbon pump in the model also increased total dissolved CO_2 in the surface oceans causing carbonate-ion concentration to drop on the 100–1,000 year time-scale. According to OPDYKE and WILKINSON (1990), this would act to reduce shallow-water carbonate accumulation and such conditions might also favour dolomitization in shallow-marine settings. Lowermost Tertiary sediments in the shallow-shelf

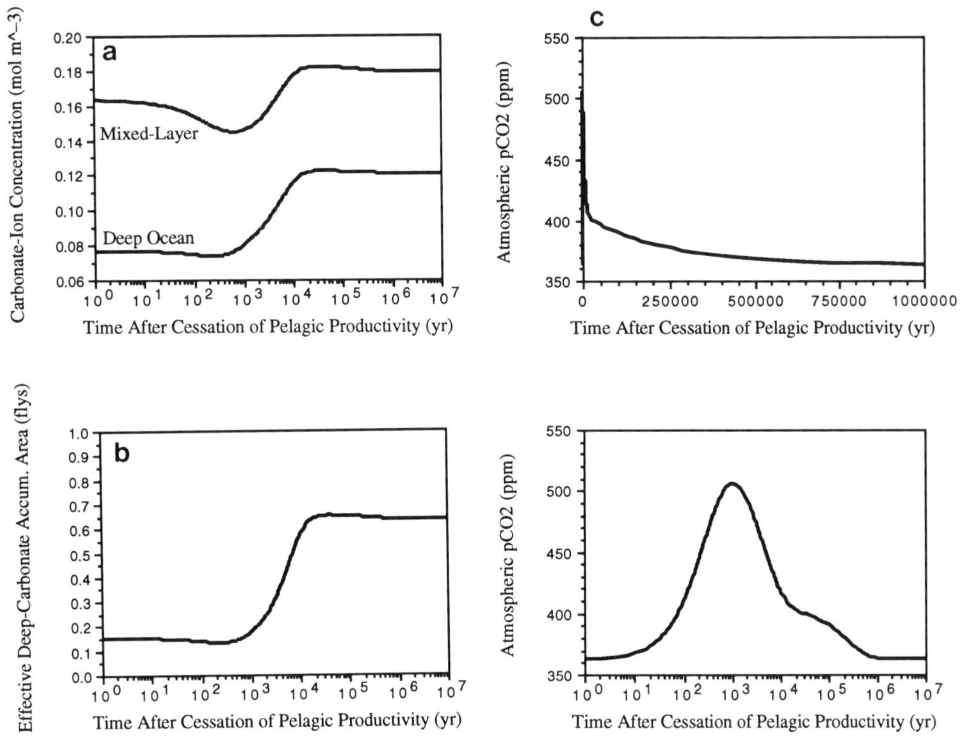

Fig. 4. (a) Model results for carbonate-ion concentration as a function of time after the K/T extinction of calcareous plankton. Note that the horizontal time axis is logarithmic. (b) Model results for effective fractional area of the deep-sea floor that was accumulating carbonate. This variable is computed based on deep-ocean carbonate-ion concentration as a function of time, ocean hypsometry, the saturation constant for calcite and aragonite as a function of pressure, and the estimated calcite to aragonite productivity (see text). The horizontal time axis is logarithmic. (c) Model results for atmospheric pCO_2 as a function of time, shown on linear and logarithmic scales (after CALDEIRA and RAMPINO, 1993).

Brazos River, Texas boundary section contain dolomite, which is absent from the uppermost Cretaceous deposits (BARRERA and KELLER, 1990).

After about 1,000 years of the model run, the ocean has largely completed its outgassing and the influx of alkalinity from rivers begins to make the oceans more alkaline, drawing atmospheric CO_2 back into the oceans. From ~1,000–10,000 years, the influx of carbon and alkalinity from rivers that is not balanced by pelagic carbonate accumulation begins to overwhelm the effects of reduced carbon pumping to the deep sea and both deep-ocean and surface-ocean carbonate-ion concentrations begin to rise (Fig. 4a), causing a near-exponential decrease in atmospheric pCO_2. From that time up until about 100,000 years, a less steep exponential drop in atmospheric pCO_2 level occurs. This may be related to the fact that shallow-water carbonate sedimentation might for a time exceed the influx of carbon and alkalinity from rivers. In the revised model, the carbonate-ion concentrations of the surface and deep ocean stabilized within about 20,000 years, as a result of the increased shallow-water carbonate accumulation. Between about 10^5 and 10^6 years, total carbonate sedimentation is again in balance with chemical weathering and the atmospheric pCO_2 level slowly decays to an equilibrium value.

Within about 1,000 years after the cessation of pelagic productivity, the carbon-isotope value of the surface-ocean layer in the model became as great as that of the deep ocean. From about 10^3 to 10^6 years after the productivity turn-off, both the deep-ocean and surface carbon-isotopic signals moved towards the model's pre-cessation mixed-layer value. KUMP (1991) came to similar conclusions regarding carbon-isotopic behaviour using a somewhat different model. Within a million years, the system approaches a new steady state and the isotopic composition of pelagic carbonate in the model (neglecting the effects of organic carbon burial) once again equals the pre-cessation value. The deep-ocean $\delta^{13}C$ values in the model actually became greater than those of the surface ocean approximately 1,000 years after the cessation of productivity. This effect lasted for roughly a million years. This trend is seen during the post-K/T boundary "Strangelove Ocean" interval in a number of ocean drilling sites (e.g. ZACHOS and ARTHUR, 1986).

Plankton extinction, DMS reduction and possible climatic warming

Another possible climatic effect resulting from the mass extinction of marine organisms was investigated by RAMPINO and VOLK (1988). Marine phytoplankton release the gas dimethyl sulphide (DMS), which oxidizes to form sulphuric acid aerosol that may be the precursor for most cloud condensation nuclei (CCN) over the oceans (see Chapter 9 by WANG). If the total liquid water content in the clouds is held constant, changes in the number density of CCN would affect the size of droplets in the cloud, which affects the cloud reflectivity or albedo. Modelling has quantified the effects of either increasing the number density of cloud droplets, which increases the cloud albedo, or decreasing the number density with its associated decrease in the cloud albedo. The extinction of almost all marine calcareous phytoplankton at the K/T boundary would have drastically reduced output of DMS and thus cloud condensation nuclei.

RAMPINO and VOLK (1988) used model results that showed the effect upon stratiform top-of-cloud albedo due to variations in the CCN number density, along with Global Climate Model (GCM) results to convert the equivalent changes in incoming solar radiation into changes in global average surface temperature. For a DMS reduction of ~80% from the reference value, global temperature was predicted to rise by more than 6°C and by nearly 10°C when the DMS is reduced by 90%. Even a 50% reduction in DMS should produce a significant global warming of 3–4°C. The thermal mass and mixing structure of the ocean would delay the full surface warming by several thousand years. These results indicate that a drastic reduction in DMS production, if maintained, could lead to a substantially warmer Earth essentially simultaneous with the phytoplankton extinction event. This issue is also discussed in Chapter 14 by PENG.

As noted above, several authors have suggested that a greenhouse warming from increased atmospheric CO_2, N_2O, CH_4 and/or H_2O could have taken place at the K/T boundary. A considerable decrease in marine cloud albedo could have provided an independent mechanism for climate warming. This global heating could have contributed to extinctions subsequent to the impact and initial plankton destruction. A severe climatic warming might also inhibit the full recovery of the marine biosphere and may have contributed to the maintenance of low productivity Strangelove Ocean conditions for 10^5 years or longer.

Evidence of environmental perturbations at the K/T boundary

In addition to the extinctions themselves, several lines of evidence point to a trauma in the global biosphere at the K/T boundary. Carbon isotope analyses provide a proxy for fluctuations of marine productivity and burial rates of organic carbon. Sedimentological/stratigraphic studies of pelagic and continental shelf sections contain data on accumulation rates of carbonate, organic carbon, amount and composition of detrital components and evidence for varying degrees of dissolution, diagenesis and precipitation of non-biogenic carbonates at the boundary.

Proxy paleotemperature analyses from oxygen-isotopic variations in marine carbonates and paleo-floral/palynological studies of terrestrial boundary sections provide a record, albeit problematical, of Late Cretaceous to Early Tertiary climate changes.

Carbon isotope and carbonate changes

Many studies report a significant negative anomaly in $\delta^{13}C$ of up to 3 parts per thousand (per mil) in the carbonate fine fraction (< 63 μm, largely nannoplankton debris) and in individual planktonic foraminifera, in K/T boundary sediments (Fig. 5). This anomaly appears to be a worldwide phenomenon and it indicates a surface ocean of very low primary productivity, the so-called Strangelove Ocean (HSU and MCKENZIE, 1990). In the South Atlantic, for example, the carbon isotope anomaly begins just above the K/T boundary (as marked by the iridium anomaly in the boundary clay) and reaches the minimum value ~50,000 years later; $\delta^{13}C$ values did not return to their pre-boundary values for at least 300,000 years and

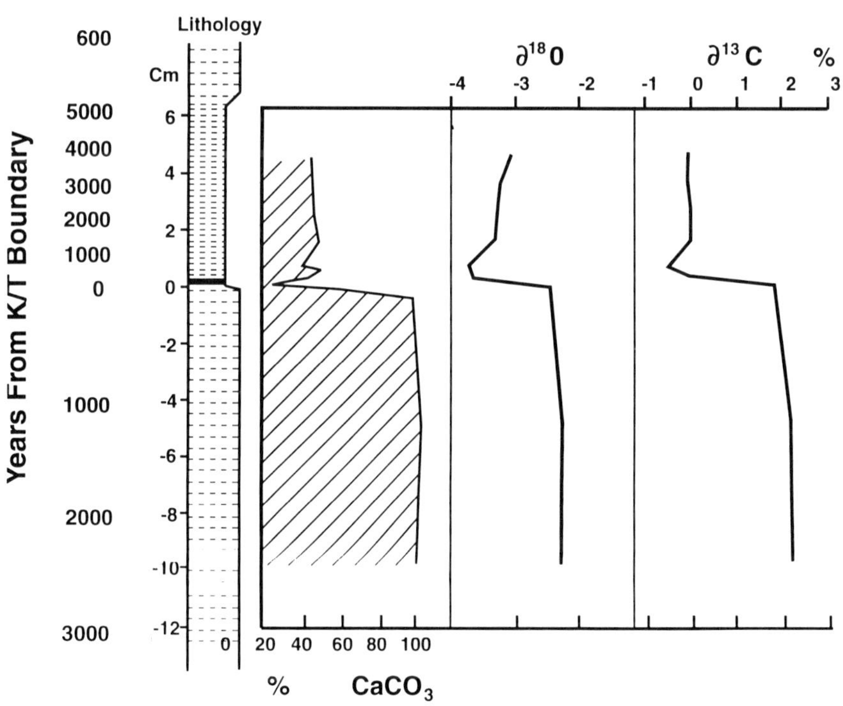

Fig. 5. Detailed K/T boundary section from Agost, Spain showing negative oxygen- and carbon-isotope shifts at the K/T boundary (as marked by an iridium anomaly), and severe reduction in pelagic biogenic calcium carbonate deposition (after SMIT, 1990).

Eocene-Oligocene

The Eocene-Oligocene boundary has been dated at 36.4 years in the most widely used recent geologic time-scale, although younger ages (33–34 years) have been suggested (MONTANARI, 1990; SWISHER and PROTHERO, 1990). Biotic changes seem to have begun in the middle Eocene and marine species apparently suffered a ~35% extinction from Late Eocene through Early Oligocene time (RAUP, 1992). This interval of transition from the Late Eocene to the Early Oligocene is marked by a number of indicators of extraterrestrial impacts, including a double iridium anomaly (GANAPATHY, 1982a,b; MONTANARI, 1990), shocked quartz, the North American tektite strewn field (GLASS, 1982) and several microtektite layers in the ocean (GLASS, 1982; D'HONDT et al., 1987; MILLER et al., 1987). The large (~100 km diameter) Popigai impact structure in Siberia is radiometrically dated at 39 ± 3 Ma, within the Late Eocene-Early Oligocene crisis interval (GRIEVE, 1991).

In pelagic environments, the Late Eocene/Early Oligocene biotic changes in plankton seem to have been stepwise over 1–2 million years (KELLER, 1986) and some steps may correlate with the microtektite and Ir layers. For example, each of the microtektite horizons identified by KELLER (1986) seems to have had some effect on microplankton communities, producing stepwise events of accelerated faunal turnover characterized by generally less than 15% species extinction. The relative species abundance changes at each stepwise extinction event, however, indicate a turnover involving >60% of the population. The final plankton extinction was selected as the formal epoch boundary (CORLISS et al., 1984).

The Eocene-Oligocene transition is marked by an abrupt global increase in $\delta^{18}O$ of 1.5–2 per mil in planktonic foraminifera that represents one of the major steps in Cenozoic

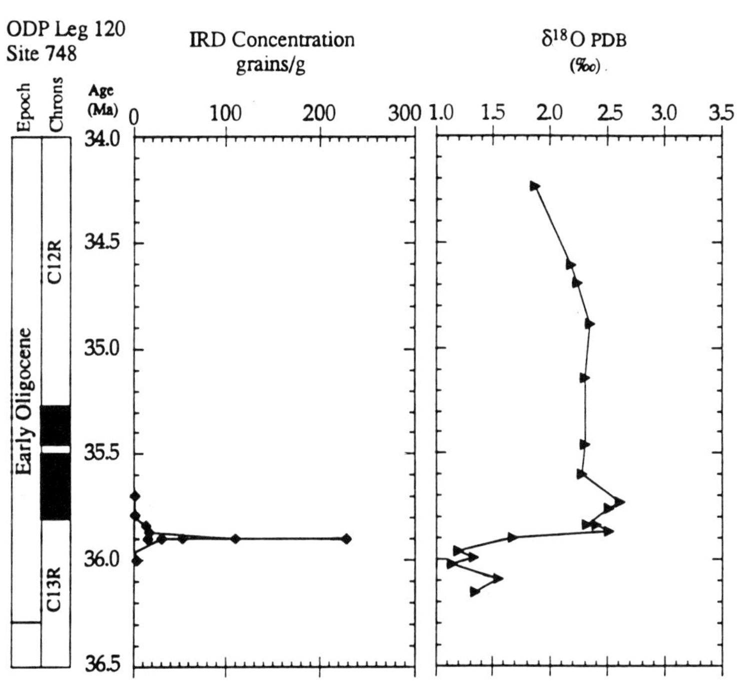

Fig. 7. $\delta^{18}O$ values from benthic foraminfera and ice-rafted debris (IRD) distribution for the 34–36 Ma interval at Ocean Drilling Project Site 748 on the Kerguelen Plateau in the Southern Ocean. Rapid glaciation and climate cooling are indicated by thc spike of ice-rafted sediment near the Eocene/Oligocene boundary. Chrons are periods of normal (black) and reversed (white) global magnetism (after ZACHOS et al., 1992).

cooling (THUNELL, 1981, MILLER et al., 1987). An abrupt increase in $\delta^{18}O$ in benthic foraminifera of >1 per mil, indicating either cooling of deep waters, ice-volume increases, or some combination of the two, also marks the E/O transition interval (ZACHOS et al., 1992). The first indications of probable Antarctic glacierization at sea level are ice-rafted sediments in Middle Eocene (~45.5 Ma) deposits on the Maud Rise and Kerguelen Plateau, but a distinct strengthening of glacial conditions and the apparent onset of continental East Antarctic glaciation is recorded by abundant ice-rafted debris on the Kerguelen Plateau in the earliest Oligocene (~36 and 35.8 Ma). The occurrence of *Nothofagus* (southern beech) and relatively warm sea-surface temperatures around Antarctica suggest temperate conditions close to the ice sheet (EHRMANN and MACKENSEN, 1992; MACKENSEN and EHRMANN, 1992).
ZACHOS et al. (1992) found that the same transition may be a very rapid climatic change, marked by a spike of ice-rafted sediment in the southern Indian Ocean (ODP Site 748 on the Kerguelen Plateau) and by rapid shifts in the carbon- and oxygen-isotope records. Although the oxygen-isotopic shift is permanent, the pulse of ice rafting apparently lasted less than 100,000 years (Fig. 7), perhaps as little as 10,000 years! They interpreted this as indicating the brief appearance of a large Antarctic ice sheet during the transition from a non-glaciated to a glaciated world.

Late Triassic

BICE et al. (1992) discovered two layers of shocked quartz at the Triassic-Jurassic (Tr/Jr) boundary (~205 Ma, but perhaps as young as 201 Ma), indicating multiple impacts, and the best estimated age of the ~100 km wide Manicouagan crater in Quebec is indistinguishable from the estimated age of the boundary (OLSEN et al., 1990). Trace metal anomalies, including Ir (~200–400 ppt), are reported to occur in two European sections (Kendelbach, Austria and Audries Bay, England) (MCLAREN and GOODFELLOW, 1990), associated with a negative $\delta^{13}C$ anomaly.
The vertebrate faunal and floral transition are apparently quite marked in the Newark deposits of eastern North America, where the pollen and spore transition, showing a fern spike followed by gradual replacement of flora, took place in <20,000 years; the vertebrate extinction may have occurred over a period ≤700,000 years, but is less well constrained (OLSEN et al., 1990). WEEMS (1992) argues that the vertebrate extinctions were gradual or perhaps stepwise (see also BENTON, 1985, 1986), but his arguments are based largely on tabulations of family diversity of vertebrates, with limited time resolution of possible abrupt killing events in the record (see MCLAREN and GOODFELLOW, 1990). Marine groups also show a rapid extinction and the few calcareous nannoplankton then in existence exhibit an almost complete turnover at the boundary (HALLAM, 1990).
The latest Triassic climatic record is, however, still rather poorly known (e.g. SIMMS, 1990; WEEMS, 1992; DICKINS, 1993). Negative shifts in oxygen- and carbon-isotope ratios have been noted in some European boundary sites (MCLAREN and GOODFELLOW, 1990). The post-Triassic disappearance of Alpine reefs was attributed to cooling, but isotopic measurements across the Tr/Jr boundary that suggest a temperature decrease may be problematic (HALLAM, 1990). A major increase in rainfall is indicated in some continental areas in the mid-Carnian, succeeded in the Norian by more arid climates (climatic cooling?) (SIMMS and RUFFELL, 1990; WEEMS, 1992), and JOHNSON and SIMMS (1989) suggested a causal con-

nection with the Late Triassic extinctions. These changes could have occurred in the wake of large-body impacts, but a causal connection between the impact event(s) and Late Triassic climatic shifts is not yet convincing.

Permian-Triassic

The Permian-Triassic (P/Tr) boundary (~250 Ma) is marked by the disappearance of ~96% of marine species (RAUP, 1979), 99% of reptile genera and a major floral extinction, including the abrupt obliteration of the diverse *Glossopteris* flora in the Southern Hemisphere (TIWARI and VIJAYA, 1992). Sudden disappearance of Late Permian palynomorphs, a global spike in fungal spores and vast numbers of acritarchs at the boundary horizon around the world (e.g. Israel, Greenland, Australia, Pakistan) indicate a sudden global catastrophe, followed by a disaster ecology (Fig. 8) (e.g. BALME, 1979; PIASECKI, 1984; VISSCHER and BRUGMAN, 1988; ESHET, 1990). Planktonic (including foraminifera and radiolaria) and nektonic groups and groups with planktonic growth stages, also show a sudden crash at the boundary (VALENTINE, 1986; LI et al., 1991). Shallow water faunas in the earliest Triassic are depauperate, whereas algal mats (another post-disaster form) suddenly become widespread (see various papers in CASSINIS, 1988).

The end-Permian extinctions are commonly regarded as gradual in nature and or composed of a series of extinction episodes from mid-Permian to early Triassic (MAXWELL, 1989, 1992). However, MCLAREN and GOODFELLOW (1990) summarized the ways in which problems related to the official definition of the P/Tr boundary (e.g. first appearances versus last appearances of different groups of organisms) and problems in correlation and sampling have tended to obscure the fact that the major extinctions and biomass loss apparently took place over a time no longer than a few times 10,000 years at the end of the Permian and they concluded that extraterrestrial impact was very probably the ultimate cause of the global Permian-Triassic killing event.

The P/Tr boundary is also marked by sudden shifts in carbon, oxygen (SWEET et al., 1992), sulphur (LI et al., 1991) and strontium isotope ratios and by europium and cerium anomalies (DAO-YI et al., 1989; BHANDARI et al., 1992; CHIFANG et al., 1992; GRUSZCZYNSKI et al., 1989, 1992). The entire carbon-isotope excursion is protracted over a period of about 2 million years in the Carnic Alps, (HOLSER et al., 1991; HOLSER and MAGARITZ, 1992) but the major negative isotope shifts of ~1.5 and 2 per mil take place rapidly in two abrupt (<50,000 years) events that coincide with abrupt negative shifts in $\delta^{18}O$ (Fig. 9). In the southern Alps, the abrupt negative shift in $\delta^{18}O$ of more than –4 per mil coincides with a sudden negative $\delta^{13}C$ excursion of –1.5 per mil (MAGARITZ et al., 1988). If the entire $\delta^{18}O$ signal is related to paleotemperature changes, then a drastic warming is indicated. A marked positive shift in $\delta^{34}S$ at the boundary indicates the development of anoxic conditions in the oceans.

Two iridium anomalies, coinciding with the two major negative $\delta^{13}C$ shifts, mark the boundary in a number of localities (e.g. CASSINIS, 1988; BHANDARI et al., 1991; HOLSER et al., 1991) (Table I) (Fig. 9). A larger Ir anomaly reported from Chinese sections (e.g. 2,000 ppt at Shangxi; XU et al., 1985) has not been reproduced (ZHOU and KYTE, 1988), but this may be a function of sample inhomogeneities. Microspherules are apparently abundant in the boundary clay (see DAO-YI et al., 1989). This is similar to the lower part of

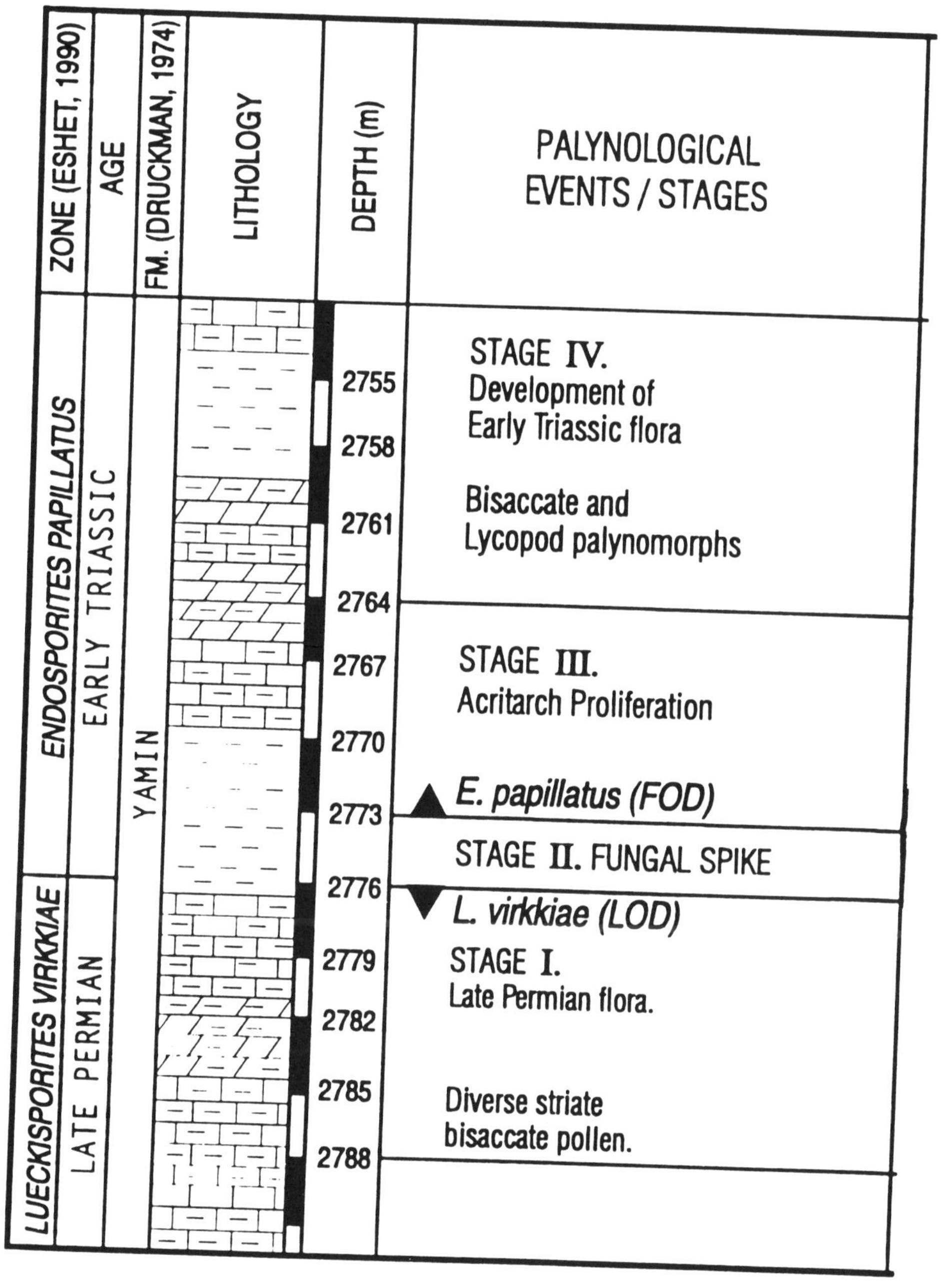

Fig. 8. Palynological changes evidenced in shallow-water sediments at the Permian/Triassic boundary in a borehole from Israel (depths in metres below surface). Sequence of palynological stages represent extinction of Late Permian palynoflora (mainly gymnosperms), proliferation of fungal spores, zone of abundant opportunistic marine plankton (acritarchs), and eventual recovery of the Early Triassic palynofiora (first primitive lycopods and then gymnosperms). The last occurrence datum (LOD) of the palynomorph *L. virkkiae* marks the disappearance of almost all Late Permian pollen. The boundary layer (a reddish clay) contains only fungal spores, evidence of an environmental disaster. Lithology: bricks, limestone; slanted bricks, dolomite; dashed lines, mudstones (after ESHET, 1990).

RAMPINO (1990a,b) and could lead to a greater greenhouse warming of up to 4°C over a few hundred thousand years, but such a warming is still of questionable significance in contributing to mass extinctions. By contrast, COURTILLOT (1990) estimated an extremely high Deccan CO_2 emission rate of 3×10^{19} g (7×10^{17} mol) of CO_2 outgassed over only ~100 years. This estimate of CO_2 outgassing from the Deccan eruptions seems much too high based on information on the CO_2 emission rates from historical basaltic eruptions.

The possible role of volcanism in mass extinctions

The wealth of physical evidence for an extraterrestrial impact at the K/T boundary and the close timing of many of the extinctions with the worldwide impact ejecta layer, suggest that the primary cause of the K/T mass extinctions was the collision of an asteroid or comet with the Earth (ALVAREZ, 1986; GLEN, 1990). Some critics of the impact interpretation (the so-called volcanists), however, have attempted to attribute all of the important components of the K/T boundary layer to volcanism, or to dismiss the impact evidence (e.g. LYONS and OFFICER, 1992). However, suggestions that the anomalous iridium at the boundary was a product of Deccan volcanism (e.g. CROCKET et al., 1988) are apparently ruled out by the generally low iridium values in the Deccan lavas, in soils and weathering profiles between flows and in K/T boundary sections on the Indian sub-continent (ROCCHIA et al., 1988b; BHANDARI et al., 1993). Furthermore, despite arguments to the contrary, experimental and observational evidence shows that shocked quartz (with multiple planar deformation features) and stishovite cannot be produced at the pressures characteristic of volcanic eruptions (GRATZ et al., 1992; DA SILVA et al., 1990) and that the K/T Haitian tektite glass has features apparently incompatible with a volcanic origin (SIGURDSSON et al., 1991).

However, several workers have argued that volcanism could have been a contributing factor in the events at the K/T boundary (e.g. COURTILLOT, 1990). This would be especially interesting if it could be clearly shown that extinctions and/or climatic changes preceded the K/T impact event (or events), as the Deccan eruptions may have begun prior to the major Ir anomaly (COURTILLOT, 1990), and/or that some of the extinctions took place over ~10^5 years after the impact. KELLER and BARRERA (1990) maintained that K/T extinctions and environmental changes were not instantaneous and may have begun prior to the deposition of the inferred impact horizon. They presented evidence that the K/T microfossil extinctions began ~300,000 years prior to the K/T iridium-rich layer and may have continued for about 50,000 years after the deposition of the Ir spike. However, even if all the extinctions were not instantaneous (and an impact model does not demand that they were), the abrupt negative shift in $\delta^{13}C$ at the boundary indicates a sudden major loss of biomass, or killing event (MCLAREN and GOODFELLOW, 1990).

As noted above, some boundary sections show evidence for significant climatic fluctuations prior to the boundary event as defined by the iridium anomaly. For example STOTT and KENNETT (1989, 1990) suggest that shifts in $\delta^{18}O$ in Antarctic deep-sea sediments evidence a 4°C warming beginning some 500,000 years before the boundary, followed by a 4°C cooling across the boundary and ZACHOS et al. (1987) presented evidence that a climatic cooling preceded the K/T boundary by about 200,000 in the North Pacific. However, other sections do not exhibit such pre-boundary fluctuations and the climate record at the K/T boundary remains problematic.

Future climates?

Astronomical observations show that the Earth lies within a swarm of Earth-crossing asteroids and comets (e.g. SHOEMAKER et al., 1990; CHAPMAN and MORRISON, 1994). Impact of objects greater than a few kilometres in diameter have predictable far-field effects that are very likely to constitute a global environmental disaster. The collision rates of asteroids and comets of various sizes with the Earth and the corresponding production of impact craters, can be calculated from the observed flux of Earth-crossing bodies of different types (SHOEMAKER et al., 1990). The estimated production rates of large craters from asteroid and comet collision on Earth during the last 100 Ma based on the recent flux is estimated as $(4.9 \pm 2.9) \times 10^{-15}$ km^{-2} $year^{-1}$, which is in excellent agreement with the rate of $(5.4 \pm 2.7) \times 10^{-15}$ km^{-2} $year^{-1}$ estimated from the geologic record of impact craters over the last 120 Ma.

The chances of a large-body impact affecting global climate in the near future have been calculated by CHAPMAN and MORRISON (1994). They find that there is a 1/10,000 chance that a large (~2 km diameter) asteroid or comet will collide with the Earth during the next hundred years. They point out that although impacts of this magnitude are so infrequent as to be beyond human experience, the long-term statistical hazard is comparable to other more familiar hazards. For example, the chances of the average American dying in a passenger aircraft crash are estimated at about 1/20,000, or in a flood, about 1/30,000 (CHAPMAN and MORRISON, 1994).

Consideration of the "impact winter" phenomenon led, in part, to the appreciation that the climatic effects of dense smoke and dust clouds produced by nuclear war might be severe (TURCO et al., 1983). Initial model calculations suggested that for most scenarios of nuclear exchange, smoke from burning cities and forest fires would cover large portions of the globe and cause a major drop in temperatures to sub-freezing levels in midlatitudes even in summer, creating a so-called nuclear winter.

There has been considerable debate as to the severity and extent of nuclear winter (e.g. COVEY et al., 1984; SCHNEIDER and THOMPSON, 1988), but most recent GCM studies predict that (for a worst-case July smoke injection) land temperatures in the 30° to 70°N latitude zone could drop from 5 to 15°C, with freezing events in mid-latitudes during the first months (TURCO et al., 1990). During the subsequent 1–3 years, significant cooling could continue in areas covered by the residual smoke cloud and ocean-surface cooling of ~2–6°C may extend for several years. Longer term (decadal) climatic cooling might arise primarily through climate feedbacks such as increased snow cover and sea-ice and perturbed sea surface temperatures.

The after-effects of large-body impacts, including massive dust clouds, increased water vapour in the atmosphere from an ocean impact, water and/or ice clouds, wildfires and soot production, production of NO_x and acid rain, heating pulses due to re-entering ejecta, increased greenhouse gases (CO_2, SO_2) and sulphuric acid aerosols from impact into carbonate or evaporite sediments together with perturbations of biogeochemical cycles resulting in fluctuations of CO_2, reduction of dimethylsulphide for cloud condensation nuclei and other effects, are predicted to produce climatic changes on various time-scales ranging from instantaneous to 10^6 years. These climatic and environmental fluctuations may have been severe enough to cause the mass extinctions on land and in the sea, which mark major geologic boundaries, such as the Cretaceous/Tertiary boundary (~65 Ma).

The K/T boundary is marked by a suite of convincing evidence of large-body impact, including a global boundary layer with elevated levels of iridium and other characteristic trace metals, shock-deformed minerals, tektites and microtektites, microspherules, impact-wave deposits and the Chicxulub impact crater. The boundary is also marked by the apparently sudden extinction of ~75% of marine species, including >95% of calcareous plankton, a major floral transition and all of the extant dinosaurs. These extinctions can be explained by the severe conditions predicted at the boundary including a brief dark and cold "impact winter", global wildfires and climatic changes resulting from changes in greenhouse gases as a result of the marine extinctions.

A review of the details of biotic changes and the environmental, isotopic and climatic indicators at times of recognized extinction events reveals some general patterns that may implicate large-body impacts and their climatic after-effects as a general cause of pulses of extinction. The major mass extinctions are accompanied by abrupt negative carbon-isotope shifts indicating sudden losses of biomass and reduction in marine productivity, abrupt negative shifts in oxygen-isotopic ratios indicating possible climatic warming, sudden crashes in global plankton and in organisms having planktonic growth stages, negative sulphur-isotope shifts suggesting development of anoxic conditions, and outbreaks of disaster fauna and flora suggesting a sudden ecological catastrophe. These conditions, combined with growing evidence of coincident large-body impacts from iridium anomalies, microtektites, shocked minerals and well-dated craters, lead to the working hypothesis that all extinction pulses are the result of the impacts of large asteroids or comets on the earth's surface.

Large impacts may trigger flood basalt volcanism, as indicated by a correspondence among mass extinctions, evidence of impact, and flood basalt eruptions over the last 250 Ma. Aerosols and carbon dioxide from massive flood basalt eruptions are predicted to produce climatic effects that could exacerbate environmental changes at mass extinction boundaries. Episodic delivery of impact energy to the Earth may even be great enough to have an effect on terrestrial geologic processes that determine long-term climate (cf. Chapter 2 by BERGER and Chapter 3 by BARRON).

Although the chances of a large asteroid or comet collision in the near future are small such an occurrence is as likely as other "accepted" hazards of modern life. When the next large-body impact occurs the climatic effects, especially on human food resources, would be massive: crop failures would almost certainly result in mass starvation. In response to the possible threat, programmes to detect large asteroids and comets that are on collision courses with the Earth and to attempt to divert these bodies if necessary, have been proposed (CHAPMAN and MORRISON, 1994). If these are enacted the climatic impacts can be reduced or removed; if they are not, a massive climatic shock equivalent to that which caused the Cretaceous/Tertiary extinctions is a possible future climate.

References

AHARON, P., SCHIDLOWSKI, M. and SINGH, I. B., 1987. Chronostratigraphic markers in the end-Precambrian carbon isotope record of the Lesser Himalaya. *Nature*, 327: 699–702.

ALVAREZ, L. W., ALVAREZ, W., ASARO, F. and MICHEL, H. V., 1980. Extraterrestrial cause for the Cretaceous-Tertiary extinction. *Science*, 208: 1095–1108.

ALVAREZ, W., 1986. Toward a theory of impact crises. *Eos, Trans. Am. Geophys. Union*, 67: 649–658.

ALVAREZ, W., ALVAREZ. L. W., ASARO, F. and MICHEL. H. V., 1982. Iridium anomaly approximately synchronous with terminal Eocene extinctions. *Science*, 216: 886–888.
ALVAREZ, W. and ASARO, F., 1990. An extraterrestrial impact. *Sci. Am.*, 263(4): 78–84.
ALVAREZ, W., ASARO, F. and MONTANARI, A., 1990. Iridium profile for 10 million years across the Cretaceous-Tertiary boundary at Gubbio (Italy). *Science*, 250: 1700–1702.
ANDERS, E., WOLBACH, W. S. and LEWIS, R. S., 1986. Cretaceous extinctions and wildfires. *Science*, 234: 264.
ANDERSON, D. L., TANIMOTO, T. and ZHANG, Y.-S., 1992. Plate tectonics and hotspots. The third dimension. *Science*, 256: 1645–1651.
ASARO, F., ALVAREZ, L. W., ALVAREZ, W. and MICHEL, H. V., 1982. Geochemical anomalies near the Eocene/Oligocene and Permian/Triassic boundaries. *Geol. Soc. Am., Spec. Pap.*, 190: 517–528.
ASKIN, R. A., 1990. Campanian to Paleocene spore and pollen assemblages of Seymour Island, Antarctica. *Rev. Palaeobot. Palynol.*, 65: 105–113.
BAKSI, A. K., 1988. Reply to comment on "Critical evaluation of the age of the Deccan traps, India. Implications for flood basalt volcanism and faunal extinctions". *Geology*, 16: 758–759.
BAKSI, A. K. and FARRAR, E., 1990. Evidence for errors in the geomagnetic polarity time scale at 17–15 Ma. $^{40}Ar/^{39}Ar$ dating of basalts from the Pacific Northwest, USA. *Geophys. Res. Lett.*, 17: 1117–1120.
BALME, B. E., 1979. Palynology of Permian-Triassic boundary beds at Kap Stosch, East Greenland. *Meddelelser Om Gronland*, Bd. 200, Nr. 6: 35.
BARLOW, N. G., 1990. Application of the inner solar system cratering record to the earth. *Geol. Soc. Am., Spec. Pap.*, 247: 189–195.
BARRERA, E, HUBER, B. T., SAVIN, S. M. and WEBB, P.-N., 1987. Antarctic marine temperatures. Late Campanian through early Paleocene. *Paleoceanography*, 2: 21–47.
BARRERA, E. and KELLER, G., 1990. Stable isotope evidence for gradual environmental changes and species survivorship across the Cretaceous/Tertiary boundary. *Paleoceanography*, 5: 867–890.
BECKER, R. T., HOUSE, M. R., KIRCHGASSER, W. T. and PLAYFORD, P. E., 1991. Sedimentary and faunal changes across the Frasnian/Famennian boundary in the Canning Basin of Australia. *Hist. Biol.*, 5: 183–196.
BENTON, M. J., 1985. Mass extinction among non-marine tetrapods. *Nature*, 316: 811–814.
BENTON, M. J., 1986. More than one event in the late Triassic mass extinction. *Nature*, 321: 857–861.
BERNER, R. A., 1990. Atmospheric carbon dioxide levels of Phanerozoic time. *Science*, 249: 1382–1396.
BERNER, R. A., LASAGA, A. C. and GARRELS, R. M., 1983. The carbonate-silicate geochemical cycle and its effect on atmospheric carbon dioxide over the past 100 million years. *Am. J. Sci.*, 283: 641–683.
BHANDARI, N., SHUKLA, P. N. and PANDEY, J., 1991. Iridium enrichment at the Cretaceous/Tertiary boundary in Meghalaya. *Curr. Sci.*, 56: 1003–1005.
BHANDARI, N., SHUKLA, P. N. and AZMI, R. J., 1992. Positive Europium anomaly at the Permo-Triassic boundary, Spiti, India. *Geophys. Res. Lett.*, 19: 1531–1534.
BHANDARI, N., SHUKLA, P. N. and CINI CASAGNOLI, G., 1993. Geochemistry of some K/T sections in India. *Palaeogeogr., Palaeoclimatol., Palaeoecol.*, 104: 199–211.
BICE, D. M., NEWTON, C. R., MCCAULEY, S., REINERS, P. W. and MCROBERTS, C. A., 1992. Shocked quartz at the Triassic-Jurassic boundary in Italy. *Science*, 259: 443–446.
BOERSMA, A. and SHACKLETON, N. J., 1981. Oxygen- and carbon-isotope variations and planktonic-foraminifer depth habitats, late Cretaceous to Paleocene, Central Pacific, Deep Sea Drilling Project Sites 463 and 465. *Initial Reports of the Deep Sea Drilling Project, Vol. 62*. Government Printing Office, Washington, DC, pp. 513–526.
BOERSMA, A., SHACKLETON, N., HALL, M. and GIVEN, Q., 1979. Carbon and oxygen isotope records at DSDP site 384 (North Atlantic) and some Paleocene paleotemperatures and carbon isotope variations in the Atlantic Ocean. *Initial Reports of the Deep Sea Drilling Project, Vol. 43*. Government Printing Office, Washington, DC, pp. 695–717.
BRAND, U., 1989. Global climatic changes during the Devonian-Mississippian. Stable isotope biogeochemistry of brachiopods. *Palaeogeog., Palaeoclimatol., Palaeoecol., Global Planet. Change Sec*, 75: 311–329.
BRENCHLEY, P. J., 1989. The Late Ordovician extinction. In: S. K. DONOVAN (Editor), *Mass Extinctions, Processes and Evidence*. Columbia University Press, New York, pp. 104–132.

BUGGISCH, W., 1991. The global Frasnian-Famennian "Kellwasser Event". *Geol. Rundsch.*, 80: 49–72.
CALDEIRA, K. and RAMPINO, M. R., 1990a. Carbon dioxide emissions from Deccan volcanism and a K/T boundary greenhouse effect. *Geophys. Res. Lett.*, 17: 1299–1302.
CALDEIRA, K. and RAMPINO, M. R., 1990b. Deccan volcanism, greenhouse warming and the Cretaceous/Tertiary boundary. *Geol. Soc. Am., Spec. Pap.*, 247: 117–123.
CALDEIRA, K. and RAMPINO, M. R., 1993. The aftermath of end-Cretaceous mass extinction. Biogeochemical stabilization of the carbon cycle and climate? *Paleoceanography*, 8: 515–525.
CALDEIRA, K., RAMPINO, M. R., VOLK, T. and ZACHOS, J. C., 1990. Biogeochemical modelling at mass extinction boundaries. Atmospheric carbon dioxide and ocean alkalinity at the K/T boundary. In: E. G. KAUFFMAN and O. H. WALLISER (Editors), *Extinction Events in Earth History*, Springer-Verlag, Berlin, pp. 333–345.
CAMPBELL, I. H., CZAMANSKE, G. K., FEDORENKO, V. A., HILL, R. I. and STEPANOV, V., 1992. Syn chronism of the Siberian Traps and the Permian-Triassic boundary. *Science*, 258: 1160–1163.
CAPUTO, M., 1985. Late Devonian glaciation in South America. *Palaeogeog., Palaeoclimatol., Palaeoecol.*, 51: 291–317.
CASSINIS, G. (Editor), 1988. Proceedings of the field conference on Permian and Permian-Triassic boundary in the South-Alpine segment of the western Tethys and additional regional reports. *Mem. Soc. Geol. Ital.*, 34: 366.
CHAPMAN, C. R. and MORRISON, D., 1994. Impacts on the Earth by asteroids and comets: assessing the hazard. *Nature*, 367: 33–40.
CHIFANG, C., YAOQI, Z. et al., 1992. Geochemical constraints on the Permo-Triassic boundary event in South China. In: W. C. SWEET et al. (Editors), *Permo-Triassic Events in the Eastern Tethys*. Cambridge University Press, Cambridge, pp. 158–168.
CLAEYS, P., CASIER J-G. and MARGOLIS, S.V., 1992. Microtektites and mass extinctions. Evidence for a Late Devonian asteroid impact. *Science*, 257: 1102–1104.
COCKS, L. R. M. and RICKARDS, R. B. (Editors), 1988. A global analysis of the Ordovician-Silurian boundary. *Bull. Br. Mus. (Nat. Hist.), Geol.* No. 43: 394.
COLBATH, G. K., 1986. Abrupt terminal Ordovician extinction in phytoplankton associations, southern Appalachians. *Geology*, 14: 943–946.
CORFIELD, R. M., CARTLIDGE, J. E., PREMOLI-SILVA, I. and HOUSLEY, R. A., 1990. Oxygen and carbon isotope stratigraphy of the Palaeogene and Cretaceous limestones in the Bottaccione Gorge and the Contessa Highway sections, Umbria, Italy. *Terra Nova*, 3: 414–422.
CORLISS, B. H., AUBRY, M.-P., BERGGREN, W. A., FENNER, J. M., KEIGWIN JR., L. D. and KELLER, G., 1984. The Eocene/Oligocene boundary event in the deep sea. *Science*, 226: 806–810.
COURTILLOT, V., 1990. A volcanic eruption. *Sci. Am.*, October, 85–92.
COURTILLOT, V., BESSE, J., VANDAMME, D., MONTIGNY, R., JAEGER, J.-J. and CAPPETTA, H., 1986. Deccan flood basalts at the Cretaceous-Tertiary boundary. *Earth Planet. Sci. Lett.*, 80: 361–374.
COVEY, C., SCHNEIDER, S. H. and THOMPSON, S. L., 1984. Global atmospheric effects of massive smoke injection from a nuclear war: results from general circulation model simulations. *Nature*, 308: 21–25.
COVEY, C., GHAN, S. J., WALTON, J. J. and WEISSMAN, P. R., 1990. Global environmental effects of impact-generated aerosols: results from a general circulation model. *Geol. Soc. Am., Spec. Pap.*, 247: 263–270.
CROCKET, J. H., OFFICER, C. B., WEZEL, F. C. and JOHNSON, G. D., 1988. Distribution of noble metals across the Cretaceous/Tertiary boundary at Gubbio, Italy. Iridium variation as a constraint on the duration and nature of Cretaceous/Tertiary boundary events. *Geology*, 16: 77–80.
CROFT, S. K., 1982. A first-order estimate of shock heating and vaporization in oceanic impacts. *Geol. Soc. Am., Spec. Pap.*, 190: 143–151.
CROWLEY, T. J. and NORTH, G. R., 1991. Environmental consequences of an asteroid impact. In: *Paleoclimatology*. Oxford University Press, New York, Ch. 9, pp. 171–182.
CRUTZEN, P. J., 1987. Acid rain at the K/T boundary. *Nature*, 330: 108–109.
DAOHAN, C., LINZHONG, L. and JIAQING, Z., 1986. Mass extinction, earth's rotating and periodic cratering. *Sci. Sin.*, 29: 531–539.
DAO-YI, X., QIN-WEN, Z., YI-YIN, S., YAN, Z., ZHI-FANG, C. and JIN-WEN, H., 1989. *Astrogeological Events in China*. Van Nostrand Reinhold, New York, p. 264.

DE SILVA, S. L., WOLFF, J. A. and SHARPTON, V. L., 1990. Explosive volcanism and associated pressures; implications for models of endogenically shocked quartz. *Geol. Soc. Am., Spec. Pap.*, 247: 139–145.

DAVENPORT, S. A. et al., 1990. Chondritic metal toxicity as a seed stock kill mechanism in impact-caused mass extinctions. *Geol. Soc. Am., Spec. Pap.*, 247: 71–76.

DELAUBENFELS, M. W., 1956. Dinosaur extinction. one more hypothesis. *J. Paleontology*, 30: 207–212.

D'HONDT, S. L., KELLER, G. and STALLARD, R. F., 1987. Major element compositional variation within and between different late Eocene microtektite strewnfields. *Meteoritics*, 22: 61–79.

DICKINS, J. M., 1993. Climate of the Late Devonian to Triassic. *Palaeogeog., Palaeoclimatol., Palaeoecol.*, 100: 89–94.

DONOVAN, S. K., 1989. Introduction. In: S. K. DONOVAN (Editor), *Mass Extinctions, Processes and Evidence*. Columbia University Press, pp. xi–xiv.

DOWSETT, H. J. and LOUBERE, P., 1992. High resolution late Pliocene sea-surface temperature record from the Northeast Atlantic Ocean. *Mar. Micropaleontol.*, 20: 91–105.

DUNNING, G. R. and HODYCH, J. P., 1990. U/Pb zircon and baddeleyite ages from the Palisades and Gettysburg sills of the northeastern United States. Implications for the age of the Triassic/Jurassic boundary. *Geology*, 18: 795–798.

EBINGER, C. J., YEMANE, T., WOLDEGABRIEL, G., ARONSON, J. L. and WALTER, R. C., 1993. Late Eocene-Recent volcanism and faulting in the southern main Ethiopian rift. *J. Geol. Soc., London*, 150: 99–108.

EHRMANN, W. U. and MACKENSEN, A., 1992. Sedimentological evidence for the formation of an East Antarctic ice sheet in Eocene/Oligocene time. *Palaeogeog., Palaeoclimatol., Palaeoecol.*, 93: 85–112.

ELSTON, W. E. and TWIST, D., 1989. Is the Rooiberg Felsite (Bushveld Complex) an impact melt? Unpublished.

EMILIANI, C., 1980. Death and renovation at the end of the Mesozoic. *Eos, Trans. Am. Geophys. Union*, 61: 504–506.

EMILIANI, C., KRAUS, E. B. and SHOEMAKER, E. M., 1982. Sudden death at the end of the Mesozoic. *Earth Planet. Sci. Lett.*, 55: 317–334.

ENGELHARDT, W. V., MATTHAI, S. K. and WALZEBUCK, J., 1992. Araguinha impact crater, Brazil. I. The interior part of the uplift. *Meteoritics*, 27: 442–457.

ERICKSON, D. J. and DICKSON, S. M., 1987. Global trace-element biogeochemistry at the K/T boundary. Oceanic and biotic response to a hypothetical meteorite impact. *Geology*, 15: 1014–1017.

ESHET, Y., 1990. The palynostratigraphy of the Permian-Triassic boundary in Israel. Two approaches to biostratigraphy. *Isr. J. Earth-Sci.*, 39: 1–15.

FAN, D., YANG, R. and HUANG, Z., 1984. The lower Cambrian black shale series and the iridium anomaly in South China. In: *Academia Sinica, Developments in Geoscience. Contribution to 27th Geological Congress, Moscow*. Science Press, Beijing, pp. 215–224.

GANAPATHY, R., 1980. A major meteorite impact on the earth 65 million years ago. Evidence from the Cretaceous-Tertiary boundary clay. *Science*, 209: 921–923.

GANAPATHY, R., 1982a. Evidence for a major meteorite impact on the Earth 34 million years ago. Implication for Eocene extinctions. *Science*, 216: 885–886.

GANAPATHY, R., 1982b. Evidence for a major meteorite impact on Earth 34 million years ago. Implication on the origin of North American tektites and Eocene extinction. *Geol. Soc. Am., Spec. Pap.*, 190: 513–516.

GELDSETZER, H. H. J., GOODFELLOW, W. D., MCLAREN, D. J. and ORCHARD, M. J., 1987. Sulfur-isotope anomaly associated with the Frasnian/Famennian extinction, Medicine Lake, Alberta, Canada. *Geology*, 15: 393–396.

GERLACH, T. M. and GRAEBER, E. J., 1985. Volatile budget of Kilauea volcano. *Nature*, 313: 273–277.

GERSTEL, J., THUNELL, R. C., ZACHOS, J. C. and ARTHUR, M. A., 1986. The Cretaceous/Tertiary boundary event in the North Pacific. Planktonic foraminiferal results from Deep Sea Drilling Project Site 577, Shatsky Rise. *Paleoceanography*, 1: 97–117.

GERSTL, S. A. and ZARDECKI, A., 1982. Reduction of photosynthetically active radiation under extreme stratospheric aerosol loads. *Geol. Soc. Am., Spec. Pap.*, 190: 201–210.

GILMOUR, I., WOLBACH, W. S. and ANDERS, E., 1989. Major wildfires at the Cretaceous-Tertiary boundary. In: S. V. M. CLUBE (Editor), *Catastrophes and Evolution, Astronomical Foundations.* Cambridge University Press, Cambridge, pp. 195–213.

GILMOUR, I., WOLBACH, W. S. and ANDERS, E., 1990. Early environmental effects of the terminal Cretaceous impact. *Geol. Soc. Am., Spec. Pap.*, 247: 383–390.

GLASS, B. P., 1982. Possible correlations between tektite events and climatic changes? *Geol. Soc. Am., Spec. Pap.*, 190: 251–256.

GLEN, W., 1990. What killed the dinosaurs? *Am. Sci.*, 78: 354–371.

GOODFELLOW, W. D., GELDSETZER, H. H. J., MCLAREN, D. J., ORCHARD, M. J. and KLAPPER, G., 1989. Geochemical and isotopic anomalies associated with the Frasnian-Famennian extinction. *Hist. Biol.*, 2: 51–72.

GOODFELLOW, W. D., NOWLAN, G. S., MCCRACKEN, A. D., LENZ, A. C. and GREGOIRE, D. C., 1992. Geochemical anomalies near the Ordovician-Silurian boundary, Northern Yukon Territory, Canada. *Hist. Biol.*, 6: 1–23.

GRATZ, A. J., NELLIS, W. J. and HINSEY, N. A., 1992. Laboratory simulations of explosive volcanic loading and implications for the cause of the K/T boundary. *Geophys. Res. Lett.*, 19: 1391–1394.

GRIEVE, R. A. F., 1991. Terrestrial impact. the record in the rocks. *Meteoritics*, 26: 175–194.

GRIFFIS, K. and CHAPMAN, D. J., 1988. Survival of phytoplankton under prolonged darkness. Implications for the Cretaceous-Tertiary boundary darkness hypothesis. *Palaeogeog., Palaeoclimatol., Palaeoecol.*, 67: 305–314.

GRUSZCZYNSKI, M., HALAS, S., HOFFMAN, A. and MALKOWSKI, K., 1989. A brachiopod calcite record of the oceanic carbon and oxygen isotope shifts at the Permian-Triassic transition. *Nature*, 337: 64–68.

GRUSZCZYNSKI, M., HOFFMAN, A., MALKOWSKI, K. and VEIZER, J., 1992. Seawater strontium isotopic perturbation at the Permian-Triassic boundary, West Spitsbergen and its implications for the interpretation of strontium isotopic data. *Geology*, 20: 779–782.

HALLAM, A., 1989. The case for sea-level change as a dominant causal factor in mass extinction of marine invertebrates. *Phil. Trans. R. Soc. London*, B325: 437–455.

HALLAM, A., 1990. The end-Triassic mass extinction event. *Geol. Soc. Am., Spec. Pap.*, 247: 577–583.

HEIMANN, A., FLEMING, T. H., ELLIOT, D. H. and FOLAND, K. A., 1994. A short interval of Jurassic continental flood basalt volcanism in Antarctica as demonstrated by $^{40}Ar/^{39}Ar$ geochronology. *Earth Planet. Sci. Lett.*, 141: 19–41.

HEUER, K., 1953. *The End of the World.* Rinehart & Co., New York, 220 pp.

HICKEY, L. J., 1980. Paleocene stratigraphy and flora of the Clark's Fork Basin. *Univ. Mich. Pap. Paleontol.*, 24: 33–49.

HILDEBRAND, A. R. et al., 1991. Chicxulub Crater. A possible Cretaceous/Tertiary boundary impact crater on the Yucatán Peninsula, Mexico. *Geology*, 19: 867–871.

HOLSER, W. T. and MAGARITZ, M., 1992. Cretaceous/Tertiary and Permian/Triassic boundary events compared. *Geochim. Cosmochim. Acta*, 56: 3297–3309.

HOLSER, W. T., SCHÖNLAUB, H. P., BOECKELMANN, K. and MAGARITZ, M., 1991. The Permian-Triassic of the Gartnerkofel-1 Core (Carnic Alps, Austria). Synthesis and conclusions. In: W. T. HOLSER and H. P. SCHÖNLAUB (Editors), *The Permian-Triassic boundary in the Carnic Alps of Austria (Gartnerkofel Region). Abhandlung Geol. Bundesanst.*, 45: 213–232.

HSU, K. J., 1980. Terrestrial catastrophe caused by cometary impact at the end of the Cretaceous. *Nature*, 285: 201–203.

HSU, K. J. and MCKENZIE J. A., 1985. A "strangelove" ocean in the earliest Tertiary. In: E. T. SUNDQUIST and W. S. BROECKER (Editors), *Carbon Cycle and Atmospheric CO_2, Natural Variations Archean to Present.* Geophysical Monograph 32, Am. Geophys. Union, Washington, DC, pp. 487–492.

HSU, K. J. and MCKENZIE, J. A., 1990. Carbon-isotope anomalies at era boundaries; global catastrophes and their ultimate cause. *Geol. Soc. Am., Spec. Pap.*, 247: 61–70.

HSU, K. J., MCKENZIE, J. A. and HE, Q. X., 1982a. Terminal Cretaceous environmental and evolutionary changes. *Geol. Soc. Am., Spec. Pap.*, 190: 317–328.

HSU, K. J., MCKENZIE, J. A., WEISSERT, H. et al., 1982b. Mass mortality and its environmental and evolutionary consequences. *Science*, 216: 249–256.

HSU, K. J., OBERHANSLI, H. et al., 1985. "Strangelove ocean" before the Cambrian explosion. *Nature*, 316: 809–811.

HUT, P., ALVAREZ, W. et al., 1987. Comet showers as a cause of mass extinctions. *Nature*, 329: 118–

126.
JABLONSKI, D., 1986. Evolutionary consequences of mass extinctions. In: D. M. RAUP and D. JABLONSKI (Editors), *Patterns and Processes in the History of Life*. Springer-Verlag, Berlin, pp. 313–329.
JAIPRAKASH, B. C., SINGH, J. and RAJU, D. S. N., 1993. Foraminiferal events across K/T boundary and age of Deccan volcanism in Palakollu area, Krishna-Godavari Basin, India. *J. Geol. Soc. India*, 41: 105–117.
JOHNSON, A. L. A. and SIMMS, M. J., 1989. The timing and cause of Late Triassic marine invertebrate extinctions. Evidence from scallops and crinoids. In: S. K. DONOVAN (Editor), *Mass Extinctions, Processes and Causes.* Columbia University Press, New York, pp. 174–194.
JOHNSON, K. R., 1992a. Leaf-fossil evidence for extensive floral extinction at the Cretaceous-Tertiary boundary, North Dakota, USA. *Cretaceous Res.*, 13: 91–117.
JOHNSON, K. R., 1992b. High latitude deciduous vegetation and muted floral response to the Cretaceous-Tertiary boundary event in New Zealand. *Geol. Soc. Am., Abstr. Prog.*, 24: A333.
JOHNSON, K. R. and HICKEY, L. J., 1990. Megafloral change across the Cretaceous/Tertiary boundary in the northern Great Plains and Rocky Mountains, U.S.A. *Geol. Soc. Am., Spec. Pap.*, 247: 433–444.
JOHNSON, K. R., NICHOLS, D. J., ATTREP JR, M. and ORTH, C. J., 1989. High-resolution leaf-fossil record spanning the Cretaceous/Tertiary boundary. *Nature*, 340: 708–711.
JONES, D. S., MUELLER, P. A. et al., 1987. Biotic, geochemical and paleomagnetic changes across the Cretaceous/Tertiary boundary at Braggs, Alabama. *Geology*, 15: 311–315.
JONES, D. S. et al., 1991. Integrated geochronology of a classic Pliocene fossil site in Florida. Linking marine and terrestrial biochronologies. *J. Geol.*, 99: 637–648.
KAIHO, K., 1992. A low extinction rate of intermediate-water benthic foraminifera at the Cretaceous/Tertiary boundary. *Mar. Micropaleont.*, 18: 229–259.
KAJIWARA, Y. and KAIHO, K., 1992. Oceanic anoxia at the Cretaceous/Tertiary boundary supported by the sulfur isotopic record. *Palaeogeog., Palaeoclimatol., Palaeoecol.*, 99: 151–162.
KASTING, J. F., RICHARDSON, S. M., POLLACK, J. B. and TOON, O. B., 1986. A hybrid model of the CO_2 geochemical cycle and its application to large impact events. *Am. J. Sci.*, 286: 361–389.
KAUFFMAN, E. G., 1988. The dynamics of marine stepwise mass extinction. *Rev. Esp. Paleontol., no. Extraordinario*, 57–71.
KELLER, G., 1986. Stepwise mass extinctions and impact events. Late Eocene to Early Oligocene. *Mar. Micropaleontol.*, 10: 267–293.
KELLER, G., 1988a. Biotic turnover in benthic foraminifera across the Cretaceous/Tertiary boundary at El Kef, Tunisia. *Palaeogeog., Palaeoclimatol., Palaeoecol.*, 66: 153–171.
KELLER, G., 1988b. Extinction, survivorship and evolution of planktic foraminifera across the Cretaceous/Tertiary boundary at El Kef, Tunisia. *Mar. Micropaleontol.*, 13: 239–263.
KELLER, G., 1989a. Extended Cretaceous/Tertiary boundary extinctions and delayed population change in planktonic foraminifera from Brazos River, Texas. *Paleoceanography*, 4: 287–332.
KELLER, G., 1989b. Extended period of extinctions across the Cretaceous/Tertiary boundary in planktonic foraminifera of continental-shelf sections. Implications for impact and volcanism theories. *Geol. Soc. Am. Bull.*, 101: 1408–1419.
KELLER, G. and BARRERA, E., 1990. The Cretaceous/Tertiary boundary impact hypothesis and the paleontological record. *Geol. Soc. Am., Spec. Pap.*, 247: 563.
KRASSILOV, V. A., 1975. Climatic changes in Eastern Asia as indicated by fossil floras. II. Late Cretaceous and Danian. *Palaeogeog., Palaeoclimatol., Palaeoecol.*, 17: 157–172.
KUMP, L., 1991. Interpreting carbon-isotope excursions. Strangelove oceans. *Geology*, 19: 299–302.
KYTE, F. T., 1988. The extraterrestrial component in marine sediments. Description and interpretation. *Paleoceanography*, 3: 235–247.
KYTE, F. T. and WASSON, J. T., 1986. Accretion rate of extraterrestrial matter. Iridium deposited 33 to 67 million years ago. *Science*, 232: 1225–1229.
KYTE, F. T., ZHOU, Z. and WASSON, J. T., 1981. High noble metal concentrations in a late Pliocene sediment. *Nature*, 292: 417–420.
KYTE, F. T., ZHOU, Z. and WASSON, J. T., 1988. New evidence on the size and possible effects of a late Pliocene oceanic asteroid impact. *Science*, 241: 63–65.
LEARY, P. N. and RAMPINO, M. R., 1990. A multi-causal model of mass extinctions. Increase in trace metals in the oceans. In: E. G. KAUFFMAN and O. H. WALLISER (Editors), *Extinction Events in*

Earth History. Springer-Verlag, Berlin, pp. 45–55.

LeCheminant, A. N. and Heaman, L. M., 1989. Mackenzie igneous events, Canada. Middle Proterozoic hotspot magmatism associated with ocean opening. *Earth Planet. Sci. Lett.*, 96: 38–48.

Lehman, T. M., 1987. Late Maastrichtian paleoenvironments and dinosaur biogeography in the western interior of North America. *Palaeogeog., Palaeoclimatol., Palaeoecol.*, 60: 189–217.

Lewis, J. S. et al., 1982. Chemical consequences of major impact events on Earth. *Geol. Soc. Am, Spec. Pap.*, 190: 215–222.

Li., Z., Zhan, L., Yao, J. and Zhou, Y., 1991. On the Permian-Triassic events in South China – Probe into the end Permian abrupt extinction and its possible causes. *Saito Ho-on Kai Special Publication* No. 3, pp. 371–385.

Lowrie, W., Alvarez, W. and Asaro, F., 1990. The origin of the White Beds below the Cretaceous-Tertiary boundary in the Gubbio section, Italy. *Earth Planet. Sci. Lett.*, 98: 303–312.

Lyell, C., 1830–33. *Principles of Geology, Being an Attempt to Explain the Former Changes of the Earth's Surface by Processes Still in Operation*, 3 Vols. John Murray, London, 1700 pp.

Lyons, J. B. and Officer, C. B., 1992. Mineralogy and petrology of the Haiti Cretaceous/Tertiary section. *Earth Planet. Sci. Lett.*, 109: 205–224.

Macdougal, J. D., 1988. Seawater strontium isotopes, acid rain and the Cretaceous-Tertiary boundary. *Science*, 239: 485–487.

Mackensen, A. and Ehrmann, W. U., 1992. Middle Eocene through Early Oligocene climate history and paleoceanography in the Southern Ocean. Stable oxygen and carbon isotopes from ODP Sites on Maud Rise and Kerguelen Plateau. *Mar. Geol.*, 108: 1–27.

MacLeod, N. and Keller G., 1991a. Hiatus distributions and mass extinctions at the Cretaceous/Tertiary boundary. *Geology*, 19: 497–501.

Macleod, N. and Keller, G., 1991b. How complete are Cretaceous/Tertiary boundary sections? A chronostratigraphic estimate based on graphic correlation. *Geol. Soc. Am. Bull.*, 103: 1439–1457.

Magaritz, M., 1989. ^{13}C minima follow extinction events. A clue to faunal radiation. *Geology*, 17: 337–340.

Magaritz, M., Bar, R., Baud, A. and Holser, W. T., 1988. The carbon-isotope shift at the Permian/Triassic boundary in the southern Alps is gradual. *Nature*, 331: 337–339.

Magaritz, M., Holser, W. T. and Kirschvink, J. L., 1986. Carbon-isotope events across the Precambrian/Cambrian boundary on the Siberian Platform. *Nature*, 320: 258–259.

Mahoney, J. J., 1988. Deccan Flood Basalts. In: J. D. Macdougall (Editor), *Continental Flood Basalts*. Kluwer, Dordrecht, pp. 151–194.

Margolis, S. V., Claeys, P. and Kyte, F. T., 1991. Microtektites, microkrystites and spinels from a Late Pliocene asteroid impact in the Southern Ocean. *Science*, 251: 1594–1597.

Margolis, S. V., Mount, J. F., Doehne, E., Showers, W. and Ward, P., 1987. Late Cretaceous-Tertiary boundary oxygen and carbon isotope stratigraphy, diagenesis and biotic extinctions at Zumaya, Spain. *Paleoceanography*, 2: 361–377.

Marvin, U. B., 1990. Impact and its revolutionary implications for geology. *Geol. Soc. Am., Spec. Pap.*, 247: 147–154.

Maxwell, W. D., 1989. The end Permian mass extinction. In: S. K. Donovan (Editor), *Mass Extinctions. Processes and Evidence.* Columbia University Press, New York, pp. 152–173.

Maxwell, W. D., 1992. Permian and Early Triassic extinction of non-marine tetrapods. *Palaeontology*, 35: 571–583.

McGhee Jr, G. R., 1988. The Late Devonian extinction event. Evidence for abrupt ecosystem collapse. *Paleobiology*, 14: 250–257.

McGhee Jr, G. R. et al., 1986. Late Devonian "Kellwasser Event" mass-extinction horizon in Germany. No geochemical evidence for a large-body impact. *Geology*, 14: 776–779.

McKay, C. P. and Thomas, G. E., 1982. Formation of noctilucent clouds by an extraterrestrial impact. *Geol. Soc. Am., Spec. Pap.*, 190: 211–214.

McKenzie, D. P. and Bickle, M. J., 1988. The volume and composition of melt generated by extension of the lithosphere. *J. Petrol.*, 29: 625–679.

McKinney, M. L., 1987. Taxonomic selectivity and continuous variation in mass and background extinctions of marine taxa. *Nature*, 325: 143–145.

McLaren, D. J., 1970. Presidential address. Time, life and boundaries. *J. Paleontol.*, 44: 801–815.

McLaren, D. J., 1985. Mass extinction and iridium anomaly in the Upper Devonian of Western Australia. A commentary. *Geology*, 13: 170–172.

McLaren, D. J. and Goodfellow, W. D., 1990. Geological and biological consequences of giant impacts. *Annu. Rev. Earth Planet. Sci.*, 18: 123–171.

McLean, D. M., 1978. A terminal Mesozoic "greenhouse": lessons from the past. *Science*, 201: 401–406.

McLean, D. M., 1985. Deccan Traps mantle degassing in the terminal Cretaceous marine extinctions. *Cretaceous Res.*, 6: 235–259.

Melosh, H. J., 1982. The mechanics of large meteoroid impacts in the Earth's oceans. *Geol. Soc. Am., Spec. Pap.*, 190: 121–128.

Melosh, H. J., 1989. *Impact Cratering, A Geologic Process.* Oxford, New York, 245 pp.

Melosh, H. J., Schneider, N. M., Zahnle, K. J. and Latham, D., 1990. Ignition of global wildfires at the Cretaceous/Tertiary boundary. *Nature*, 343: 251–254.

Miller, K. G., Fairbanks, R. G. and Mountain, G. S., 1987. Tertiary oxygen isotope synthesis, sea level history and continental margin erosion. *Paleoceanography*, 2: 1–19.

Milne, D. H. and McKay, C. P., 1982. Response of marine plankton communities to a global atmospheric darkening. *Geol. Soc. Am., Spec. Pap.*, 190: 297–304.

Montanari, A., 1990. Geochronology of terminal Eocene impacts; an update. *Geol. Soc. Am., Spec. Pap.*, 247: 607–616.

Napier, W. M. and Clube, S. V. M., 1979. A theory of terrestrial catastrophism. *Nature*, 282: 455–459.

Nichols, D. J., Jarzen, D. M., Orth, C. J. and Oliver, P. Q., 1986. Palynological and iridium anomalies at Cretaceous-Tertiary boundary, south-central Saskatchewan. *Science*, 231: 714–717.

O'Keefe, J. D. and Ahrens, T. J., 1982a. Impact mechanics of the Cretaceous-Tertiary extinction bolide. *Nature*, 298: 123–127.

O'Keefe, J. D. and Ahrens, T. J., 1982b. The interaction of the Cretaceous/Tertiary extinction bolide with the atmosphere, ocean and solid earth. *Geol. Soc. Am., Spec. Pap.*, 190: 103–120.

O'Keefe, J. D. and Ahrens, T. J., 1989. Impact production of CO_2 by the Cretaceous/Tertiary extinction bolide and resultant heating of the Earth. *Nature*, 338: 247–249.

Olsen, P. E., Fowell, S. J. and Cornet, B., 1990. The Triassic/Jurassic boundary in continental rocks of eastern North America; a progress report. *Geol. Soc. Am., Spec. Pap.*, 247: 585–593.

Opdyke, B. N. and Wilkinson, B. H., 1988. Surface area control of shallow cratonic and deep marine carbonate accumulation. *Paleoceanography*, 3: 685–703.

Opdyke, B. N. and Wilkinson, B. H., 1990. Ocean temperature, saturation state and Phanerozoic/Holocene carbonate accumulation rates. *Geol. Soc. Am. Abstr. Prog.*, 22: A131.

Orth, C. J., 1989. Geochemistry of the bio-event horizons. In: S. K. Donovan (Editor), *Mass Extinctions. Processes and Evidence.* Columbia University Press, New York, pp. 37–72.

Orth, C. J., Attrep Jr., M. and Quintana, L. R., 1990. Iridium abundance patterns across bio-event horizons in the fossil record. *Geol. Soc. Am., Spec. Pap.*, 247: 45–59.

Orth, C. J., Gilmore, J. S., Quintana, L. R. and Sheehan, P. M., 1986. Terminal Ordovician extinction. Geochemical analysis of the Ordovician/Silurian boundary, Anticosti Island, Quebec. *Geology*, 14: 433–436.

Piasecki, S., 1984. Preliminary palynostratigraphy of the Permian-Lower Triassic sediments in Jameson Land and Scoresby Land, East Greenland. *Bull. Geol. Soc. Den.*, 32: 139–144.

Pollack, J. B. et al., 1982. Environmental effects of an impact generated dust cloud. Implications for the Cretaceous-Tertiary extinctions. *Science*, 219: 287–290.

Prinn, R. G. and Fegley, B., 1987. Bolide impacts, acid rain and biospheric traumas at the Cretaceous/Tertiary boundary. *Earth Planet. Sci. Lett.*, 88: 1–15.

Rampino, M. R., 1987. Impact cratering and flood basalt volcanism. *Nature*, 327: 468.

Rampino, M. R., 1992. A major Late Permian impact event on the Falkland Plateau. *Eos, Trans. Am. Geophys. Union*, 73: 336.

Rampino, M. R. and Caldeira, K., 1992. Antipodal hotspot pairs on the earth. *Geophys. Res. Lett.*, 19: 2011–2014.

Rampino, M. R. and Caldeira, K., 1993. Major episodes of geologic change. Correlations, time structure and possible causes. *Earth Planet. Sci. Lett.*, 114: 215–227.

Rampino, M. R. and Haggerty, B. M., 1994. Extraterrestrial impacts and mass extinctions. In: T. Gehrels (Editor), *Hazards Due to Comets and Asteroids.* University of Arizona Press, Tucson,

AZ.
RAMPINO, M. R. and SELF, S., 1984. Sulphur-rich volcanic eruptions and stratospheric aerosols. *Nature*, 310: 677–679.
RAMPINO, M. R., SELF, S. and STOTHERS, R. B., 1988. Volcanic winters. *Annu. Rev. Earth Planet. Sci.*, 16: 73–99.
RAMPINO, M. R. and STOTHERS, R. B., 1984a. Terrestrial mass extinctions, cometary impacts and the Sun's motion perpendicular to the galactic plane. *Nature*, 308: 709–712.
RAMPINO, M. R. and STOTHERS, R. B., 1984b. Geologic rhythms and cometary impacts. *Science*, 226: 1427–1430.
RAMPINO, M. R. and STOTHERS, R. B., 1986. Geologic periodicities and the Galaxy. In: R. SMOLUCHOWSKI, J. N. BAHCALL and M. S. MATTHEWS (Editors), *The Galaxy and the Solar System.* University of Arizona Press, Tucson, AZ, pp. 241–259.
RAMPINO, M. R. and STOTHERS, R. B., 1988. Flood basalt volcanism during the past 250 million years. *Science*, 241: 663–668.
RAMPINO, M. R. and VOLK, T., 1988. Mass extinctions, atmospheric sulphur and climatic warming at the K/T boundary. *Nature*, 332: 63–65.
RAUP, D. M., 1979. Size of the Permo-Triassic bottleneck and its evolutionary implications. *Science*, 206: 217–218.
RAUP, D. M., 1990. Impact as a general cause of extinction; a feasibility test. *Geol. Soc. Am., Spec. Pap.*, 247: 27–32.
RAUP, D. M., 1991. A kill curve for Phanerozoic marine species. *Paleobiology*, 17: 37–48.
RAUP, D. M., 1992. *Extinctions. Bad Genes or Bad Luck*? Norton, New York, 210 pp.
RAYMO, M. E. et al., 1989. Late Pliocene variation in northern hemisphere ice sheets and North Atlantic deep water circulation. *Paleoceanography*, 4: 413–446.
RENNE, P. R. and BASU, A. R., 1991. Rapid eruption of the Siberian Traps flood basalts at the Permo-Triassic boundary. *Science*, 253: 176–179.
RENNE, P. R. et al., 1992. The age of Parana flood volcanism, rifting of Gondwanaland and the Jurassic-Cretaceous boundary. *Science*, 258: 975–981.
RETALLACK, G. J., LEAHY, G. D. and SPOON, M. D., 1987. Evidence from paleosols for ecosystem changes across the Cretaceous/Tertiary boundary in eastern Montana. *Geology*, 15: 1090–1093.
ROBERT, C. and CHAMLEY, H., 1990. Paleoenvironmental significance of clay mineral associations at the Cretaceous-Tertiary passage. *Palaeogeog., Palaeoclimatol., Palaeoecol.*, 79: 205–219.
ROCCHIA, R., BOCLET, D., BONTE, P., BUFFETAUT, E., ORUE-ETXEBARRIA, JAEGER, J-J. and JEHANNO, C., 1988a. Structure of the iridium anomaly at the K-T boundary of the Sopelana section (Spanish Basque country). *C. R. Acad. Sci. Paris*, 307, serie II: 1217–1223.
ROCCHIA, R., BOCLET, D., COURTILLOT, V. and JAEGER, J.-J., 1988b. A search for iridium in the Deccan traps and inter-traps. *Geophys. Res. Lett.*, 15: 812–815.
SAITO, T., YAMANOI, T. and KAIHO, K., 1986. End-Cretaceous devastation of terrestrial flora in the boreal Far East. *Nature*, 323: 253–255.
SANCETTA, C., HEUSSER, L. and HALL, M. A., 1992. Late Pliocene climate in the Southeast Atlantic. Preliminary results from a multi-disciplinary study of DSDP Site 532. *Mar. Micropaleontol.*, 20: 59–75.
SANDBERG, C. A., ZIEGLER, W., DREESEN, R. and BUTLER, J. L., 1988. Late Frasnian mass extinction. Conodont event stratigraphy, global changes and possible causes. *Cour. Forsch.-Inst. Senckenberg*, 102: 263–307.
SCHNEIDER, S. H. and THOMPSON, S. L., 1988. Simulating the climatic effects of nuclear war. *Nature*, 333: 221–227.
SCHUBERT, J. K. and BOTTJER, D. J., 1992. Early Triassic stromatolites as post-mass extinction disaster forms. *Geology*, 20: 883–886.
SCHULTZ, P. H. and GAULT, D. E., 1990. Prolonged global catastrophes from oblique impacts. *Geol. Soc. Am., Spec. Pap.*, 247: 239–261.
SEARS, J. W. and ALT, D., 1992. Impact origin of large intracratonic basins, the stationary Proterozoic crust and the transition to modern plate tectonics. In: M. J. BARTHOLOMEW et al. (Editors), *Basement Tectonics 8*, Kluwer, Dordrecht, pp. 385–392.
SEBAI, A., FERAUD, G., BERTRAND, H. and HANES, J., 1991. $^{40}Ar/^{39}Ar$ dating and geochemistry of tholeiitic magmatism related to the early opening of the Central Atlantic rift. *Earth Planet. Sci. Lett.*, 104: 455–472.

SEPKOSKI JR., J. J., 1990. The taxonomic structure of periodic extinction. *Geol. Soc. Am., Spec. Pap.*, 247: 33–44.

SHACKLETON, N. J., BACKMAN, J. et al., 1984. Oxygen isotope calibration of the onset of ice-rafting and history of glaciation in the North Atlantic region. *Nature*, 307: 620–623.

SHARPTON, V. L., DALRYMPLE, G. B. et al., 1992. New links between the Chicxulub impact structure and the Cretaceous/Tertiary boundary. *Nature*, 359: 819–821.

SHEEHAN, P. M. et al., 1991. Sudden extinction of the dinosaurs. Latest Cretaceous, Upper Great Plains, U.S.A. *Science*, 254: 835–839.

SHOEMAKER, E. M., WOLFE, R. F. and SHOEMAKER, C. S., 1990. Asteroid and comet flux in the neighborhood of Earth. *Geol. Soc. Am., Spec. Pap.*, 247: 155–170.

SIGNOR III, P. W. and LIPPS, J. H., 1982. Sampling bias, gradual extinction patterns and catastrophes in the fossil record. *Geol. Soc. Am., Spec. Pap.*, 190: 291–296.

SIGURDSSON, H., 1982. Volcanic pollution and climate: the 1783 Laki eruption. *Eos, Trans. Am. Geophys. Union*, 63: 601–602.

SIGURDSSON, H., 1990. Assessment of the atmospheric impact of volcanic eruptions. *Geol. Soc. Am., Spec. Pap.*, 247: 99–110.

SIGURDSSON, H., D'HONDT, S., ARTHUR, M. A. et al., 1991. Glass from the Cretaceous/Tertiary boundary in Haiti. *Nature*, 349: 482–487.

SIGURDSSON, H., D'HONDT, S. and CAREY, S., 1993. The impact of the Cretaceous/Tertiary bolide on evaporite terrane and generation of major sulfuric acid aerosol. *Earth Planet. Sci. Lett.*, 109: 543–559.

SIMMS, M. J., 1990. Climatic and biotic change in the Late Triassic. *J. Geol. Soc., London*, 147: 321–327.

SIMMS, M. J. and RUFFELL, A., 1990. Synchroneity of climatic change and extinctions in the Late Triassic. *Geology*, 17: 265–268.

SMIT, J., 1990. Meteorite impact, extinctions and the Cretaceous-Tertiary boundary. *Geol. Mijnbouw*, 69: 187–204.

SMIT, J. and HERTOGEN, J., 1980. An extraterrestrial event at the Cretaceous-Tertiary boundary. *Nature*, 285: 198–200.

SMIT, J., MONTANARI, A. et al., 1992. Tektite-bearing, deep-water clastic unit at the Cretaceous-Tertiary boundary in northeastern Mexico. *Geology*, 20: 99–103.

STANLEY, S. M., 1984. Marine mass extinctions. A dominant role for temperature. In: M. H. NITECKI (Editor), *Extinctions*. University of Chicago Press, Chicago, IL, pp. 69–117.

STANLEY, S. M., 1988. Paleozoic mass extinctions. Shared patterns suggest global cooling as a common cause. *Am. J. Sci.*, 288: 334–352.

STANLEY, S. M., 1992. An ecological theory for the origin of *Homo*. *Paleobiology*, 18: 237–257.

STANLEY, S. M. and CAMPBELL, L. D., 1985. Neogene mass extinction of western Atlantic molluscs. *Nature*, 293: 457–459.

STORETVEDT, K. M., MITCHELL, J. G., ABRANCHES, M. C., MAALOE, S. and ROBUIN, G., 1992. The coast-parallel dolerite dykes of east Madagascar; age of intrusion, remagnetization and tectonic aspects. *J. Afr. Earth Sci.*, 15: 237–249.

STOTHERS, R. B., 1993. Flood basalts and extinction events. *Geophys. Res. Lett.*, 20: 1399–1402.

STOTHERS, R. B., WOLFE, J. A., SELF, S. and RAMPINO, M. R., 1986. Basaltic fissure eruptions, plume heights and atmospheric aerosols. *Geophys. Res. Lett.*, 13: 725–728.

STOTT, L. D. and KENNETT, J. P., 1989. New constraints on early Tertiary palaeoproductivity from carbon isotopes in foraminifera. *Nature*, 342: 526–529.

STOTT, L. D. and KENNETT, J. P., 1990. The paleooceanographic and paleoclimatic signature of the Cretaceous/Paleogene boundary in the Antarctic. Stable isotope results from ODP Leg 113. *Proc. Ocean Drilling Prog., Sci. Results*, 113: 829–848.

SWISHER III, C. C. and PROTHERO, D. R., 1990. Single-crystal $^{40}Ar/^{39}Ar$ dating of the Eocene-Oligocene transition in North America. *Science*, 249: 760–762.

THIERSTEIN, H. R., 1980. Selective dissolution of late Cretaceous and earliest Tertiary calcareous nannofossils. Experimental evidence. *Cretaceous Res.*, 2: 165–176.

THIERSTEIN, H. R., 1981. Late Cretaceous nannoplankton and the change at the Cretaceous-Tertiary boundary. *Soc. Econ. Paleont. Mineral., Spec. Publ.*, 32: 355–394.

THIERSTEIN, H. R., 1982a. The terminal Cretaceous extinction event and climatic stability. In: W. H. BERGER and J. C. CROWELL (Editors), *Climate in Earth History*. National Academy Press, Wash-

Chapter 5

Observations from the surface: projection from traditional meteorological observations

P. D. JONES

Introduction

Although meteorological recording began earlier, the period since the middle of the 19th century is that traditionally associated with instrumental records. This chapter examines in detail the reliability of temperature, precipitation and sea-level pressure data, concentrating on the construction of the hemispheric and global average temperature series. The key piece of observational evidence in the "global warming debate" is the "global" temperature series (JONES and WIGLEY, 1990). How much has the temperature risen and how certain are we of the causes? The remainder of the chapter considers these issues, together with an assessment of the record in a longer paleoclimatic context (see also Chapter 6 by DIAZ and KILADIS) and the representativeness of the longer surface record with respect to free-atmosphere observations since the late 1950s. The chapter concludes by considering possible explanations of the instrumental record (for more discussion of the possible explanations see also Chapter 4 by RAMPINO, Chapter 6 by DIAZ and KILADIS, Chapter 9 by WANG et al., Chapter 10 by ANDREAE, Chapter 11 by BRASSEUR et al. and Chapter 12 by HENDERSON-SELLERS) and indicates what each suggests concerning the course of climate change in the next century.

Development of instrumentation and meteorological observational networks

The instrumental period of weather recording began in the late 17th century with the development of barometers and thermometers. Up until the late 18th century, the early observers were dedicated amateurs, mostly physicians, parsons and country squires and their equivalents in countries other than England. Careful analyses are necessary to enable the early records to be used. Calibration scales and the exposure of the early instruments must be scrutinized with, in some cases, experiments performed to duplicate the conditions (CHENOWETH, 1992, 1993). Long records have been developed for some European regions (e.g. MANLEY (1974) and PARKER et al. (1992) for Central England; DETTWILLER (1981) for Paris and SCHAAKE (1982) for Berlin).

There was some correspondence and liaison between the early observers through societies such as the Royal Society in London, but little central organisation. Many of the original data have been lost and are only available in society proceedings. Sometimes the entire records of early observers have been published (e.g. KINGTON (1988a) for Lyndon in England). The first real weather observing network was set up in 1778 by the *Société Royale de*

Médicine in France under the patronage of Louis XVI. The French meteorologist Louis Cotte was appointed to establish and maintain the network of observers. Meanwhile in Mannheim, then the capital of the Palatinate of the Rhine, the Elector Karl Theodor founded the *Societas Meteorologica Palatina* in 1780. In both cases detailed instructions on how and when to record information were sent to observers. In the German case, standard instruments were also issued. The networks, however, did not survive the change in the political climate in Europe in the late 18th and early 19th centuries and both ceased in the 1790s. KINGTON (1988b) has produced, from the station data, weather maps for most of Europe for the 1780s.

Further expansion of the meteorological recording network began following the Vienna Meteorological Congress of 1873. This meeting provided the impetus for the expansion of recording networks worldwide and for the international exchange of data that continues today under the auspices of the World Meteorological Organization (WMO). When national meteorological centres were established in the late 19th century, observing networks expanded, instruments and instrument shelters were standardised, and uniform instructions to observers were issued. As a result, a reasonably comprehensive global network of temperature, precipitation and pressure recording stations emerged, many of which have continued in operation to the present. Unfortunately, standardisation of instruments and exposure did not extend to a universally adopted policy so that, even today, vastly different protocols are followed. The effects of this are discussed in the remainder of this section. Further discussion on the history of instrumentation has been presented by LAMB and JOHNSON (1966), VON RUDLOFF (1967) and MIDDLETON (1966, 1969).

Land areas represent only 30% of the Earth's surface and it is important to consider marine weather observations. Both merchant and naval vessels have taken observations of weather, including measurements of the temperature of the sea surface (SST), since the 1850s. Data were stored in log books by the major maritime nations following an international agreement signed in Brussels in 1853. The force behind this was an American naval captain, Matthew Fontaine Maury. In the last 20 years, major international efforts have been made to put all this log-book information into computer data banks. Two such compilations exist, the Comprehensive Ocean-Atmosphere Data Set (COADS) produced at Boulder, Colorado (WOODRUFF et al., 1987) and the UK Meteorological Office marine data bank (BOTTOMLEY et al., 1990). About 80 million individual observations of SST exist between 1854 and 1990.

Homogeneity of temperature records

In this section we define homogeneity according to CONRAD and POLLAK (1962), who call a numerical series representing the variations of a climatological element homogeneous if the variations are caused only by variations of weather and climate. It is vital that all meteorological observations contain as few non-meteorological variations as possible. Although observers may take readings with meticulous care, other influences can easily affect the readings. The following sections discuss the most important causes of non-homogeneity.

- changes in instruments, exposure and measurement technique;
- changes in station location;
- changes in observation times and the methods used to calculate monthly averages;
- changes in the station environment with particular reference to urbanisation.

Changes in instrumentation, exposure and measuring techniques

For land-based temperature, the effects of changes in thermometers have been slight so far this century. More important to the long-term homogeneity is the exposure of the thermometers. Readings now are generally from thermometers located in louvered screened enclosures, which are generally painted white. Earlier readings often were from shaded wall locations, sometimes under roof awnings. These need corrections which can only be estimated by parallel measurements or by reconstructing past locations (CHENOWETH, 1992, 1993).

The greatest potential discontinuity in temperature time series is currently being induced in first order stations in the United States. Here mercury-in-glass thermometers have been replaced by a thermistor installed in a small louvered housing (model HO-83). The change was made by the National Weather Service in the USA at most first order sites during early 1982. There appears to have been little thought given to effects on the homogeneity of the record. The continuity of the record for climatic purposes was only a minor consideration of the weather service. Parallel measurements were not made as is generally recommended.

The first signs of a problem were noticed by the frequent breaking of records for extreme daily maximum temperatures. A warm bias was clearly evident, possibly being indicative of an aspiration problem in the replacement instrumentation (GALL et al., 1992; KESSLER et al., 1993). As more developed countries might follow the lead of the United States, the problem is likely to recur. However, with good design and adequate testing through parallel observations, the problems induced by any changes should be surmountable.

Marine temperature data are seriously affected by homogeneity problems. SST data were generally taken before World War II by collecting some seawater in an uninsulated canvas bucket and measuring the temperature. As there was a short time between sampling and reading, the water in the bucket cooled slightly due to evaporation. Since World War II most readings are made in the intake pipes which take water on to ships to cool the engines. The change appears to have been quite abrupt, occurring within a month during December 1941. The sources of ships' observations in the two marine data banks (COADS and UKMO) changed from principally European to mainly American at this time (see also FOLLAND and PARKER, 1995).

Studies of the differences between the two methods indicate bucket temperatures are cooler than intake ones by 0.3–0.7°C (JAMES and FOX, 1972). A correction technique for bucket measurements has been derived based on the physical principles related to the causes of the cooling (BOTTOMLEY et al., 1990; FOLLAND and PARKER, 1995). The cooling depends on the prevailing meteorological conditions, and so depends on the time of year and on location. Although a day-to-day phenomenon, the various influences on the bucket are basically linear, so cooling amounts can be calculated on a monthly basis. Assuming that there has been no major change in the seasonal cycle of SSTs over the last 100 years, the time between sampling and reading can be estimated as that which best minimises the residual seasonal cycle in the pre-World War II data. The estimate of between 3 and 4 min is both quite consistent spatially and agrees with instructions given to marine observers (BOTTOMLEY et al., 1990; FOLLAND and PARKER, 1991, 1995).

Coverage of SST data is generally dictated by shipping routes and is, therefore, poor over the Southern and tropical oceans. From the late 1970s, it has been possible to include infor-

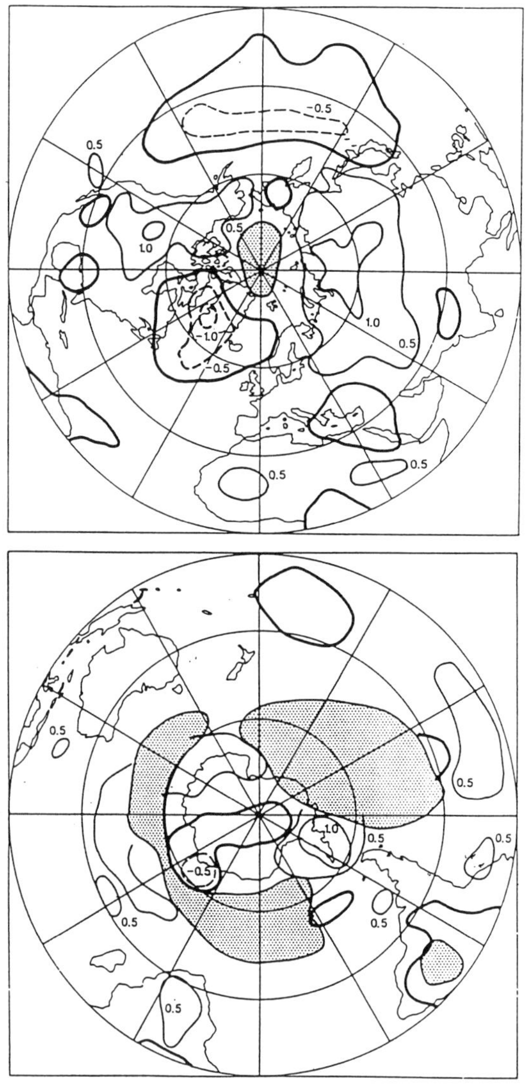

Fig. 6. Annual temperature anomalies for the Northern and Southern Hemispheres. The anomalies are based on the 1950–1979 period. The contour interval is 0.5°C, with negative anomalies dashed. Shaded areas indicate regions with insufficient data.

major exception to this was a part of eastern Antarctica during MAM and SON. In tropical and temperate latitudes Australia, the Indian Ocean and the eastern equatorial Pacific were warmer than normal. The regions with the greatest warming over Antarctica were the Australian sector in JJA, SON and DJF and the Antarctic Peninsula in MAM and JJA.

The annual anomaly maps for the two hemispheres (Fig. 6) obviously encompass the average features of the four seasons. Over the Northern Hemisphere relative warmth prevailed over most of Eurasia, especially Siberia, the western half of North America and western Africa. Temperatures were cooler during the decade over the North Pacific Ocean and the northern Atlantic region surrounding Greenland and Iceland. Most regions of the Southern Hemisphere have experienced warmth during the decade except for parts of eastern Antarctica, the Amazon Basin and part of the southwestern Pacific. Warmth during the decade was greatest over the Indian Ocean, southern Africa, the eastern equatorial Pacific Ocean, the Antarctic Peninsula and the Australian sector of Antarctica.

Although this decade has clearly been the warmest recorded since the beginning of the instrumental era, it is clear that the warming has not been global in extent. The term "global warming" therefore, is something of a misnomer. The warmth has also varied seasonally, being more significant in the boreal spring and winter and less in the boreal autumn and summer. Assessment of the spatial patterns of change is particularly important when considering possible causal factors for the changes.

Diurnal temperature change

Most of the focus of past and future temperature change has been concerned with mean temperature. For past changes, this has resulted from a lack of readily available databases of monthly mean maximum and minimum temperatures. These data, since November 1994, are routinely exchanged between countries. For models, the diurnal cycle in the control runs is often poorly simulated. Few GCM experiments have so far considered maximum and minimum temperatures and the associated diurnal range. The issue is, however, particularly important with regard to climate change impact studies. The impacts of a change will, for example, be very different if the change affects only minimum temperatures rather than affecting both minimum and maximum equally.

Recently developed data sets (KARL et al., 1993) have permitted analyses of maximum and minimum temperatures to be made for 37% of the global land mass (encompassing the contiguous United States, Canada, Alaska, the former Soviet Union, China, Japan, the Sudan, South Africa and eastern Australia). The analyses indicate (Fig.7) that over the 1952–1989 period, minimum temperatures have risen at a rate three times that of maximum temperatures. The reduction in the diurnal temperature range is approximately equal to the tempera-

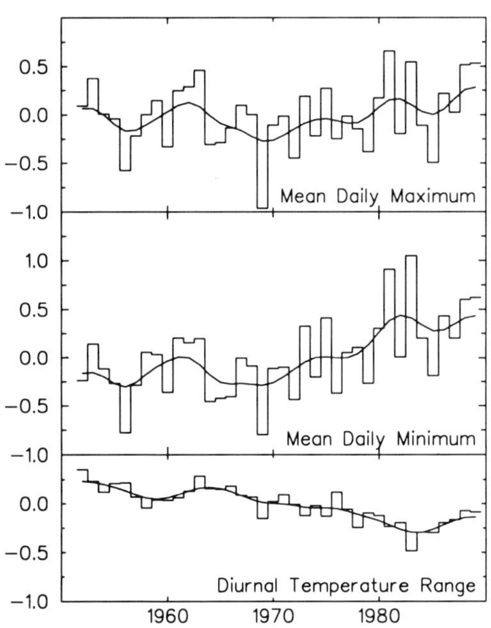

Fig. 7. Annual maximum, minimum temperatures and the diurnal temperature range for 1952–1989.

period changes the appearance of the time series from the "Sahel drought" of the last 20 years (relative to 1931–1960) to the "Sahel wet period" of the 1920s to the 1960s (relative to 1961–1990). The difference between the two analyses is not a simple change in level because of the use of standardised anomalies. Interannual variability is greater at most stations during the 1961–1990 period compared to 1931–1960. The reasons for the decline in rainfall have been widely discussed and a variety of causes have been postulated (LAMB, 1978; LOUGH, 1986; NICHOLSON, 1989). Rainy season (June–September) totals for the region appear to be partially predictable from the previous seasons' (March–May) sea surface temperatures in the Atlantic Ocean (FOLLAND et al., 1986; ROWELL et al., 1992).

Global precipitation data sets have been developed recently, principally to assess the performance of GCMs (see section on Precipitation and pressure homogeneity). Assessment is achieved by comparing spatial patterns of model (control) and real-world precipitation climatologies (HULME, 1991). GCM climatologies, however, with 2× CO_2 levels (perturbed) indicate an enhanced hydrological cycle with between a 7 and 15% increase in global total precipitation (MITCHELL et al., 1990). A reason for analysing global-scale precipitation, therefore, is to see if any increase is occurring. Detection of future climate change is developed in more detail later. Although large-scale precipitation series are a poor detection variable, because of greater uncertainties in both observed and model data, the impacts of climate change will be principally felt through changes in precipitation. Runoff, for example, is much more sensitive to precipitation than temperature (WIGLEY and JONES, 1985).

Hemispheric- and global-scale precipitation have been studied in two papers (BRADLEY et al., 1987a; DIAZ et al., 1989). In both studies, the non-normal distribution of precipitation and its large spatial variability were accounted for by transforming all seasonal and annual precipitation totals using the gamma distribution. With this, precipitation totals are scaled between 0 and 1, median values that occur half the time having a value of 0.5. The distribution was fitted over the 1921–1960 period, and for each station, seasonal and annual estimates of the scale and shape parameters of the distribution were made. Large scale averages were calculated after the transformed precipitation data were interpolated onto an equal-area grid.

For the Northern Hemisphere (BRADLEY et al., 1987a), the most interesting finding was a decline in precipitation during the last 40 years over subtropical areas (5–35°N) and an increase in mid-latitudes (35–70°N). These trends are consistent with many other analyses and are partially accounted for by the Sahel and the former Soviet Union trends discussed earlier. For the NH as a whole, there is considerable decade-to-decade variability but little overall trend. Northern summer and autumn seasons show a slight decline whereas winter and spring seasons show an increase. These trends are in accord with hemispheric temperature changes, providing evidence for a more intense hydrological cycle when average temperatures are higher.

The Southern Hemisphere analyses (DIAZ et al., 1989) indicate increases in precipitation in both subtropical and temperate latitudes. The rise in precipitation is primarily due to increases in autumn and spring over South America. A slight increase is evident over Australasia and no change over southern Africa. The changes in precipitation in both hemispheres are broadly consistent with a number of GCM results with doubled CO_2 levels (MITCHELL et al., 1990). The consistencies, however, are only on the largest of scales and do not survive scrutiny on small spatial and seasonal scales.

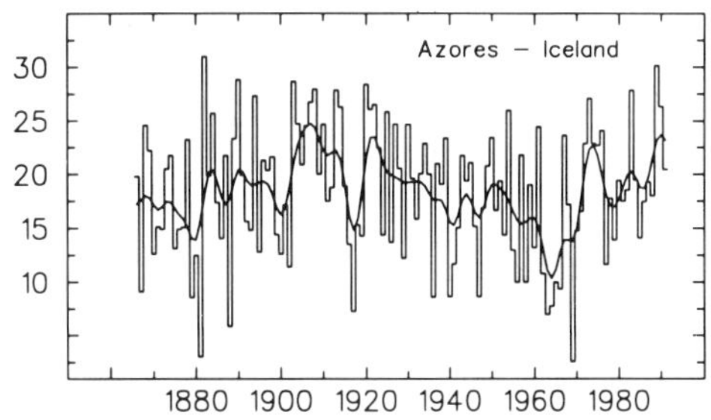

Fig. 10. Winter (November–March) pressure difference (mb) between Ponta Delgada, Azores and Stykkisholmur, Iceland. Year dated by the January. Higher values indicate stronger westerlies over the North Atlantic.

Pressure

The interrelationships of the climate system mean that past changes in the patterns of temperature and precipitation patterns will undoubtedly also be accompanied by changes in circulation throughout the troposphere. Changes that have been noted range from local-scale studies such as the decline in westerlies over the British Isles (LAMB, 1972) to much larger regional scale indices such as the North Atlantic Oscillation (VAN LOON and ROGERS, 1978), North Pacific winter pressure (TRENBERTH, 1990) and various Southern Hemisphere indices (including the Southern Oscillation) (KAROLY, 1995).

The North Atlantic/European region has probably been studied more than other parts of the world, almost certainly because of the availability of long (>150 years) records. The well-known out-of-phase relationship between winter temperatures over northern Europe (especially Fennoscandia) and Greenland has long been recognised (VAN LOON and ROGERS, 1978). The link between surface temperature anomalies in the region and the circulation of the North Atlantic region is evident in indices such as the North Atlantic Oscillation (NAO). Figure 10 shows seasonal (November–March) values of the NAO (difference in pressure between Iceland and the Azores). Stronger westerlies prevailed during the first two decades of the 20th century and again during the 1970s and 1980s. The weakest westerlies occurred during the 19th century and the 1960s.

In the North Pacific region a major change in the strength and position of the Aleutian Low took place during the mid-1970s. After about 1977 the strength of the low pressure centre

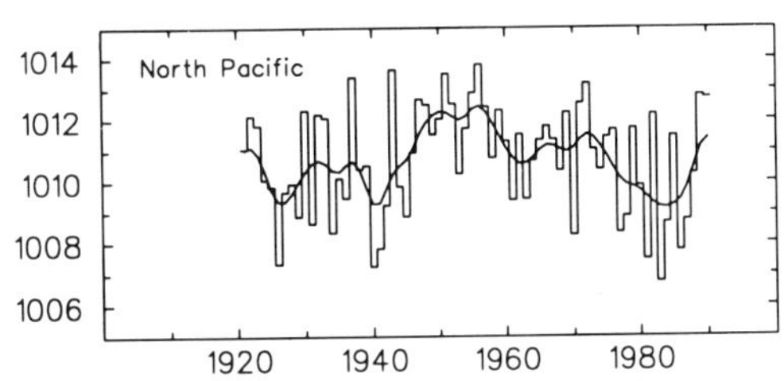

Fig. 11. Winter (November–March) pressure average (mb) for the North Pacific 30–70°N, 150°E-150°W. Year dated by the January.

increased and the centre moved slightly south particularly in winter (Fig. 11). In many respects the change may have been a return to the conditions that prevailed during the 1920s, 1930s and the early 1940s. Some of the change since the 1970s has reversed in the most recent years. The effects of the change in the mid-1970s in the climate of the whole N. Pacific region have been discussed by numerous workers (e.g. DOUGLAS et al., 1982; NITTA and YAMADA, 1989; TRENBERTH, 1990). The decreased pressures led to greater advection of warm maritime air on to the northwest of North America leading to enhanced precipitation and warmer temperatures. Alaska has experienced the greatest regional warmth over the Northern Hemisphere with average decadal temperatures 1°C warmer than the 1951–1980 period during the 1980s. In contrast sea temperatures have been much cooler than normal in the central North Pacific and cooler air temperatures have been experienced over Japan and the Okhotsk/Kamchatka region. Although the region only explains part of the hemispheric increase, the change in climate of the region, particularly in the winter, has been abrupt. Some of the cause of the extratropical circulation change may be tropical in origin due to a lack of any significant La Niña (see PHILANDER 1985 and Chapter 6 by DIAZ and KILADIS) events (TRENBERTH, 1990). Modelling studies support the tropical Pacific link (GRAHAM, 1995).

The last 140 years in a longer term context

The purpose of this section is to address, partly, the question of how much of the rise of temperatures can be explained by natural fluctuations. In addressing this issue we need to consider what is known about climatic variations of the last 1,000 years. Prior to the development of instruments, climatic information must be inferred from various naturally recording phenomena as well as from written historical records. The reconstruction of past climates is generally referred to as paleoclimatology. Numerous indicators exist, e.g. trees, ice cores, historical records, corals, varves, glacier movements, etc. (see BRADLEY, 1985 for a complete discussion of the methods and their strengths and weaknesses). In the following sections the changes in climate inferred, from the proxy climate sources, for the last 1,000 years are surveyed (see WILLIAMS and WIGLEY, 1983; JONES and BRADLEY 1992; BRADLEY and JONES, 1992a, 1993; see Chapter 6 by DIAZ and KILADIS for more details).

Over the last 1,000 years three main episodes are evident in many, but not all, proxy records from the Northern Hemisphere. Although less well studied, Southern Hemisphere evidence from South America, Australia and New Zealand suggests that these main events also occurred there. The first episode is a cool period spanning approximately the 9th and 10th centuries. Following this a warm period, the "Medieval Warm Period", occurred which was at its maximum during the early 12th century A.D. After this period, climate entered the period known as the "Little Ice Age". This period was marked by cooler conditions between the 14th and 19th centuries, although not always. It was probably cool from 1300 to 1450 and again from 1550 to 1850 (WILLIAMS and WIGLEY, 1983).

A general warming of the climate from the 10th to the 12th century is evident from both North America and Europe. By the 12th century, trees were living beyond their present limits in Alaska and the Yukon. In Europe winter temperatures were generally mild. Conditions in Greenland were best for human exploitation during the 10th and 11th centuries and this

allowed the establishment of two main Norse settlements in western and southern Greenland with combined populations of over 5,000 people (McGOVERN, 1981). The ensuing cooler conditions are confirmed by glacier advances in North America and Europe during the 14th, 16th and 17th centuries. The advances in different regions were never synchronous, but they point to cooler conditions for about 300 years interspersed with a few warmer episodes in different regions at different times (GROVE, 1988). Paintings of glaciers in Switzerland (ZUMBUHL, 1976) show evidence of cooler summers, with the glaciers reaching their maximum extent in the late 17th century. Tree ring evidence in Switzerland and Scandinavia indicates similar timings. Hill farms were abandoned in many areas of upland Europe.
It is impossible to say when the coldest periods were globally, because of different seasonal timings of the coldest phases in different parts of the world. From an analysis of 16 "summer-responding" proxy records in the middle-to-high latitudes of the Northern Hemisphere BRADLEY and JONES (1993) conclude that the coldest conditions of the last 560 years were between 1570 and 1730 and in the 19th century. Longer single-site and composite instrumental records confirm the annual warming from the late 19th century. The long European records, however, show that the 1880s were the coldest decade, at least since 1700, so part of the warming since then may reflect this unusually low starting point.

Upper air data during the last 40 years

There are two methods for measuring temperatures at different levels in the atmosphere. These are the conventional radiosonde network (ANGELL, 1988) and the microwave-sounding unit (MSU) on the TIROS-N series of satellites (SPENCER and CHRISTY, 1990 and see also Chapter 7 by DICKINSON). The conventional network extends back to 1958 and the MSU data to 1979. Before comparing the two sets of analyses with each other and with the surface, their advantages and limitations are discussed. Both sets of information are used routinely in global scale analyses undertaken at several operation centres (e.g. ECMWF, NMC, World Meteorological Centre-C at Melbourne), but changes to the operational models and data assimilation schemes mean that the homogeneity of long-time series (as for surface pressures also, see above) must be questionable (TRENBERTH, 1992). In the following it is preferable to use the original station and satellite observations rather than the analysed fields. There are many potential problems even with the original station measurements due to changes in sonde design and the timing of observations (see GAFFEN, 1994; PARKER and COX, 1994)
The conventional radiosonde network began in the late 1940s, but an important instrument alteration in the mid-1950s and a change of observation timing means that the most consistent long series begin in 1958 (ANGELL, 1988). Extension back prior to 1958 may be possible but would be limited spatially to the mid-to-high latitudes of the Northern Hemisphere because of data availability. The most widely studied data set is that produced by Angell and Korshover (see ANGELL, 1988) using a network of 63 stations spread as evenly over the world as data availability allows. The data are available seasonally from 1958 for seven zones of the world, each the average of the stations in each zone. Using up to four times as many stations as ANGELL (1988) for the northern polar zone, KAHL et al. (1993) have confirmed the earlier results. The spatial representativeness of the Angell network on a global

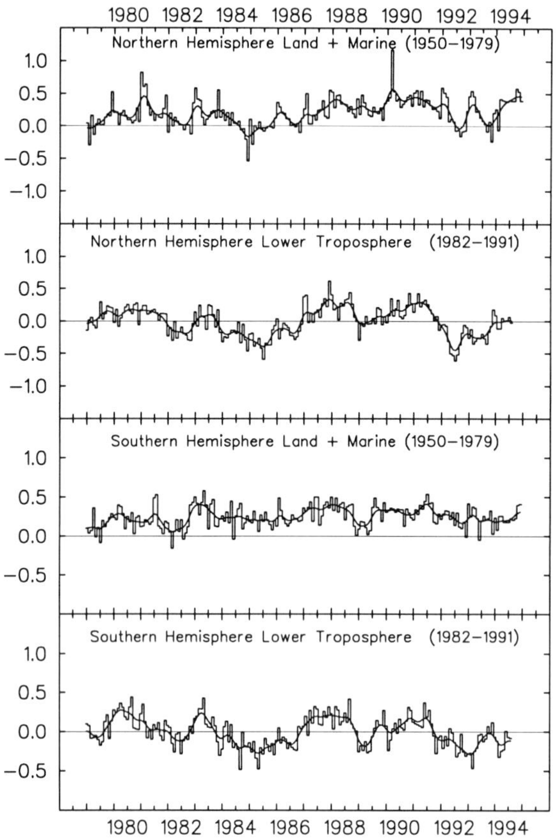

Fig. 12. Hemispheric monthly averages of land and marine temperatures (surface; data from Figs. 1 and 2) and MSU2R data (lower troposphere; SPENCER and CHRISTY, 1992b). The time intervals given in each panel are the reference intervals from which anomalies have been calculated.

Explanations of the instrumental temperature record

Explanations of changes in hemispheric- and global-mean climate over the past 100 years or so have concentrated on temperature variations. Temperature is the most reliably measured variable, in terms of accuracy of regional averages, and the most widely studied. Changes in precipitation and pressure, particularly on the regional scale, have undoubtedly occurred and variations in all three are strongly interrelated. It might be suspected that precipitation totals over global land-areas have increased over the 20th century but we are not as confident of the data as we are for the 0.5°C rise in global temperatures. Despite the fact that precipitation variations have a greater impact on human activities than variations in other variables,

TABLE III

INTERCORRELATIONS BETWEEN RADIOSONDE (100–50MB) AND MSU4 DATA (ABBREVIATIONS AS IN TABLE I)

Time period	Correlation pair	NH	SH	Global
Seasonal (52)	MSU4 v 100–50	0.89	0.78	0.86
Annual (13)	MSU4 v 100–50	0.93	0.97	0.99

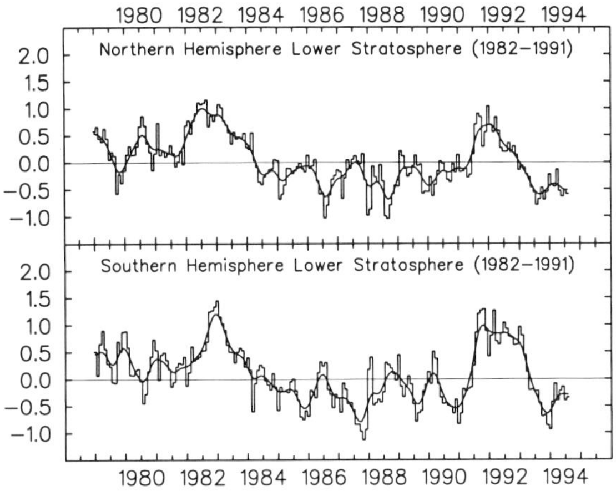

Fig. 13. Hemispheric monthly averages of MSU4 data (lower stratosphere; SPENCER and CHRISTY, 1992b). The time intervals given in each panel are the reference intervals from which anomalies have been calculated.

we concentrate here on global-mean temperature, which is a fundamental measure of the state of the climate system. All variations in the historic and geologic past are considered as cold or warm epochs.

The possible causal factors affecting global-mean temperature are conveniently grouped as either "internal" or "external" (ROBOCK, 1978). Internal factors are those which must be considered even in the absence of potentially larger changes in external forcing. Factors include natural changes in planetary albedo resulting from changes in cloudiness or in surface characteristics and changes in the circulation of the atmosphere and/or the ocean. The latter factor determines the lower boundary conditions for the atmosphere and the rates of heat exchange between the ocean and atmosphere. The former factors determine the spatial distribution of heat sources and sinks, as well as the horizontal and vertical fluxes of sensible and latent heat. Man-induced surface albedo changes should strictly be classified as external but, although their effects may be important at the regional scale, they are unlikely to be at the global scale.

The main external factors are changes in the Sun's output or luminosity, changes in the fraction of shortwave solar radiation reaching the troposphere (due to injection of particulate matter and sulphate aerosols in the stratosphere following volcanic eruptions), changes in the reflectivity of clouds due to industrial activity and changes in the vertical distribution of outgoing longwave radiation flux in the troposphere due to increasing concentrations of greenhouse gases.

Internal forcing of the climate

It is likely that most of the interannual variability shown in Figs. 1–3 is the result of natural variability, even though the precise causes cannot be quantified. On slightly longer time-scales, such as 2–8 years (but still classed as natural variability as it results from internal forcing), the ocean circulation and related changes in sea surface temperatures in the tropical Pacific Ocean associated with the El Niño/Southern Oscillation (ENSO) phenomenon (RASMUSSEN and CARPENTER, 1982; PHILANDER, 1983; see also Chapter 6 by DIAZ and

KILADIS) have a noticeable effect on hemispheric and global temperatures. This phenomenon, along with the similar long-term trends, is responsible for much of the high annual correlation ($r = 0.79$, 1901–1990) between the two hemispheres. Many warm years relative to the filtered curves have been shown to be El Niño or "warm event" years (BRADLEY et al., 1987b; VAN LOON and SHEA, 1985). In contrast many cold years relative to the filtered curve have been previously classified as La Niña (PHILANDER, 1985) or "cold event" years. Even during the last 12 years the warm and cold event years can be seen in the MSU2R record (see Fig. 12). Using the Southern Oscillation Index (SOI), the pressure difference between Tahiti and Darwin, (see ROPELEWSKI and JONES, 1987; ALLAN et al., 1991; and Chapter 6 by DIAZ and KILADIS for more details), JONES (1989) has removed the ENSO influence from the annual hemispheric and global temperature series. The SOI explains about 25–30% of the high frequency (<20 years) variations in hemispheric temperature anomalies (see Fig. 14). Removal of the influence may assist in the early detection of greenhouse gas-related trends (NICHOLLS and KATZ, 1991). When the ENSO effect is factored out, the warming in recent years becomes more obvious and the long-term warming more significant.

In explaining the century time-scale temperature history we need to consider natural variability on decadal and longer scales. It is not possible to estimate the scale of natural internal variability from the observational record because there is only a single realisation of global temperature change on this time-scale and this record is almost certainly contaminated by many external factors. We do know, however, that considerable decadal and longer time-scale variability should occur as a result of the way the ocean modulates all high-frequency forcing. Because of the ocean's large thermal inertia, the system tends to behave like a first

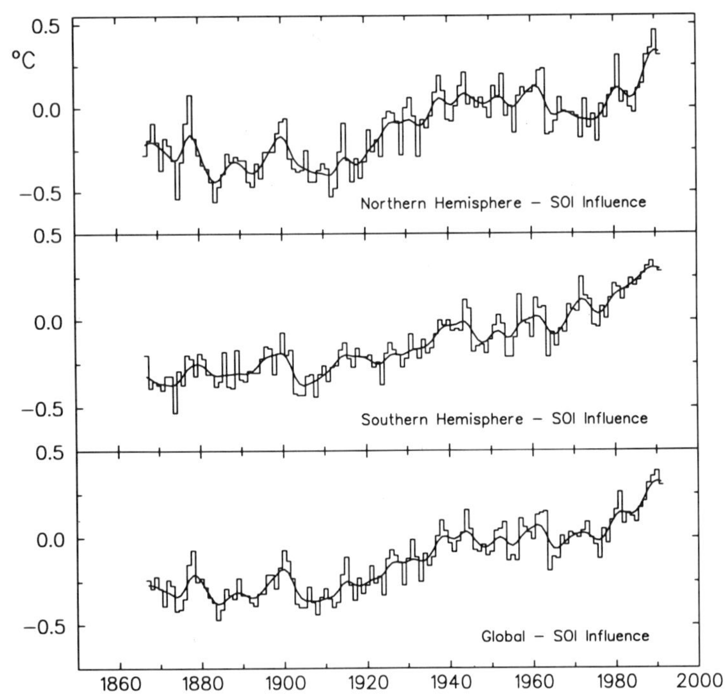

Fig. 14. Hemispheric and global surface air temperatures for the 1867–1992 period after subtracting the "SOI explained" part of the temperature record.

order autoregressive process in which white-noise atmospheric forcing is converted to a red-noise response (HASSELMANN, 1976). In this way even if there are only purely random changes in the atmospheric part of the system, the large heat capacity will produce a shift towards lower frequencies. This effect can be quantified by using an appropriate energy balance model, forcing it with white noise of an amplitude sufficient to reproduce the observed high frequency variations of global mean temperature. WIGLEY and RAPER (1990, 1991) have run such a model and have shown that low-frequency changes in global mean temperature similar to those that occur can be produced. Similar effects have been seen in some GCM results (STOUFFER et al., 1989; HANSEN et al., 1988). On the century time-scale, these variations could easily be as much as 0.2–0.3°C. Thus, up to half of the observed warming since the 19th century could be a natural, internal fluctuation. Equally well, however, a much larger externally forced warming of 0.7–0.8°C could have been partially offset by a natural, internally generated cooling.

External forcing of the climate

The most widely considered factors are changes in solar irradiance, the effects of explosive volcanic eruptions and the enhanced greenhouse effect. Solar irradiance changes and their effect on weather and climate are one of the most studied areas of climatology. Most theories have been shown to fall down when statistically tested with independent data (PITTOCK, 1979, 1983). Recently solar weather relationships have received a resurgence of interest, following work by VAN LOON and LABITZKE (1988) (see next section).

It is only recently that we have obtained, through satellite observations, evidence that the Sun's output does vary in parallel with the 11-year sunspot cycle (WILLSON and HUDSON, 1991; see also Chapter 7 by DICKINSON). These variations, however, are small, amounting to at most about 0.1%. Because only 12% of this is absorbed at the Earth's surface (the intercepted radiation covers only a quarter of the Earth's surface and 30% is reflected back to space) the change in forcing of 0.24 W m^{-2} would be virtually impossible to detect. If the climate system were able to reach a new equilibrium state after the forcing, a change of 0.08–0.24°C could occur at the surface. However, as the climate system takes decades to reach a new equilibrium and the forcing is approximately cyclic with a period of about 11 years the response is less than 0.03°C in global-mean terms. Periods of prolonged sunspot minima such as the Maunder, Spörer and Wolf minima have occurred in the historic past (EDDY, 1976; STUIVER and BRAZIUNAS, 1989). These were long enough for the climate to reach a new equilibrium and they may be the cause of the Little Ice Age events (WIGLEY and RAPER, 1995).

Solar radius variations have been suggested as a cause of variations over the past 100 years (GILLILAND, 1981, 1982). These should affect solar output, according to astrophysical studies, with a cycle of about 80 years, and there is evidence of quasi-cyclic radius changes over the past few centuries. A new approach (FRIIS-CHRISTENSEN and LASSEN, 1991) uses the length of the sunspot cycle which is known to be related to solar activity. The authors achieve a high correlation (~0.95) between highly smoothed cycle length data and similarly smoothed hemispheric average temperatures for the Northern Hemisphere. Further discussion of this relationship is given by KELLY and WIGLEY (1992) and SCHLESINGER and RAMANKUTTY (1992). Both ideas, however, suffer from the same problem – namely that the forcing is not great enough and the nature of the mechanism, and presumably also any required amplification, is not known. Only time, and satellite observations of solar irradiance (measurements above the atmosphere) and other solar parameters such as the radius, will eventually resolve the uncertainties surrounding solar forcing.

The situation with regard to volcanic influences is more clear cut, at least on short time-scales. It was Benjamin Franklin, in the 1780s, who first associated unusual weather in Europe with an Icelandic volcanic eruption in 1783. Curiously, this particular eruption (Skaftáreldar, often referred to as Laki) was not of the explosive type now associated with effects on climate. Hemispheric climatic effects generally only occur if the ejected dust and gases reach the stratosphere, since it is only here that their residence time is sufficient to have a noticeable cooling effect. Normally, however, such a layer is only likely to occur through the violence of an explosive eruption and the rapid and direct injection of dust and sulphur dioxide into this part of the atmosphere.

There is now a considerable body of modelling and observational evidence for the cooling effect of individual volcanic eruptions (BRADLEY, 1988; BRADLEY and JONES, 1992b, 1993; see also Chapter 4 by RAMPINO). Major eruptions, of the size, for example, of Krakatoa (1883) appear to have caused short time-scale cooling of the lower atmosphere of a few tenths of a degree Celsius, beginning within a few months of the eruption and noticeable for 1 or 2 years subsequently (KELLY and SEAR 1984; SEAR et al., 1987). Less violent eruptions, if they have high sulphur dioxide emissions, may also have noticeable effects: the best example here is the eruption of Agung in 1963 (ANGELL and KORSHOVER, 1985). In recent centuries, the eruption of Tambora in 1815 appears to have been the most violent. The following summer was so cool and wet in Europe and eastern North America that 1816 came to be known as "the year without a summer" (HARRINGTON, 1992). Unfortunately, Tambora occurred too early for us to be able to quantify its cooling effect reliably in global terms.

When the record of historical information on major eruptions is compared with hemispheric temperatures in Figs. 3 and 14, short term cooling for a year or two can be seen to be associated with the eruptions of Krakatoa (1883), Pelee/Soufrière/Santa Maria (1902) and Agung (1963). The effect of the recent eruption of Mt. St. Helens (1980) was negligible because most of the explosion was directed more sideways than vertically, while the effects of El Chichón (1982) and Mt. Pinatubo (1991) may have been, to some extent, confounded by El Niño events which occurred at about the same time. Clear cooling, however, is evident during the period April–November 1992 and again during the same period in 1993 in both surface and MSU2R data (Fig. 12). The stratospheric warming following the event is apparent in Fig. 13, despite being superimposed on the compounding effects of other sources of natural variability. Modelling studies (HANSEN et al., 1992) indicated a cooling after the Mt. Pinatubo eruption of about 0.5°C in global terms. The effects of the eruption may not be finished yet, but the magnitude of the observed cooling is at most 0.3°C for annual means. The tropospheric/stratospheric contrast is less clear after the El Chichón eruption. When just the annual mean values are examined (Figs. 3 and 14), however, there are many short term cooling events that are *not* associated with volcanic activity. It is necessary, therefore, to go to the monthly time-scale to identify volcanic effects clearly (SEAR et al., 1987).

The decadal-to-century time-scale effects of volcanic eruptions are much more debatable, although sound arguments for their importance can be made (BRADLEY and JONES, 1993). One of the problems here is that we have no convincing continuous record of the volcanic aerosol forcing, and so no way to say with any confidence what the longer time-scale climatic response might be. There are numerous independent indicator time series which might be (and have been) used. These include the historical record of eruptions together with estimates of their strength, the Volcanic Explosivity Index (SIMKIN et al., 1981) and LAMB's (1970) record of the Dust Veil Index. This latter index uses a variety of sources including tephra volume and unusual sunsets to assign an index value to each eruption. For a few eruptions in the 17th century, Lamb uses the presumed temperature cooling following the eruption. Another historic record is the record of atmospheric transparency which, unfortunately, comes from a rather limited network of observation sites (PIVOVAROVA, 1977). There are also records of sulphate concentration and acidity in Greenland and Antarctic ice cores (HAMMER et al., 1980; MOORE et al., 1991). These show strong parallels with the eruption history and are thought to be an indicator of the history of stratospheric sulphate aerosol loading. Unfortunately, when these different forcing "proxies" are compared, they correlate rather poorly (BRADLEY and JONES, 1985). Furthermore, none of the indices is clearly a measure of the global amount and duration of sulphur dioxide in the stratosphere following an eruption.

If one accepts the volcanic influence on short time-scales, then, even if the aerosol is removed from the stratosphere within a few years, as seems likely, it still could be argued that there may be more prolonged climatic effects because of oceanic thermal inertia (ROBOCK, 1979). The warming between 1920 and 1940 may, thereby, be partly attributable to the dearth of large eruptions over this period (ROBOCK, 1991). Unfortunately, modelling studies do not offer any strong support for this idea. In any event, explaining the 1920–1940 warming in this way is a rather selective interpretation of the data as cooling began in the early 1940s and the first major eruption was not until 1957.

What then of the enhanced greenhouse effect – the term for the build-up of radiatively active gases in the atmosphere due to anthropogenic activity? At least in this case (see also Chapter 9 by WANG et al.) we have a good record of the forcing changes over the past few centuries (WATSON et al., 1990, 1992). During this time carbon dioxide (CO_2) concentrations have increased from around 280 ppmv to more than 350 ppmv today, methane (CH_4) concentrations have more than doubled (from 800 ppbv to around 1700 ppbv), nitrous oxide has increased by some 10% and, over the last 30 years or so, the concentrations of a large number of chlorofluorocarbons (CFCs) have increased at alarming rates. These are all important greenhouse gases. Molecule for molecule, CH_4 is some 30 times more powerful than CO_2 while some CFCs are more than 10,000 times as efficient as CO_2. The combined effect of these gases amounts to a radiative forcing change at the top of the troposphere of about 2.5 W m^{-2} over the past few centuries, i.e. roughly equivalent to increasing solar output by 1% (HOUGHTON et al., 1990). The corresponding equilibrium global-mean temperature change would be a warming of between 0.7°C and 2.5°C (the uncertainty is due to a lack of quantitative understanding of various feedbacks in the climate system's response to external forcing, in particularly the climate sensitivity). However, because the thermal inertia of the oceans causes a lag in the response, the expected observed warming is substantially less than this – probably somewhere between 0.5°C and 1.3°C over the period 1880–1992 (HOUGHTON et al., 1990, 1992).

Thus, the observed warming of around 0.5°C appears to be consistent with that expected from the greenhouse effect, but only if the climate sensitivity (best guess IPCC value in

1992 is 2.5°C) is at the low end (~1.4°C) of the range predicted by General Circulation Models (WIGLEY and BARNETT, 1990; WIGLEY and RAPER, 1990). This does not, however, mean that we can claim to have positively detected the enhanced greenhouse effect, nor to have demonstrated that it is relatively small. Given the possible magnitude of natural, internally generated variability, and the possibility that changes may have occurred as a result of other external forcing factors, the observed warming could still be attributed to factors other than the greenhouse effect. Alternatively, a larger greenhouse effect may have been offset by these other factors.

There are a number of characteristics of the observed record that appear to be in conflict with the greenhouse hypothesis (WIGLEY and BARNETT, 1990). The warming between 1920 and 1940 was too rapid, while the cooling between 1940 and the mid-1970s occurred at a time when all greenhouse gas concentrations were increasing rapidly. Furthermore, the records from the individual hemispheres do not accord with expectations; more ocean in the Southern Hemisphere should lead to a damped response there, yet it seems more as if the Southern Hemisphere has been leading slightly in the warming race.

Of course, all of these discrepancies can be explained away, at least qualitatively. The rapid early 20th century warming could be a manifestation of internal variability or it may partly reflect lessened volcanic activity and/or solar irradiance changes. The cooling from 1940 may also just be natural variability superimposed on (and offsetting) the greenhouse effect. This cooling appears to have been forced in the Northern Hemisphere and communicated in a reduced form to the Southern Hemisphere by interhemispheric transport mechanisms.

One possible hemispherically specific forcing mechanism is changes in ocean circulation, specifically changes in the rate of North Atlantic Deep Water formation. Alternatively, this cooling (and NH/SH differences in general) could be the result of anthropogenic sulphur dioxide emissions, which, it has been suggested, could lead to noticeable increases in the albedo of oceanic cloud masses (CHARLSON et al., 1991). Since the dominant sources of sulphur dioxide are in the Northern Hemisphere, and since this gas has a short atmospheric residence time, the implied cloud albedo changes would be mainly in the Northern Hemisphere. Invoking this effect enables a higher climate sensitivity (2.5° compared to 1.4°C) to be compatible with the observational record (WIGLEY and RAPER, 1992). Incorporating changes in solar irradiance (using the solar cycle length series as a proxy) further alters the estimation of the climate sensitivity made from the observational record (SCHLESINGER and RAMANKUTTY, 1992).

Given these uncertainties, most of which, because of the absence of suitable data, can never be resolved, it is impossible to give any firm interpretations of the undeniable global-scale warming that has occurred this century. We can note that the warming is at the lower end of the range predicted by the models of greenhouse conditions, but this does not preclude the possibility that the greenhouse effect is even smaller than current models suggest. Neither, however, does it preclude the possibility of a very strong greenhouse effect which, up until now, has been partially offset either by natural variations in climate or by other anthropogenic influences.

Climatologists, therefore, are keeping a close watch on global temperatures and on possible forcing factors. Together with advances in modelling, this will eventually lead to reduced uncertainties in the magnitudes of all the factors which are thought to influence the climate (including the enhanced greenhouse effect) and to better predictions of future climatic change. In the meantime, attempts to explain past variations in global-mean temperature, particularly those using a single external factor, are bound to be unsatisfactory. Quite probably, we will have to wait until future observations have clarified the magnitude of the enhanced greenhouse effect before we can say much more about the causes of past climatic change. Almost certainly, we will find that the past record (which is only known in global terms with some certainty since the middle of the 19th century) will be explained by an as yet unknown combination of the various factors discussed here, some of whose histories we now know well and some which we do not. Determining the combination is, unfortunately, embroiled in the pro- and anti-greenhouse debate.

Projections to the future

The near future (<30 years)

On this time-scale any change due to the enhanced greenhouse effect is likely to be at most 0.5°C. It is likely that any greenhouse warming should be detectable in this time frame despite the scale of natural variability of the climate system (see the next section for a more complete discussion of greenhouse effect detection). We cannot adequately explain the changes that have taken place over the last 140 years. IPCC estimates an increase of temperature relative to 1990 of about 0.5°C by 2010 (GREGORY and MITCHELL, 1992; IPCC emissions scenario IS92a).

Prediction of climate trends and patterns has been one of the goals of climatology since the birth of scientific instrumentation. The most widely tried forecasting method (predictions on time-scales of 5–10 years) has used presumed solar weather relationships, all of which have been found to fail when better statistics and/or more data have been utilised (PITTOCK, 1979, 1983). The FRIIS-CHRISTENSEN and LASSEN (1991) approach (see above) is the latest in a long line of speculative ideas. A more convincing hypothesis, from a statistical point of view, has been postulated (VAN LOON and LABITZKE, 1988) which links changes in tropospheric circulation with the 11-year solar cycle, the exact relationship dependent on the phase of the wind direction in the stratosphere at the equator (the quasi-biennial oscillation, QBO). The relationships have been shown to be strong in certain regions and it has even been used in seasonal forecasting (BARNSTON and LIVESEY, 1989). Such work is, however, still somewhat premature, as no mechanism has been proposed. Although sunspot data are available since 1700, data for the QBO are only available since the late 1940s. If the relationship is found to continue to hold during the next 20 years, greater emphasis should be placed on finding the link through which the process operates.

Another means of climate prediction that has been developed recently is Singular Spectrum Analysis (SSA) (GHIL and VAUTARD, 1991). The method is an eigenvector decomposition of a time series, concentrating on variations with periods greater than 5 years. Projection of the leading components into the future gives the forecast. GHIL and VAUTARD (1991) used the method to project global mean temperatures into the future, forecasting a fall in temperatures during the early 1990s, followed by a later rise. The method is somewhat controversial, with ELSNER and TSONIS (1991b) suggesting that the principal oscillations used by the technique depend critically on the length of the data record used (see also ALLEN et al.,

1992). Whoever is right, the method warrants further research. One means of ascertaining whether the method is useful is to use it with longer time series. Some of the recent years can then be kept back for verification of the method (as is the common practice in paleoclimatology).

The second half of the next century

Whether people are capable of altering the climate of the Earth through the enhanced greenhouse effect, by past and continuing emissions of greenhouse gases, is the most important issue in climatology today. The issue, popularly termed "global warming", could have a significant impact on society. On the time-scale of the next century, if our current understanding of the climate system is correct, the average global temperature by 2100 should be about 3°C warmer than present values (GREGORY and MITCHELL, 1992; IPCC emissions scenario IS92a). A temperature rise of this order, coupled with the associated changes in the hydrological components (especially precipitation, cloudiness and evaporation) will undoubtedly change the climate of every location on the planet. Regionally, the warming and the disturbances in the hydrology might be large; a few regions might even cool. Impacts and policy responses will be determined by local-to-regional scale effects. The enhanced greenhouse effect, however, will only be detected at the global scale, and is to some extent independent of what happens at smaller scales. In this final section, we discuss this aspect of the problem: detection of the effect in the global observational record.

Much of the climatological research undertaken today, both directly in atmospheric sciences and in many related disciplines, is focussed towards improvements in our understanding of this one issue. There have been a number of international assessments over the last 10 years of which the two IPCC reports (HOUGHTON et al., 1990, 1992) are the most recent. The important aspects of the issue are therefore constantly changing, as research solves some questions, but generally asks many more. Modelling is a case in point. Although dramatic improvements have been made, principally through General Circulation Models (see Chapter 9 by WANG et al.) over the last 25 years, it is likely to be several decades before modellers will have much faith in their scenarios (cf. Chapter 7 by DICKINSON and Chapter 3 by BARRON). In other words, predictions made now for some future time frame will change countless times before their veracity can be assessed, possibly with reference to satellite data (Chapter 7 by DICKINSON).

How should policymakers respond to the projections? Given uncertainties in other areas, a typical response strategy is to spread the risk by insurance. In the global warming issue, the "no regrets" response would be to incorporate all energy saving options into new designs, cutting back on all wasteful energy usages while funding research that it is hoped will narrow scientific uncertainty. Even this response is too radical for many, who state that nothing should be done until we have categorical proof that we can see the effect in the climatic record. The importance of detecting the effect in the observations is therefore clear.

In attempts to detect the enhanced greenhouse effect, the key step is to be able to attribute an observed change in climate to this specific cause. A rise in temperature over the last 130 years has occurred, but it might have occurred without any increase in greenhouse gases. Attribution almost certainly requires the identification in the observational record of a multivariate signal characteristic of greenhouse gas-induced climatic change. The signal should ideally be unique to this specific cause. This approach to detection has been termed the fingerprint method (MADDEN and RAMANATHAN, 1980; MACCRACKEN and MOSES, 1982). In most applications tried so far, the detection variable has been a vector, whose components have been different scalar variables (e.g. temperature, precipitation, etc.) or the same variable measured at different places (e.g. temperatures measured at different locations on the Earth's surface or at different levels in the atmosphere). The strategy of detection is to compare the observed time series of the vector with an estimate from GCM results (e.g. the difference in the equilibrium response experiments for $2\times CO_2$ and $1\times CO_2$ or the difference between a transient model run with time-dependent forcing and the $1\times CO_2$ equilibrium response) using pattern correlation techniques.

Some thought should be given to the choice of detection vector. All the individual components should have high signal-to-noise ratios (SNRs). Poor choices in this regard are precipitation and sea-level pressure, while a better choice would be surface temperature. The detection vector should not be model specific. If GCMs give different results this would lower confidence in the model signal which is being sought. The detection vector should be easily distinguishable from other signals. Important in this regard is that it should, ideally, be very different from the pattern of natural climatic variability. Finally, suitable long observation records should exist. Long records are needed because it is the decadal-to-century time-scale that is of most interest.

At present all detection attempts have been unsuccessful. BARNETT (1986) and BARNETT and SCHLESINGER (1987) failed to find the $2\times CO_2$ signal from the Oregon State University GCM in surface temperature data. More recently, SANTER et al. (1993) failed to find the spatial pattern of the signal from a number of GCMs in surface temperature data, while KAROLY et al. (1994) failed to find the elevational signal (temperature change at different levels in the atmosphere) from GCMs in tropospheric and stratospheric radiosonde data.

The lack of success can be interpreted in a number of different ways. If the model results are correct, our lack of success could be due to the signal being obscured by the background noise of natural climatic variability or by the signals due to other forcing factors. Another possibility is that the climatic sensitivity is much less than the models suggest. In the same vein, the lack of success may be because the signal patterns, from the current generation of GCMs, are wrong.

Other than waiting for the models to improve sufficiently and/or for the accumulation of more observational data in which the signal might be more readily apparent, what can be done to speed up the time by which detection might be achieved? First, we should experiment with other pattern correlation or optimum detection techniques, addressing the issue of quantifying the signal. It is not adequate to claim detection per se, it is also necessary to attempt to quantify the strength of the signal (i.e. to estimate the climate sensitivity and/or how much of the rise in temperatures is due to this specific cause). This assessment is vital to strategies to mitigate the build-up of greenhouse gases in the atmosphere. Second, attempts should be made to detect the signals of other forcing factors such as sulphate aerosols and volcanoes, as these may be easier to detect than greenhouse gases. If these can be detected it may be possible to remove their influence and thus reduce noise levels. Addressing the attribution question is crucial. To do this we need to improve our understanding of both low-frequency natural variability and the climatic signals due to other forcing factors.

The evidence from traditional meteorological observations cannot confirm the existence of an enhanced greenhouse signal commensurate with the predictions of state-of-the-art Climate Models. Thus, if the models are to be believed (cf. Chapter 9 by WANG et al.) there must be other factors which, individually or in combination, are moderating or slowing the observed warming (see e.g. Chapter 10 by ANDREAE). If these factors persist the model predictions will be overestimated but if they "switch off" (cf. Chapter 14 by PENG), the final warming by, say, 2100 might even be greater than predicted.

Acknowledgements

Mike Hulme provided the precipitation data used in Figs. 7 and 8 and John Christy provided the MSU2R and MSU4 data used in Figs. 12 and 13. Many of the data sets and time series discussed in the chapter have been developed with the support of the United States Department of Energy, Atmospheric and Climate Research Division, under grant No. DE-FG02-86ER60397.

References

ALEXANDERSSON, H., 1986. A homogeneity test applied to precipitation data. *J. Climatol.,* 6: 661–675.

ALLAN, R. J., NICHOLLS, N., JONES, P. D. and BUTTERWORTH, I. J., 1991. A further extension of the Tahiti-Darwin SOI, early ENSO events and Darwin pressure. *J. Climate*, 4: 743–749.

ALLEN, M. R., READ, P. L. and SMITH L. A., 1992. Scientific correspondence. *Nature*, 355: 686.

ANGELL, J. K., 1988. Variations and trends in tropospheric and stratospheric global temperatures, 1958–87. *J. Climate*, 1: 1296–1313.

ANGELL, J. K. and KORSHOVER, J., 1985. Surface temperature changes following the six major volcanic episodes between 1780 and 1980. *J. Climate Appl. Meteorol.*, 24: 937–951.

ARKIN, P. A. and ARDANUY, P. E., 1989. Estimating climatic-scale precipitation from space: a review. *J. Climate*, 2: 1229–1238.

BAKER, D. G., 1975. Effect of observation time on mean temperature estimation. *J. Appl. Meteorol.*, 14: 471–476.

BARNETT, T. P., 1978. Estimating variability of surface air temperature in the Northern Hemisphere. *Mon. Weather Rev.*, 112: 303–312.

BARNETT, T. P., 1986. Detection of changes in global tropospheric temperature field induced by greenhouse gases. *J. Geophys. Res.*, 91: 6659–6667.

BARNETT, T. P. and SCHLESINGER, M. E., 1987. Detecting changes in global climate induced by greenhouse gases. *J. Geophys. Res.*, 92: 14772–14780.

BARNSTON, A. G. and LIVESEY, R. E., 1989. A closer look at the effect of the 11-year solar cycle and the quasi-biennial oscillation on Northern Hemisphere 700 mb height and extratropical North American surface temperature. *J. Climate*, 2: 1295–1313.

BLACKBURN, T., 1983. A practical method of correcting monthly average temperature biases resulting from differing times of observations. *J. Climate Appl. Meteorol.*, 22: 328–330.

BOTTOMLEY, M., FOLLAND, C. K., HSIUNG, J., NEWELL, R. E. and PARKER, D. E., 1990. *Global Ocean Surface Temperature Atlas (GOSTA).* HMSO, London, 20 pp., 313 plates.

BRADLEY, R. S., 1985. *Quaternary Paleoclimatology: Methods of Paleoclimatic Reconstruction.* Allen and Unwin, London and Boston, 472 pp.

BRADLEY, R. S., 1988. The explosive volcanic eruption signal in Northern Hemisphere continental temperature records. *Clim. Change*, 12: 221–243.

BRADLEY, R. S. and JONES, P. D., 1985. Data bases for detecting CO_2-induced climatic change. In: M. C. MACCRACKEN and F. M. LUTHER (Editors), *Detecting the Climatic Effects of Increasing Carbon Dioxide.* US Dept. of Energy, Carbon Dioxide Research Division, Washington, DC, pp. 29–53.

efforts aimed at detecting human-induced changes in our climate. In general, most studies that have examined the connection between temperature and precipitation trends, and corresponding changes in the variance of these two key climate variables, have found no clear evidence to indicate that increases or decreases in either field are associated with systematic changes in variance.

In the foregoing discussion we have provided an overview of what is known about climatic (primarily temperature) variations over the past several centuries based on a variety of historical and paleoclimatic sources. In the following section, we consider some of the key phenomena that can lead to decadal to century time-scale variability in the climate record.

Possible causes of climatic variability on decadal to century time-scales

Solar variability

A direct connection between solar variability and climatic variation at interannual to decadal time-scales has long been sought. However, such a link has proved difficult to find. Precise measurements of solar irradiance from space are less than a couple of decades in length. However, satellite measurements of the sun's energy output over the course of a full solar cycle have shown that the solar "constant" undergoes changes on the order of 0.1% of its mean value, or about 1 W m^{-2} (WILLSON and HUDSON, 1988, 1991). Observations of other stars similar to the Sun indicate that the Sun's radiative output may vary in response to its 11-year magnetic activity or sunspot cycle (BALIUNAS and JASTROW, 1990; RADICK et al., 1990). Such variations in the solar "constant" may suffice to perturb global climate on decade to century time-scales (EDDY et al., 1982; FOUKAL and LEAN, 1990). The reader is referred to PITTOCK (1978, 1983) for critical reviews of postulated solar effects on weather and climate, and for an overview of the efforts made over the past couple of decades to develop quantitative measures to explain those effects.

Since the 11-year sunspot cycle is so regular, along with the 22-year cycle of solar magnetic polarity reversals, there have been many attempts at finding similar "cycles" in climatic records (CURRIE, 1984; CURRIE and FAIRBRIDGE, 1985; CURRIE and O'BRIEN, 1988 among others). We consider two main areas of solar-climate relationships at the time-scales of interest here. The first deals with drought (or flood) recurrence, usually associated with either the 11-year or 22-year sunspot cycles (MITCHELL et al., 1979; BHALME and MOOLEY, 1981; DIAZ, 1983; STOCKTON and MEKO, 1983). In most cases, studies have concentrated on establishing a statistical association between surface temperature variations and solar (i.e. sunspot) variability. More recently, numerical models have been used to look at the effects of solar variability on the Earth's climate. One line of research has focused on changes in ocean temperatures and its coupling via energy exchanges with the deeper ocean and with the upper troposphere (e.g. GAGE and REID, 1981; REID, 1991). Another area of research has concentrated on the effects of solar variability on internal oscillations of the Earth's climate system such as the Quasi-Biennial Oscillation (QBO) (LABITZKE and VAN LOON, 1990), and the El Niño/Southern Oscillation (ENSO) phenomenon (ENFIELD and CID, 1991). Plausible candidates for coupling mechanisms linking changes in the solar forcing to climatic changes at the surface have centred on resonant excitation of atmospheric phenomena through wave coupling between the troposphere and the stratosphere (GELLER and ALPERT, 1980; SCHMIDT, 1986), and feedback mechanisms between changes in sea surface (or mixed layer) temperature and atmospheric water vapour (e.g. FLOHN et al., 1992).

At time-scales of ~100–1000 years, solar variability may be an important source of variability in the climate system. Evidence for solar variability on this time-scale comes primarily from carbon isotopic records (DAMON et al., 1978; STUIVER and QUAY, 1980; JIRIKOVIC and DAMON, 1994). Historical observations of solar activity also suggest that solar activity may have varied over the centuries (EDDY, 1977; EDDY et al., 1977). Recent work suggests that solar irradiance during the so-called Maunder Minimum in sunspot activity in the 17th century may have been lower than present values by about 0.25%, or about 2 W m^{-2} (LEAN et al., 1992). Estimates of the climatic effect of such a reduction in solar irradiance based on GCM experiments indicate a global cooling of about 0.5°C (LEAN and RIND, 1994; RIND and OVERPECK, 1994).

Although changes in the Earth's orbital parameters on decadal to century time-scales are rather small (of the order of 0.2 W m^{-2}, see LOUTRE et al., 1992), they nevertheless add an additional forcing mechanism contributing to climatic variability on the century time-scale. We refer the reader to Chapter 2 by BERGER, which examines the response of the climate system to orbital variations.

Recent studies have found some observational support for the hypothesis that solar variability has been a major controlling factor in the observed behaviour of hemispheric and global scale surface temperature variations over the last ~130 years. An interesting study by FRIIS-CHRISTENSEN and LASSEN (1991) obtained a very high correlation ($r \approx 0.95$) between a record of Northern Hemisphere land surface temperature changes and variations in the length of the sunspot cycle, which is known to be related to solar activity. The sunspot cycle length record, however, was highly smoothed, and, because the strength of the solar cycle perturbation is relatively weak, an amplification mechanism is needed to translate these solar signals into the observed surface temperature fluctuations. Based on a study with a one-dimensional two-layer ocean model, REID (1991) concluded that solar irradiance variations could have accounted for much of the observed globally averaged sea surface temperature variations. SCHLESINGER and RAMANKUTTY (1992) performed some experiments with a simplified version of a coupled climate model to estimate the magnitude of the contribution of solar cycle radiative changes to the observed "global" temperature record. They concluded, based on best-fit estimates of a number of model parameters, and the available observations, that intercycle solar irradiance changes could have accounted for perhaps up to one-half of the observed surface temperature variations during the past few centuries. In a similar study of the relative effects of solar irradiance variations versus greenhouse gas radiative forcing of the observed "global" surface temperature record, KELLY and WIGLEY (1992) also concluded that solar irradiance variability may have contributed to the observed variability in the instrumental temperature record, although their estimate of the solar effect was somewhat lower (on the order of 20% of the observed temperature variance) than that of SCHLESINGER and RAMANKUTTY (1992).

Among the first to perform a careful analysis of available data with regard to solar-climate connections were MITCHELL et al. (1979) who considered the time-scale of drought recurrence in the western United States. In that study the authors used several long-term reconstructions of the Palmer Drought Severity Index (PDSI) (see DIAZ, 1983; KARL, 1986) using precipitation-sensitive tree-ring records from different sites. It was shown, through the use of various statistical tests, that there was a significant tendency, "for the amplitude of the 22-year drought rhythm to vary systematically in parallel with the amplitude (envelope) of the Hale sunspot cycle" (MITCHELL et al., 1979, p. 125). The variable in question was not a time series of PDSI at a given site, but rather, an index which measured the fractional area of the western United States experiencing a particular magnitude of drought. In a subsequent re-evaluation of their earlier results (STOCKTON et al., 1983), the authors restated their finding with regard to the statistical significance of the 22-year drought rhythm in the western United States, but they also found evidence of an 18.6-year variance component in the data, which they indicated might be related to the lunar nodal cycle variation (see CURRIE, 1984). We have calculated the fractional drought area index using US climate division PDSI values from 1895 to 1990. These are labelled as the Palmer Drought Area Index, or PDAI. Two sets of seasonal indices are shown in Fig. 4. One is for the states comprising the Great Plains region (from North Dakota to Texas) and the second is for the region extending from the Rocky Mountains to the Pacific coast (Montana to New Mexico and west).

The technique of singular spectrum analysis (SSA) (see VAUTARD and GHIL, 1989; GHIL and VAUTARD, 1991) was used to identify the characteristic temporal scales of variability in the PDAI time series, and made use of maximum entropy spectrum analysis (MESA) to estimate the period associated with oscillatory SSA principal component modes. In particular, we were interested in seeing if any of the leading (statistically significant) principal compo-

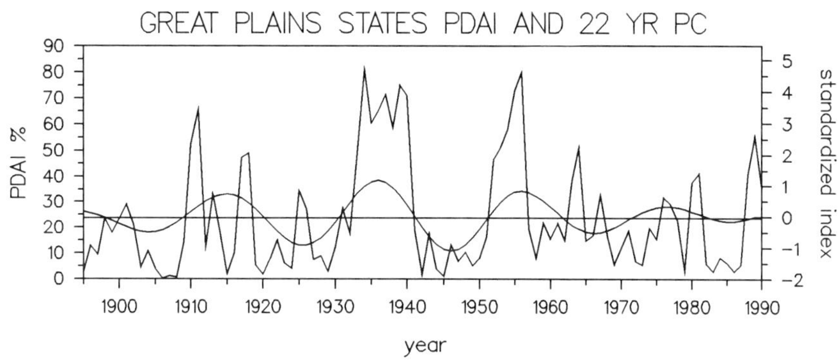

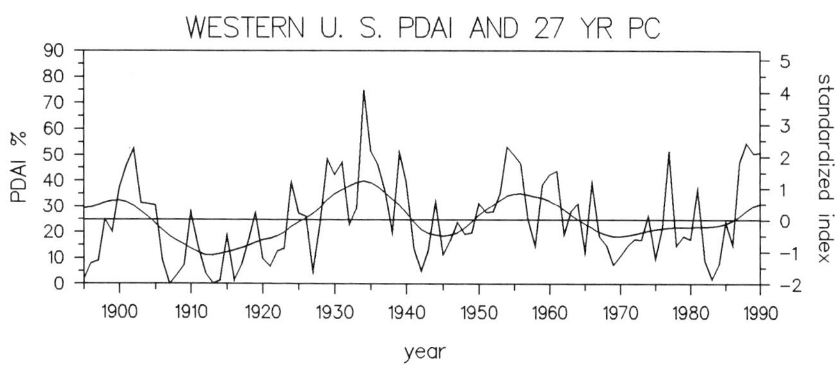

Fig. 4. Index showing the fractional drought area for the US Great Plains (top) and Rocky Mountain West (bottom) from 1895–1990 (see text). Smoothed curve (plotted in standardised units) is derived from a singular spectrum analysis of the annual mean values and corresponds to a principal component pair whose spectrum contains a significant peak near the period shown.

nents exhibited periodicities near 22 years. For annual mean values of the PDAI in the time interval from 1895 to 1990 (top curve of Fig. 4), the SSA filtering indicates that approximately 43% of the variance in the Great Plains PDAI is associated with oscillations at ~22 years. The Rocky Mountain West index contains longer term components which stretch the characteristic time-scale of drought area variability to something closer to 30 years (bottom curve of Fig. 4). Approximately 33% of the variance of annual mean western US PDAI is associated with oscillations having a time-scale of ~27 years. We note, however, that although a definite pattern of temporal variability in drought extent, tuned to a time-scale of ~20 to 30 years, is evident in the western United States, there are clearly other factors at play which modulate the occurrence and recurrence of drought in that part of the United States.

An important consideration with regard to climatic variability-solar relationships at the time-scales of interest here, is whether these relationships withstand the test of time (see RAMAGE, 1983). PITTOCK (1983) points out that three of the thirteen droughts documented in the drought study of MITCHELL et al. (1979) were clearly out of phase with respect to the 22-year Hale sunspot cycle. Although a number of hypothesis have been advanced to offer an explanation of such phase drift (e.g. CURRIE, 1981), replication of the effects of solar variability on regional climatic variability has yet to be reproduced with general circulation models of the Earth's climate system.

One of the more interesting associations between solar phenomena and climatic variations is an apparent connection between the 11-year solar cycle and atmospheric circulation anomalies modulated by the phase of the quasi-biennial oscillation (QBO) in the stratosphere (LABITZKE and VAN LOON, 1988, 1989, 1990; VAN LOON and LABITZKE, 1988). The QBO itself is unique in that it is by far the most regular atmospheric oscillation not directly related to a periodic external forcing mechanism, such as the diurnal or annual cycle (see HOLTON, 1992). The oscillation affects primarily the low latitude winds in the lower stratosphere, and is associated with a reversal in the sign of the zonal wind at a given level every 24–30 months. The signal of zonal wind of one sign propagates downward, starting at about 50 km, at a rate of about 1 km per month, until a height of about 20 km is reached. This is followed by the downward phase propagation of a signal of opposite sign, the oscillation having an amplitude of about 20 m s^{-1}. This signal is zonally symmetric and present around the globe with an approximately Gaussian distribution of about 12° latitude half-width centred on the equator.

The QBO was first documented by REED (1965) and has been reconstructed back to 1953 by NAUJOKAT (1986). It is seen that, although its period varies somewhat, the oscillation is extremely regular and possesses a high degree of predictability. Thus, any observable association between the QBO and the reasonably regular solar cycle has potential to account for regular oscillations of the climate system on longer time-scales. An important feature of the association between the QBO and the solar cycle is that correlations between any of a number of atmospheric elements with the 11-year solar cycle display opposite signs depending on the QBO phase. Hence, correlations between these atmospheric variables with solar activity without regards for the phase of the QBO nearly always yield negligible values.

Among the critical observations highlighting the above-noted relationship is the high correlation ($r > 0.75$) between 30 mb January–February temperatures throughout the Arctic and the 10.7 cm solar flux (a good measure of solar activity). To illustrate the changes, Fig. 5,

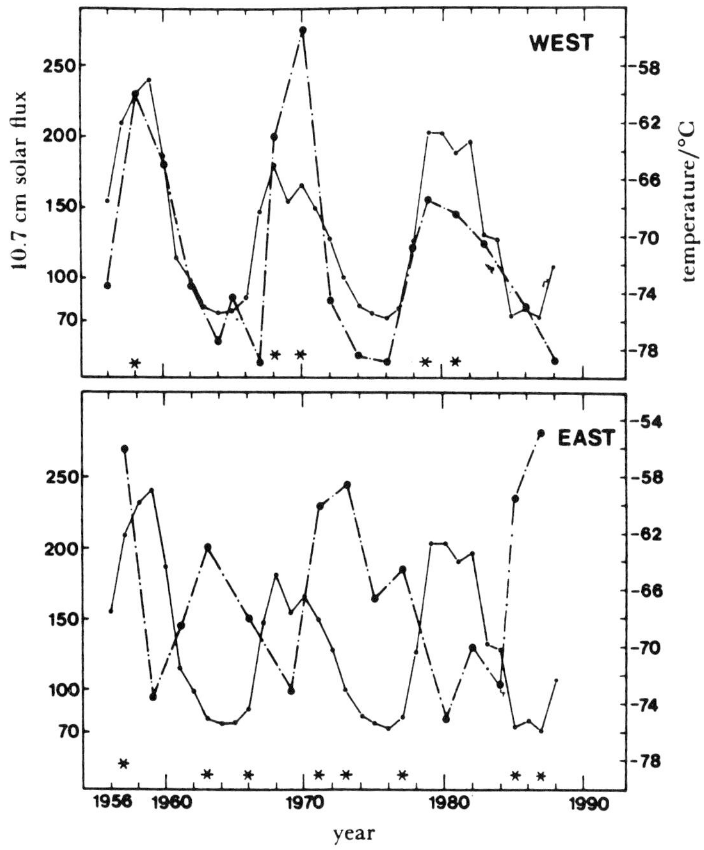

Fig. 5. The association between stratospheric temperature, solar activity and the phase of the QBO: (a) Time series of temperature (°C) at 30 mb at the North Pole (dot-dashed curve) versus the 10.7 cm solar flux (in units of 10^{-22} W m^{-2} Hz^{-1}) in the west wind phase of the QBO; (b) same as (a) but for the east wind phase. Taken from LABITZKE and VAN LOON (1990).

taken from LABITZKE and VAN LOON (1988, 1990) shows the 30 mb winter temperature at the North Pole (obtained from hand-analysed charts prepared at the Free University of Berlin) plotted against the 10.7 cm solar flux for years with west phase of the QBO, together with the corresponding years with east QBO phase. Similar differences are also evident with January–February 30 mb geopotential height field. Interestingly, the association between 50 mb temperature at the South Pole (there are insufficient 30 mb radiosonde data available from the South Pole) and the 10.7 cm solar flux shows a similarly high correlation ($r \approx 0.8$) but with the east phase of the QBO.

Since the stratosphere is in close radiative equilibrium with variations in solar irradiance, it is perhaps not surprising that we can detect a strong coupling of stratospheric climatic features with solar energy variability. Dynamical mechanisms may be the source of the different response features in the stratosphere during the different QBO phases. Finding clear solar signals in the denser and more dynamically complex troposphere is more remarkable.

VAN LOON and LABITZKE (1988) showed that sea-level pressure (SLP) correlates on large spatial scales to solar flux changes, if proper accounting of the QBO phase is made. In particular, when solar activity is high coinciding with the west QBO phase, SLP is relatively high in January–February over interior continental regions of North America, with lower than normal SLP in much of the North Atlantic and in the eastern North Pacific. During the

Fig. 6. Association between surface air temperature (°C) and 10.7 cm solar flux during west QBO phase. Map shows lines of equal correlation coefficient between local January–February mean surface temperature in the Northern Hemisphere and the 10.7 cm solar flux. Period of record varies between 16 and 19 years. Taken from VAN LOON and LABITZKE (1988).

east QBO phase, sub-tropical and mid-latitude SLP tend to be low in the North Pacific in January–February, and SLP in sub-polar latitudes tends to be higher than normal. The changes in atmospheric circulation associated with these different SLP patterns are reflected in January–February surface temperature variations in areas experiencing the most pronounced circulation changes. Figure 6, taken from VAN LOON and LABITZKE (1988) illustrates the association between solar flux changes and surface temperature modulated according to the phase of the QBO.

BARNSTON and LIVEZEY (1989) have carefully evaluated the association between solar flux variation, the QBO and Northern Hemisphere 700 mb and extratropical surface temperature in North America. They found statistically significant correlations (measured against strict field significance tests) during the fall and winter seasons for both North American temperature and the 700 mb geopotential height field during years of west QBO phase. The principal 700 mb height anomaly patterns associated with the solar activity–tropospheric circulation relationship appear to be the Tropical/Northern Hemisphere (TNH) described in MO and LIVEZEY (1986) and BARNSTON and LIVEZEY (1987), and the North Atlantic Oscillation (NAO) modes (VAN LOON and ROGERS, 1978; WALLACE and GUTZLER, 1981). The NAO pattern represents a seesaw in pressure between northern and central portions of the North Atlantic Ocean, whereas the TNH pattern, as well as the Pacific-North American or PNA pattern described by HOREL and WALLACE (1981), are part of a quasi-stationary wave-train configuration linked to tropical variations.

In the absence of a physical hypothesis that explains these statistical relationships, and given the fact that the observed record of stratospheric wind reversal which comprises the QBO is less than 40 years long, investigators have been cautious in accepting the apparent relationship as real. For instance, HAMILTON (1990) performed a series of experiments using a very large number of plausible sequences of the QBO phase in the stratosphere to test the stability of the statistical relationships present between solar activity and selected station time series of barometric pressure and temperature in the Northern Hemisphere. The results indicate the possibility that the associations evident in the modern observational record could have been substantially weaker in the past. Clearly, confirmation of a cause-and-effect linkage must await improved knowledge of the physical basis of these associations. In the absence of testable hypotheses, additional years of observation will continue to provide a means to verify the relationship.

Volcanism and climate

The notion that explosive volcanic eruptions had an impact on the Earth's climate has been around for some time. For example, an account of the 1783 Laki eruption by SIGURDSSON (1982) contains an excerpt from the writings of Benjamin Franklin, in which he conjectures that "volcanic fogs" may be responsible for unusually cold weather in Europe during that time. The 1883 eruption of Krakatoa in present-day Indonesia was noted by many observers of the time due to its spectacular sunsets and by the relative coolness of the mid-1880s which followed (see LAMB, 1970).

The volcanic explosivity index (VEI) (NEWHALL and SELF, 1982) is now used as an index of eruption strength. A VEI of 5 (considered a major eruption) is defined as an eruption with a volume of ejecta exceeding 10^9 m^3, and an eruption column height greater than 25 km. Early work on a connection between large volcanic eruptions and climatic cooling at the surface (for instance, LAMB, 1970; TAYLOR et al., 1980; SELF et al., 1981) generally agreed that a hemispheric-scale cooling on the order of 0.5°C was plausible in the 12–36 months following a large volcanic eruption.

More recent evaluations of the surface temperature effect of large volcanic eruptions (e.g. BRADLEY, 1988; MASS and PORTMAN, 1989) have suggested that the magnitude of the response of hemispheric-scale surface temperature to volcanic eruptions with a VEI of 5 or greater is of the order of 0.2–0.4°C, and that the climate response is fairly rapid, with maximum cooling occurring within 6 months to 2 years following the eruption.

Early theoretical studies (e.g. POLLACK et al., 1976) emphasised that it was the magnitude of the sulphur dioxide (SO_2) loading of the eruption that was important from the point of view of its radiative effects, and hence, its potential impact on climate. Recent modelling and observational work has improved our estimate of the climatic impact of volcanic aerosols (DUTTON and CHRISTY, 1992; HANSEN et al., 1992; LACIS et al., 1992). The 1991 eruption of Mount Pinatubo in the Philippines, which qualifies as at least a VEI-5 volcanic eruption, provided an unprecedented opportunity to observe global scale climatic changes associated with a large volume of stratospheric volcanic aerosol. Using the microwave sounding units (MSU 2R) aboard National Oceanic and Atmospheric Administration (NOAA) satellites (SPENCER and CHRISTY, 1990, 1992), Dutton and Christy estimated that a temperature drop in the lower troposphere of ~0.7°C for the Northern Hemisphere and 0.5°C for the globe had been observed by September 1992, approximately 15 months after the eruption. This value appears to be somewhat larger than the above cited estimates of volcanic effects on surface temperatures for the last century or so, possibly because stratospheric loading of SO_2 is known accurately only for the most recent eruptions. The estimated reduction in clear-sky daily integral solar irradiance at the surface for various sites is of the order of 3–5% (~6–10 W m^{-2}) within about 10 months of the eruption (DUTTON and CHRISTY, 1992).

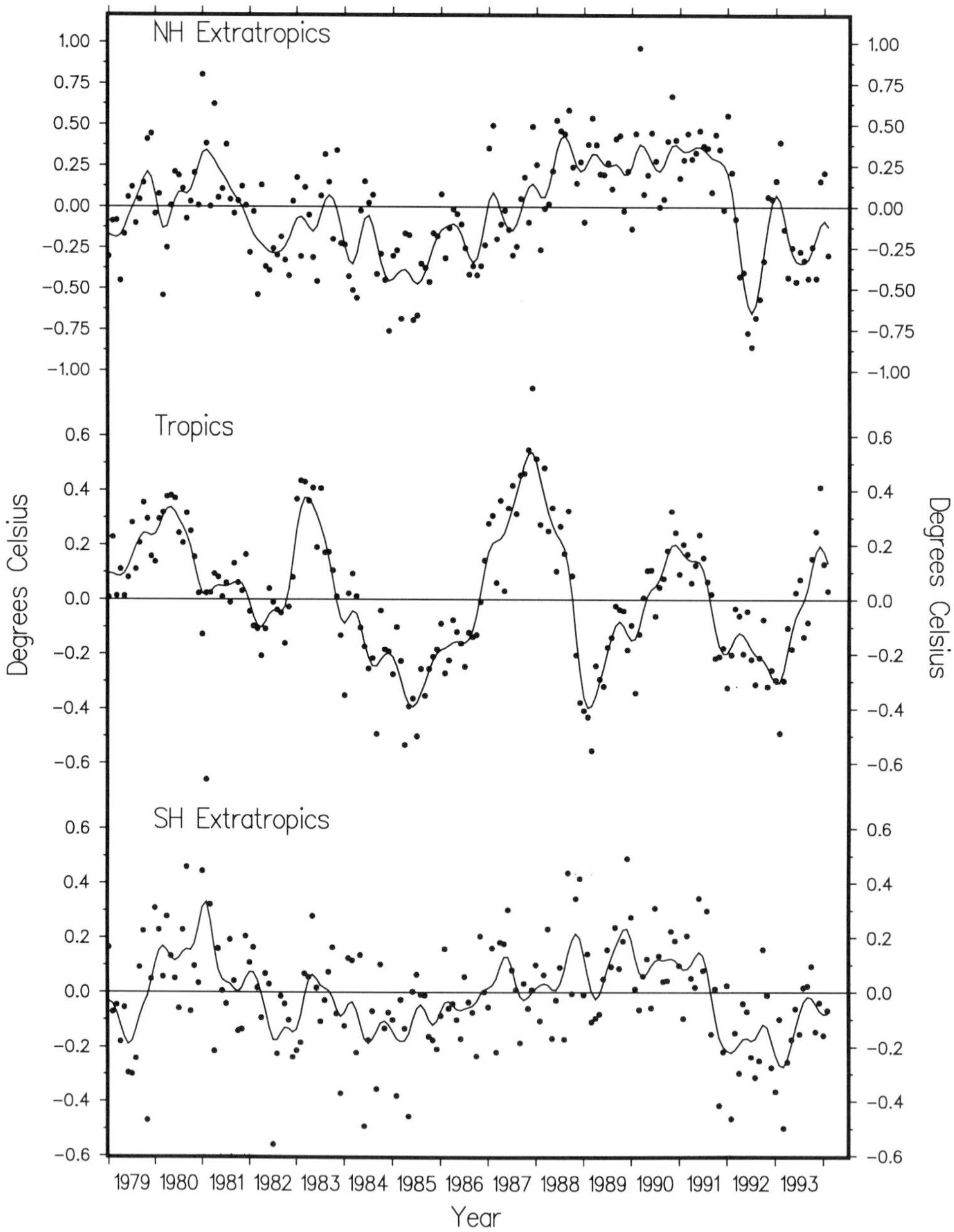

Fig. 7. Lower tropospheric temperature anomalies (°C), 1979–1994. Values are derived from the microwave sounding units (version MSU-2R) aboard NOAA polar-orbiting satellites (see text for references). Smoothed curve is obtained by applying a 9-pt. Gaussian low-pass filter to the monthly anomalies. Reference period is 1982–1992.

A global climate model simulation of the possible effect of Mount Pinatubo aerosols on global surface temperature anomalies using radiative forcing profiles consistent with the above measurements projected a global cooling of about 0.5°C (HANSEN et al., 1992). This agrees with the MSU-2R measurements (Fig. 7). The model results indicated that the maximum cumulative temperature effect would peak in late 1992, with a slow decay in the response for another couple of years. Hence, the response of global surface temperature in the model to a volcanic aerosol perturbation of the size of Mt. Pinatubo extended over approximately a 3-year period, in rough agreement with the empirical studies.

It is difficult to separate the direct volcanic effect from the other factors that affect global temperature, such as ENSO, internal feedbacks, and cryospheric changes. Figure 7 illustrates the available record of monthly lower tropospheric temperature anomalies derived from the MSU-2R data. A smoothed curve has been fitted to the data to highlight lower frequency fluctuations. We have split the globe into a tropical band and the two extratropical regions for comparison. The ENSO time-scale is strongly represented in the tropical record (see section on ENSO below), whereas temperature variability in the extratropics is dominated by longer time-scale changes. The temperature falls during 1992 as a result of the Mt. Pinatubo eruption, and to a lesser extent after the 1982 El Chichón eruption, are clearly evident for the Northern and Southern Hemispheres.

The occurrence of several large volcanic eruptions within a short period of time may result in substantial temperature declines, lasting for a decade or two, depending on the timing of the eruptions. Such was apparently the case in the early part of this century, when several large eruptions (Pelée/Soufrière/Santa María during 1902, Kamchatka in 1907, and Katmai in 1912) may have perturbed the climate enough to produce a relatively cool period prior to the rapid rise in Northern Hemisphere surface temperature evident in the observational record (JONES et al., 1986).

It was noted earlier that internal variability of the climate system, meaning the non-linear interactions in the energy exchanges between the oceans, the cryosphere and the land surface, might be largely responsible for large-scale (~ 10^7–10^8 km^2) climatic variability at decadal to century time-scales. In the following two sections we consider some of these aspects, paying special attention to such important large-scale features of the climate system as the El Niño/Southern Oscillation, followed by a consideration of cryospheric variability (snowcover and sea-ice variations) and extratropical air-sea interaction mechanisms.

El Niño and the Southern Oscillation

The El Niño/Southern Oscillation (ENSO) phenomenon is perhaps, next to the stratospheric QBO discussed in the section on solar influences, the most obvious example of quasi-periodic internal climate variability on interannual time-scales. The term El Niño (or "the child") was originally used by Peruvian fishermen in the 19th century (see ENFIELD, 1988) to refer to a Christmas-time warming of coastal sea surface temperature (SST), often associated with an abrupt decrease in productivity of the local fisheries (JORDÁN, 1991). The Southern Oscillation (SO) portion of ENSO describes the global-scale surface pressure oscillation documented by workers around the turn of the century and first studied in detail by Sir Gilbert Walker (see RASMUSSON and CARPENTER, 1982; various chapters in GLANTZ et al., 1991; DIAZ and MARKGRAF, 1992, for historical reviews). It was not until the 1960s that heat content, to develop into an alternate state. The strong seasonal dependence of the success of both dynamical and statistical models in forecasting ENSO (CANE and ZEBIAK, 1985; GRAHAM et al., 1987; CANE, 1991) is another indication that the fluctuations are tied to the base state and time-scale of ocean dynamics.

Teleconnections

Much of Walker's early work was geared towards prediction of Indian rainfall, and the SO was of great interest since it was known that the strength of the monsoon was inversely proportional to the surface pressure over southern Asia. In his investigations WALKER (1923) established that high pressure over the Australasian region was accompanied by drought over India and Australia and cool, wet winters over the southeastern United States. Surprisingly, little work on atmospheric teleconnections was undertaken until nearly a half a century later, when BJERKNES (1966) uncovered evidence that El Niño conditions were associated with a strengthening of the storm track over the North Pacific. Since then, a wealth of research has been done on the temperature and precipitation signals associated with ENSO. Much of the synopsis to follow has been taken from the work on large scale ENSO signals of RASMUSSON and CARPENTER (1982), ROPELEWSKI and HALPERT (1986, 1987), LAU and SHEU (1988), KILADIS and VAN LOON (1988) and KILADIS and DIAZ (1986, 1989). The signals to be discussed here are those which are the most highly statistically significant results of these studies. Other studies relating to specific signals are given in reference lists from these papers, and also in the discussion below for individual regions.

The most pronounced signals occur during the year that an ENSO extreme first develops (here called "Year 0") into the following year (Year + 1), following the convention of RASMUSSON and CARPENTER (1982). Figures 10 and 11, taken from KILADIS and DIAZ (1989), show regions of statistically significant differences in temperature and precipitation between cold and warm events for selected seasons, and are used as a basis for the following discussions.

Precipitation

Although the term "El Niño" was first used to describe the annual warming of the waters along the Peruvian coast, in the 1960s it became synonymous with the concurrence of abnormally high SST and heavy rains in the usually hyperarid Peruvian coastal plains. BJERKNES (1969) noted that anomalously heavy rains during El Niño also occurred at Pacific island stations such as Canton and Christmas in the equatorial dry zone. As mentioned above, this signal is one manifestation of the equator-ward shift of the convergence zones during equatorial warming of the SST associated with ENSO. As a result, the regions normally under the influence of these convergence zones, such as the Caroline and Marshall Islands, Fiji and New Caledonia experience drier than normal conditions (Fig. 10c).

ENSO appears closely tied to the Asian and Australian monsoon circulations, which are notably weaker during warm events (see WEBSTER and YANG, 1992). Figures 10 and 12 illustrate the effect of ENSO on precipitation in the region of India and around northern Australia. While precipitation increases in the central and equatorial Pacific during warm events, it becomes markedly drier over regions bordering the western Pacific and eastern Indian

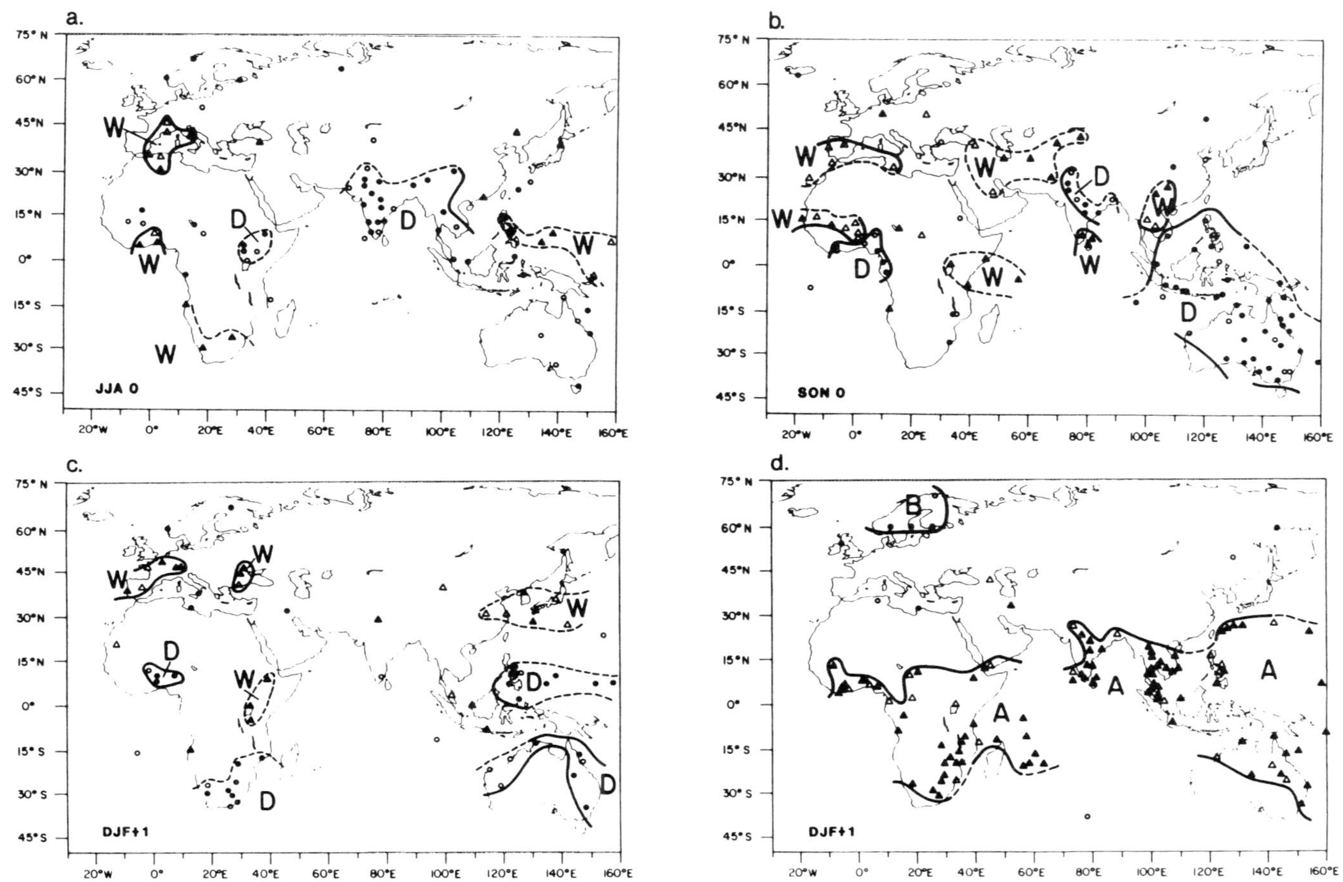

Fig. 10. Map illustrating the areas in the eastern hemisphere where the composite precipitation anomaly difference between warm and cold ENSO events (warm minus cold) is statistically significant at the 5% (solid symbols) or 10% (open symbols) level. The anomaly sign indicated in these panels is representative of warm events ("W" for wetter and "D" for drier than normal conditions); the opposite sign is representative of cold events. Panels: (a) JJA (b) SON and (c) DJF + 1 seasons. Panel (d) illustrates the corresponding surface temperature anomalies of DJF + 1 season. Significance symbols as above, "A" stands for above normal, "B" for below normal. Taken from KILADIS and DIAZ (1989).

tal Panel on Climate Changes Reports (IPCC, 1990, 1992) for a comprehensive review of these issues.

Climates of the future

Changes from anthropogenic sources

It has become increasingly apparent that human beings have the capacity to change the Earth's climate. Evidence to that effect is found in the observation that carbon dioxide (CO_2) concentration in the atmosphere has been rising inexorably since regular measurements were begun at Mauna Loa Observatory, Hawaii in the late 1950s (BOLIN and BISCHOF, 1970; KEELING et al., 1976). Since about the mid-1970s, first with the application of simple energy-balance models (EBMs) (SELLERS, 1974; BUDYKO, 1977) and then through the use of more complex general circulation models, the study of the so-called "greenhouse effect" has grown greatly (see reviews by SCHLESINGER, 1984, 1991 and Chapter 9 by WANG et al.). The use of these sophisticated tools, relying on ever-expanding computational power, have led to a global awareness that human beings are likely to modify not just regional-scale climates, but the global environmental "commons" as well. The publication of *Understanding Climatic Change* (NRC, 1975) provided an early rationale for the development of coherent national and international climate research programs. In the ensuing two decades, the World Climate Research Program sponsored by the World Meteorological Organization (WMO), the International Oceanographic Commission (IOC) and the International Council of Scientific Unions (ICSU), together with a host of national climate programs (e.g. COMMITTEE ON EARTH SCIENCES, 1989) have been established to address a multitude of issues related to climate change with emphasis on the understanding of both natural and human-induced variations and their impact on human society (see also IPCC, 1990). Most scientific assessments of the potential impact of growing concentrations of CO_2 and other radiatively active trace gases in the atmosphere (primarily methane, nitrous oxide and chlorofluorocarbons) have been consistent in projecting a global temperature rise of ~2–3°C in the next century, with greater warming in the polar regions relative to the tropics (NRC, 1979, 1982, 1983; IPCC, 1990, 1992 and many others).

There is considerable paleoclimatic evidence which indicates that relatively large variations in CO_2 accompanied the growth and decay of the latest Pleistocene glaciation (e.g. STAUFFER et al., 1984; BARNOLA et al., 1987). The most reliable information on past carbon dioxide concentrations in the atmosphere is obtained by the analysis of air bubbles trapped in Antarctic and Greenland ice sheets. Ice-core samples for about the last 160,000 years have been obtained, which show that, in general, there is a remarkable similarity between polar temperature changes, as deduced from the $\delta^{18}O$ ratio (see earlier), and CO_2 concentrations. However, the interpretation of causal mechanisms between the recorded changes in CO_2 and global temperature changes remains unclear at present. Methane (CH_4) also appears to have undergone large oscillations in connection with the build-up and decay of continental ice sheets. It is now widely accepted, based on a variety of paleoenvironmental climate indicators, that the principal forcing mechanisms associated with glacial/interglacial cycles in the last million years are varying amounts of solar insolation with latitude received at the top of the atmosphere due to the Earth's orbital variations – the so-called Milankovitch mechanisms (BERGER, 1978; IMBRIE and IMBRIE, 1980; see also Chapter 2 by BERGER).

Increasing the atmospheric concentration of "greenhouse gases" is not the only climatic impact of human activities. Two other important perturbations of the global environment have taken place which are relevant to the subject of anthropogenic climate change – increases in the tropospheric aerosol loading (e.g. CHARLSON et al., 1992; PENNER et al., 1992; see also Chapter 10 by ANDRAEA) and modifications of the land surface cover (HENDERSON-SELLERS, 1987; HENDERSON-SELLERS et al., 1993; see also Chapter 12 by HENDERSON-SELLERS). The possibility that atmospheric particulates of human origin may be capable of affecting the climate at regional and larger space scales has been a consideration for at least two decades (SCHNEIDER, 1972; BRYSON, 1974; MITCHELL, 1975; SAGAN et al., 1979; CHARLSON et al., 1992). Recent calculations of the effect of anthropogenic sulphate aerosols in the troposphere on the surface radiation balance (KIEHL and BRIEGLEB, 1993) indicate a net cooling effect, which, averaged globally, is about one-seventh the magnitude of the net greenhouse gas forcing. In some regions, such as the eastern United States and central Europe during summer, these authors estimate that the sulphate aerosol cooling effect is about equal to the radiative warming contributed by the increase in greenhouse gas concentrations. Another feature associated with sulphate aerosol screening of shortwave solar radiation is the hemispheric asymmetry introduced by the much greater industrial activity in the Northern compared to the Southern Hemisphere. KIEHL and BRIEGLEB (1993) calculate more than a factor of three difference in the anthropogenic sulphate radiative forcing for the two hemispheres.

The difficulty in evaluating a greenhouse gas climate signal in the observational record stems from the fact that, whatever the greenhouse signal, it will be embedded in a background of high natural variability which will tend to mask and distort the anthropogenic signal (KARL, 1988; KARL et al., 1991a; DIAZ and BRADLEY, 1994). Control simulations by various coupled ocean-atmosphere GCMs show multidecadal internal variations on the order of ~0.4°C for hemispherically averaged surface temperature and ~0.3°C for global means (IPCC, 1992). Various statistical techniques are being applied to the observed climate record in attempts to separate greenhouse gas-induced climate change signals from "climatic noise" or its natural variability (see BARNETT et al., 1991b; BLOOMFIELD, 1992; RICHARDS, 1993). Some of these tests are being framed in the context of both transient (where greenhouse gas concentrations in the atmosphere are allowed to increase slowly, as observed in the real world), as well as equilibrium (where the atmospheric concentration of carbon-dioxide, for instance, is suddenly doubled) coupled GCM experiments, which provide a three-dimensional picture of expected climate changes. Nevertheless, the spatial patterns of observed climatic variation at decadal time-scales, including those inferred from paleoenvironmental climate indicators, such as tree-rings and ice-cores, may also resemble the characteristic patterns of variations obtained from the climate models. If so, it may take longer to separate a greenhouse climate change signal from the background of natural variability.

An important consideration with regard to the impact of increasing oceanic and tropospheric temperatures in response to increasing greenhouse gas forcing deals with rising sea-levels, both in terms of the absolute changes, as well as the rate of change. Over about the past century, global mean sea-levels have risen about 10 cm, at a more or less uniform rate of ~1–2 mm/year (BARNETT, 1988; WARRICK and OERLEMANS, 1990; GORNITZ and SOLOW,

1991). This represents primarily a eustatic (or volumetric) change, after having taken into account the effects due to land movements in response to both glacial rebound and land subsidence or emergence. The IPCC (1990) report concludes that there is "no firm evidence of accelerations in sea level rise during this century... ." Hence, the future evolution of this important variable in relation to global warming is unclear at the present.

Complicating the assessments of the climatic responses to increasing atmospheric concentrations of "greenhouse gases" is the presence in the instrumental, historical and proxy record of strong decadal scale temperature and precipitation variability (see previous sections). There is considerable observational and theoretical background to suggest that climate behaviour in the 20th century may not differ appreciably from that in past centuries and that "natural fluctuations", intrinsic to the coupled atmosphere–ocean system, could have accounted for the bulk of the changes in the observational record (see LORENZ, 1991 and Chapter 5 by JONES).

Statistical techniques have recently been developed which attempt to establish, in a cause-and-effect manner, the recent evolution of the observed temperature field to greenhouse gas forcing as predicted by GCMs, and to develop a monitoring strategy aimed at early climate change detection (BARNETT et al., 1991b). Two of the critical variables identified on the basis of the compatibility of model versus observational signal-to-noise ratios are surface and lower tropospheric temperature and water vapour mixing ratios in the upper troposphere. The three-dimensional structural changes in the temperature field, compared with transient and equilibrium GCM climate change simulations, may offer the best hope of achieving some degree of consensus in the shorter term about the causes of global and regional scale climatic behaviour. Some recent studies which show that day and night time temperatures have changed at different rates in most of the large continental regions of the globe (KARL et al., 1991b) perhaps point towards some unexpected consequences of global climatic change. Questions regarding changes in cloudiness remain unanswered, partly because of the difficulty in assessing what spurious effects exist in cloudcover data (see, for instance, HENDERSON-SELLERS and MCGUFFIE, 1989; KARL and STEURER, 1990; PLANTICO et al., 1990), and partly because truly global coverage of cloud amount is available only for a relatively short period of time (a couple of decades).

New observing platforms promise to increase our knowledge and understanding of climatic processes and variability in the oceans and how such changes affect atmospheric circulation patterns. These new observing systems include microwave sounders on board satellites (SPENCER and CHRISTY, 1990), long-term monitoring of the oceans from unmanned buoys and through such cooperative efforts as the Expendable Bathythermograph (XBT) program aboard merchant ships, and improved ocean models run in near-real-time to assimilate these data and provide a dynamically consistent four-dimensional picture of the world oceans (LEETMAA and JI, 1989). Certainly, some tantalising, but equally disconcerting, results are appearing in the literature regarding possible major changes in the oceanic circulation that may arise, a few centuries from now, in response to climatic changes forced by the ever increasing greenhouse gas loading of the atmosphere (e.g. MANABE and STOUFFER, 1993 and Chapter 14 by PENG). Such changes could be irreversible on the time-scale of many human generations, and may have incalculable impacts on society.

The level of "natural" climatic variability still remains to be fully determined. This lack of a complete understanding of long-term variability complicates the assessment of the major factors responsible for a lack of convergence between the historical record of climate and the predictions of climate models (see Chapter 5 by JONES). Natural variability could be responsible for most or all of the observed climatic record on decade-to-century time-scales, and could well be the principal factor for future climates.

References

AAGAARD, K. and CARMACK, E. C., 1989. The role of sea-ice and other fresh water in the Arctic circulation. *J. Geophys. Res.*, 94: 14485–14498.

ACEITUNO, P., 1989. On the functioning of the Southern Oscillation in the South American sector. Part II: upper-air circulation. *J. Climate*, 2: 341–355.

ACTA ARCHAEOLOGICA, 1991. *The Norse of the North Atlantic.* Volume 61 (Special Issue), Institute of Archeology, Copenhagen, Denmark.

ALLAN, R. J., 1988. El Niño-Southern Oscillation influences in the Australasian region. *Prog. Phys. Geogr.*, 12: 4–40.

ALLAN, R. J., 1991. Australasia. In: M. H. GLANTZ, R. W. KATZ and N. NICHOLLS (Editors), *Teleconnections Linking Worldwide Climate Anomalies.* Cambridge University Press, Cambridge, pp. 73–120.

ANDERSON, R. Y., 1992. Long-term changes in the frequency of occurrence of El Niño events. In: H. F. DIAZ and V. MARKGRAF (Editors), *El Niño: Historical and Paleoclimatic Aspects of the Southern Oscillation.* Cambridge University Press, Cambridge, pp. 193–200.

ANGELL, J. K., 1981. Comparison of variations in atmospheric quantities with sea surface temperature variations in the equatorial eastern Pacific. *Mon. Weather Rev.*, 109: 230–243.

ANGELL, J. K., 1991. Changes in tropospheric and stratospheric global temperatures, 1958–1988. In: M. E. SCHLESINGER (Editor), *Greenhouse-Gas-Induced Climatic Change: A Critical Appraisal of Simulations and Observations.* Elsevier, Amsterdam, pp. 231–247.

BAERREIS, D. A. and BRYSON, R. A. (Editors), 1967. Climatic change and the Mill Creek culture of Iowa. *Arch. Archeol.,* 29: 1–673.

BALIUNAS, S. and JASTROW, R., 1990. *Nature*, 348: 520–523.

BARNETT, T. P., 1985. Variations in near-global sea level pressure. *J. Atmos. Sci.*, 42: 478–501.

BARNETT, T. P., 1988. The estimation of "global" sea-level change: a problem of uniqueness. *J. Geophys. Res.,* 89: 7980–7988.

BARNETT, T. P., 1991. The interaction of multiple time scales in the tropical climate system. *J. Climate*, 4: 269–285.

BARNETT, T. P., DUMEIL, L., SCHLESE, U., ROECKNER, E. and LATIF, M., 1989. The effect of Eurasian snow cover on regional and global climate variations. *J. Atmos. Sci.*, 46: 661–685.

BARNETT, T. P., LATIF, M., KIRK, E. and ROECKNER, E., 1991a. On ENSO physics. *J. Climate*, 4: 487–515.

BARNETT, T. P., SCHLESINGER, M. E. and JIANG, X., 1991b. On greenhouse gas signal detection strategies. In: M. E. SCHLESINGER (Editor), *Greenhouse-Gas-Induced Climatic Change: A Critical Appraisal of Simulations and Observations.* Elsevier, Amsterdam, pp. 537–558.

BARNOLA, J. M., RAYNAUD, D., KOROTKEVITCH, Y. S. and LORIUS, C., 1987. Vostok ice core: a 160,000 year record of atmospheric CO_2. *Nature*, 329: 408–414.

BARNSTON, A. G. and LIVEZEY, R. E., 1987. Classification, seasonality and persistence of low-frequency atmospheric circulation patterns. *Mon. Weather Rev.*, 115: 1083–1126.

BARNSTON, A. G. and LIVEZEY, R. E., 1989. A closer look at the effect of the 11-year solar cycle and the Quasi-biennial Oscillation on Northern Hemisphere 700 mb height and extratropical North American surface temperature. *J. Climate*, 2: 1295–1313.

BATTISTI, D. S. and HIRST, A. C., 1989. Interannual variability in the tropical atmosphere/ocean system: influence of the basic state and ocean geometry. *J. Atmos. Sci.*, 45: 1687–1712.

BERGER, A., 1978. Long-term variations of caloric insolation resulting from the earth's orbital elements. *Quat. Res.*, 9: 139–167.

BERGTHORSSON, P., 1969. An estimate of drift ice and temperature in Iceland in 1000 years. *Jökull*, 19: 94–101.

WEARE, B. C., 1983. Interannual variation in net heating at the surface of the tropical Pacific Ocean, *J. Phys. Oceanogr.*, 11: 705–717.

WEBSTER, P. J., 1982. Seasonality in the local and remote atmospheric response to sea surface temperature anomalies. *J. Atmos. Sci.*, 39: 41–52.

WEBSTER, P. J., 1994. The role of hydrological processes in ocean-atmosphere interactions. *Rev. Geophys.*, 32: 427–476.

WEBSTER, P. J. and YANG, S., 1992. Monsoon and ENSO: selectively interactive systems. *Q. J. R. Meteorol. Soc.*, 118: 877–926.

WELLS, L. E., 1987. An alluvial record of El Niño events from northern coastal Peru. *J. Geophys. Res.*, 92: 14463–14470.

WELLS, L. E., 1990. Holocene history of the El Niño phenomenon as recorded in flood sediments of northern coastal Peru. In: *Proceedings Sixth Annual Pacific Climate (PACLIM) Workshop, March 5–8, 1989.* California Department of Water Resources, Interagency Ecological Studies Program Technical Report 23, pp. 141–144.

WENDLAND, W. M. and BRYSON, R. A., 1974. Dating climatic episodes of the Holocene. *Quat. Res.*, 4: 9–24.

WILLSON, R. C. and HUDSON, H. S., 1988. Solar luminosity variations in solar cycle 21. *Nature*, 332: 810–812.

WILLSON, R. C. and HUDSON, H. S., 1991. The Sun's luminosity over a complete solar cycle. *Nature*, 351: 42–44.

WOLTER, K., 1989. Modes of tropical circulation, Southern Oscillation and Sahel rainfall anomalies. *J. Climate*, 2: 149–172.

WRIGHT, H. E., 1984. Late Glacial and Late Holocene moraines in the Cerros Cuchpanga, central Peru. *Quat. Res.*, 21: 275–285.

WRIGHT, P. B., WALLACE, J. M., MITCHELL, T. P. and DESER, C., 1988. Correlation structure of the El Niño/Southern Oscillation phenomenon. *J. Climate*, 1: 609–625.

WYRTKI, K., 1965. The average annual heat balance of the North Pacific Ocean and its relationship to the ocean circulation. *J. Geophys. Res.*, 70: 4547–4559.

WYRTKI, K., 1975. El Niño-the dynamic response of the equatorial Pacific Ocean to atmospheric forcing. *J. Phys. Oceanogr.*, 5: 572–584.

YASUNARI, T., 1987. Global structure of the El Niño-Southern Oscillation. Part II: time evolution. *J. Meteorol. Soc. Jpn.*, 65: 81–102.

ZEBIAK, S. E., 1986. Atmospheric convergence feedback in a simple model for El Niño. *Mon. Weather Rev.*, 114: 1263–1271.

Chapter 7

Satellite systems and models for future climate change

R. E. DICKINSON

Introduction

Relationship of satellite observations to the climate system

Both modelling and observing future changes of climate require global observations of uniform coverage and quality, unrestricted by national boundaries or the technological limitations of individual countries. Orbiting satellite systems have such a capability. Satellite remote sensing detects fluxes of electromagnetic radiation that are also involved in some of the more crucial physical processes of the climate system.

Satellite instruments can now provide precision measures of the energy output of the sun (WILLSON, 1993). Other instruments show the distribution by season and geography of the reflected solar radiation and terrestrial thermal emission (e.g. BARKSTROM and SMITH, 1986). Emissions at thermal and microwave wavelengths give considerable information regarding distributions of atmospheric and surface temperatures and composition. Precise measures of time of travel and amplitude of reflected signals by various satellite instruments reveal heights of the ocean surface and its wave state.

Exploitation of the various technological opportunities available to take quantitative observations of the climate system from space will increase our knowledge of the climate system and consequently improve models of the climate system and its changes. Global observations are also required for the detection of climate trends. This chapter primarily addresses the application of satellite observations to improving model projections of climate change. For that purpose, data sets of limited accuracy and duration may still be useful if they are improvements over what would otherwise be available. Trend detection, on the other hand, requires considerable attention to precision and long-term continuity of the data. Inferences as to long-term future climates from present trends require uniform climate records over decades to centuries. Satellite records, on the other hand, normally provide records of a few years in length or, with large efforts, records of up to a decade (cf. Chapter 5 by JONES and Chapter 6 by DIAZ and KILADIS). Thus satellite data alone are best used to characterize interannual variations in climate. However, they need to be merged with many other kinds of data to examine trends of longer than a decade.

Some of the key satellites and their instruments

Although satellite systems have been in place for several decades, they have shortcomings that have limited their usefulness for quantitative applications to climate modelling. Weather satellites have used visible and infrared radiances operationally to obtain imagery of clouds and vertical temperature profiles. Other products have been developed recently, including microwave estimates of atmospheric humidity and rainfall, ocean surface temperatures and indices of land vegetation. Earth resource satellites have examined, in fine detail but infrequently in time, selected areas over the land surfaces (e.g. to study Amazon deforestation; SKOLE and TUCKER, 1993). Until recently, attempts to derive precise averages of surface radiation and cloud properties from satellite radiances have been confounded by errors and drifts in instrumental calibration and by the complicated geometries of satellite viewing. Besides the more obvious dependencies on season, latitude and longitude, significant dependencies on time of day, sun angle and viewing angle require measurements to sample in multiple angular dimensions and in time. Typical low-altitude Sun-synchronous orbits take satellites over the poles every two hours or so as the Earth rotates beneath them (Fig. 1a). With such orbits, instruments see only a limited range of Sun angles, times of day and viewing angles.

Geostationary orbits (in which the satellite rotates with the Earth, Fig. 1b) provide diurnal sampling, which is especially important for clouds and precipitation. However, because of their high altitude, geostationary orbits give reduced instrumental spatial resolution and, because of their lack of global coverage, at least several satellites are needed to see most of the Earth. Tables I and II (OHRING et al., 1989) summarize the instruments on the National Oceanic and Atmospheric Administration (NOAA) polar orbiting and geostationary satellites. Satellite research has begun to provide major dividends for improvement of climate models with recent efforts to calibrate the satellite radiances and to model diurnal cycles and angular distributions of the satellite radiances. Important observations include inference of cloud properties, precipitation rates and surface solar radiation.

The most widely used polar orbiter systems have been the US NOAA weather satellites. These have temperature sounding units that measure outward fluxes of infrared and microwave thermal emissions over narrow spectral channels from specific atmospheric layers. Vertical profiles of temperature and humidity are determined by the inversion of the emissions from thick layers centred at various altitudes.

While sounding emphasizes vertical profiles, imaging refers to instruments whose emphasis is on spatial coverage, like a TV image. The NOAA satellites have provided the Advanced Very High Resolution Radiometer (AVHRR), an imaging system that looks at 1 km pixels and covers most of the Earth every day. It measures upward radiation in three channels for reflected solar radiation and in two channels for thermal emission. Originally intended to provide primarily qualitative pictures of cloud cover, this instrument has been expanded to provide much of the quantitative information on clouds, land properties, and ocean temperatures.

The Land Remote-Sensing Satellite (LANDSAT) series of satellites (originally developed by the National Aeronautics and Space Administration (NASA)) also does imagery, but with higher spatial resolution (30–80 m) and with further spectral channels. These capabilities have made possible detailed classifications of land covers for mineral surveys and descriptions of vegetation cover and land use. However, individual scenes are repeated only every few weeks. After the loss of land imagery because of clouds, this infrequent repeat cycle means that a given land scene may be sampled as little as a few times a year. Regular cover-

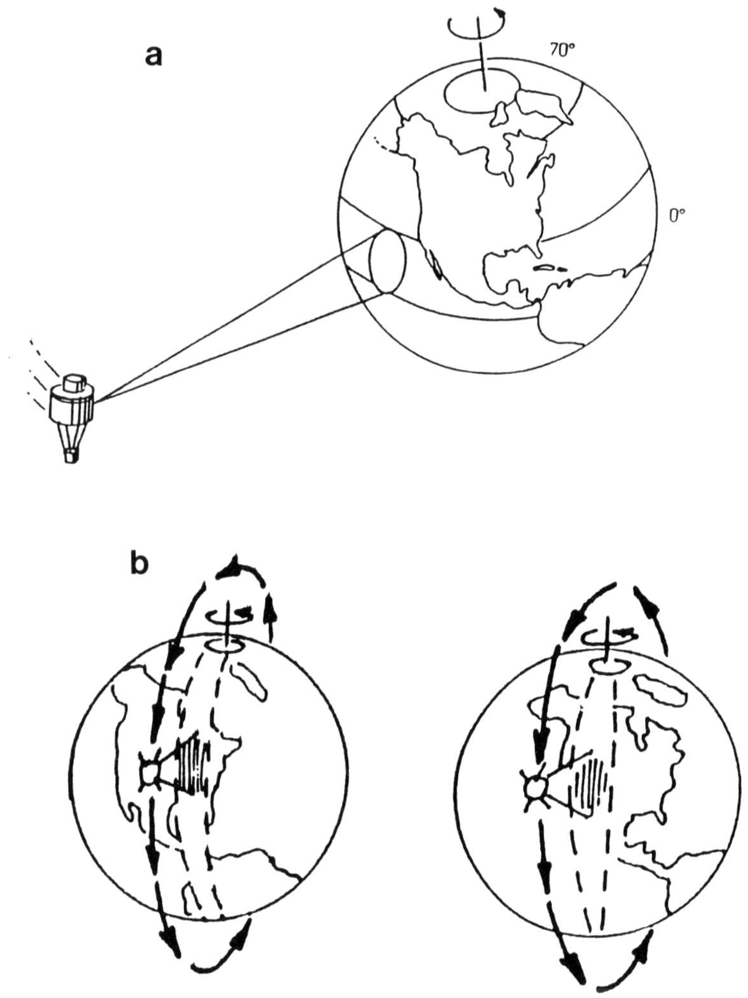

Fig. 1. (a) The geostationary satellite moves through space at the same rate that the Earth rotates, so it remains above a fixed spot on the equator and monitors one area constantly. (b) Polar orbiting satellites scan from north to south and on each successive orbit the satellite scans an area further to the west. From AHRENS, 1991.

age of a particular scene would be available only by special arrangement. LANDSAT is currently operated commercially in the US through the Earth Observation Satellite Company (EOSAT). The United States Geological Survey (USGS) at its Earth Resources Observation System (EROS) data centre provides archiving and dissemination capabilities for historical LANDSAT data.

Recently Japan and Europe have been developing large satellite programmes to complement and supplement US capabilities. Of the new systems already employed, the commercial French Système pour l'Observation de la Terre (SPOT) satellite provides similar, but somewhat higher resolution, capabilities to those of LANDSAT. European and Japanese (and soon Canadian) Synthetic Aperture Radars (SARs) provide detailed imagery of sea ice and, potentially, some land properties.

In addition to the operational satellites, the US flies Department of Defense (DOD) military satellites and NASA research satellites. The DOD satellites overlap considerably in the row swaths seen by the radar, rainfall estimation will be considerably improved from past approaches and the infrared and microwave estimates available from other instruments on TRMM or from other satellites also will be improved through calibration with this orbiting radar.

Measurements of precipitation are required for validation of climate model precipitation. How good are the absolute amounts, the spatial patterns, and the interannual variations? Interannual variability of precipitation, especially in the tropics, is closely linked to variations in sea surface temperatures (SSTs). Hence, climate models using measured time series of SSTs should produce the measured time series of precipitation, at least over large enough time- and space-scales. The El Niño Southern Oscillation (ENSO) phenomenon (e.g. LAU and BUSALACCHI, 1993) provides much of this variability.

Observations from satellite and other means are also needed to better establish an understanding of the mechanisms producing the precipitation. Current questions include the following: How do surface boundary conditions (land and ocean) affect the presence and intensity of precipitation? What are the effects of cloud radiative processes and cloud microphysical processes on precipitation? Hypotheses relating tropical SSTs to extratropical climate anomalies include a suggested explanation for the US drought of 1988 (TRENBERTH et al., 1988).

Atmospheric dynamic and thermodynamic fields

Atmospheric temperatures, winds and water vapour concentrations have traditionally been provided through radiosondes (vertically ascending instrumented balloons). These observations are required on a daily basis for initialization of weather prediction models. Assimilation procedures are used that add information from earlier observations to the current data. Archives of the results are maintained as climate data. Over the past two decades, satellite instruments have been developed to provide additional temperature and water vapour soundings to those obtained with balloons (as reviewed by SUSSKIND, 1993). These measurements depend on thermal (infrared and microwave) emissions coming from different atmospheric layers at different wavelengths. Infrared provides the best vertical resolution, but microwave is less affected by clouds. Use of non-vertical paths to the satellite, especially viewing of the limb, improves the vertical resolution; limbscanning is most successful in the upper troposphere and above, where cloud obscuration is minimal.

Where available, radiosonde temperatures from balloons are still substantially more accurate than are present satellite soundings, but the satellite temperature soundings provide data over the oceans, where no other temperature measurements are made. The data for temperatures have been obtained primarily from the NOAA sounders. SPENCER and CHRISTY (1992) have introduced climatological analyses directly using radiances from the microwave unit on the NOAA satellite. Considerable improvements in accuracy and vertical resolution of temperature are expected with the Atmospheric Infrared Sounder (AIRS) instrument being developed for EOS and for water vapour from this and microwave instruments. Geostationary satellites now add to balloon measurement of wind a limited amount of additional information by tracking of cloud movements (e.g. RUTLEDGE et al., 1991).

Additional measurement of water vapour has been made possible from microwave data (e.g. CHANG and WILHEIT, 1979; TJEMKES et al., 1991). SODEN and BRETHERTON (1994) compare direct satellite estimates of water vapour with the spatial distributions of water vapour in two climate models.

The largest source of error in temperature measurement may be not in the satellite data but in the algorithms used to invert them for temperature. For example, HURRELL and TRENBERTH (1992) correlate monthly means of Channel 2 of the Microwave Sounding Unit (MSU) of the NOAA satellites with monthly means from the European Center for Medium-Range Weather Forecasting (ECMWF) archives for optimally assimilated data. Strangely enough, the MSU fields are more highly correlated with the ECMWF analyses over regions where the model is initialized by radiosonde data than over regions where the primary temperature measurement is inversion from the NOAA satellite (which includes the MSU data). They find correlations of greater than 0.95 over extratropical continents of the Northern Hemisphere, whereas, at comparable latitudes in the Southern Hemisphere, correlations are between 0.8 and 0.9. Current research is working towards more direct methods of assimilating the satellite radiance data (e.g. EYRE, 1989; EYRE et al., 1993).

An alternative approach to use of model forecasts as a first-guess input to the inversion of temperature and humidity profiles from the NOAA satellites has been reported by SUSSKIND et al. (1984) and CHAHINE and SUSSKIND (1991). Their approach is evolving into algorithms for analysis of the AIRS data on future EOS platforms.

Data assimilation issues

Atmospheric fields are continuous in time and space, whereas satellite and radiosonde measurement systems provide limited samples. This limitation is largely overcome through use of 4-dimensional data assimilation approaches using dynamical models (e.g. HARMS et al., 1992; SCHUBERT et al., 1993). This approach has been developed as part of numerical weather prediction (NWP). Weather prediction models start with present observed conditions and are integrated forward in time for up to several weeks to derive future weather. Because the equations of atmospheric dynamics are chaotic, the initial conditions are eventually "forgotten", and a model simply begins to provide a realization of its climate (random departures from mean conditions). In principle, NWP models describe the same physical system as climate models. In practice, NWP models are integrated at much higher spatial resolutions, whereas climate models use additional physical parameterizations and system components, important on longer time-scales.

Over the first day or two of a weather forecast, the predicted winds and temperatures are largely determined by the observationally based initial conditions. Hence, these predicted fields provide an observationally based estimate of the actual winds and temperatures at that time. Direct measurements made concurrently provide another estimate.

An improved estimation can be made by combining these independent estimates. If observations can be characterized statistically as an actual value plus random noise, then multiple estimates can be optimally combined. This approach is used at a point to combine direct observations with model forecasts derived from previous observations. Observations are extended spatially by use of information on spatial "influence" functions, derived from auto-

RICHARDS, F. and ARKIN, P., 1981. On the relationship between satellite-observed cloud cover and precipitation. *Mon. Weather Rev.*, 109: 1081–1093.
ROBINSON, D. A., SERREZE, M. C., BARRY, R. G., SCHARFEN, G. and KUKLA, G., 1992. Large-scale patterns and variability of snow melt and parameterized surface albedo in the Arctic Basin. *J. Climate*, 5: 1109–1119.
ROBINSON, D. A., DEWEY, K. F. and HEIM JR., R. R., 1993. Global snow cover monitoring: an update. *Bull. Am. Meteorol. Soc.*, 74: 1689–1696.
ROSSOW, W. B., 1993. Clouds. In: R. J. GURNEY, J. L. FOSTER and C. L. PARKINSON (Editors), *Atlas of Satellite Observations Related to Global Change*. Cambridge University Press, Cambridge, pp. 141–163.
ROSSOW, W. B. and SCHIFFER, R. A., 1991. ISCCP cloud data products. *Bull. Am. Meteorol. Soc.*, 72: 2–20.
RUNNING, S. W., JUSTICE, C. O., SALOMONSON, V., HALL, D., BARKER, J., KAUFMANN, Y. J., STRAHLER, A. H., HUETE, A. R., MULLER, J.-P., VANDERBILT, V., WAN, Z. M., TEILLET, P. and CARNEGGIE, D., 1994a. Terrestrial remote sensing science and algorithms planned for EOS/MODIS. *Int. J. Remote Sensing, Special Edition*, 15: 3587–3620.
RUNNING, S. W., LOVELAND, T. R., PIERCE, L.L. and HUNT JR., E. R., 1994b. A remote sensing based vegetation classification logic for global land cover analysis. *Remote Sensing Environ., ISLSCP Special Issue*, 51: 39–48.
RUTLEDGE, G., LEGG, E. and MENZEL, P., 1991. Operational production of winds from cloud motions. *Global Planet. Change*, 4: 141–150.
SCHUBERT, S. D., ROOD, R. B. and PFAENDTNER, J., 1993. An assimilated dataset for Earth Science applications. *Bull. Am. Meteorol. Soc.*, 74: 2331–2342.
SCHWEIGER, A. J. and KEY, J. R., 1992. Arctic cloudiness: comparison of ISCCP-C2 and Nimbus-7 satellite-derived cloud products with a surface-based cloud climatology. *J. Climate*, 5: 1514–1527.
SELLERS, P. J., HALL, F. G., ASRAR, G., STREBEL, D. E. and MURPHY, R. E., 1992a. An overview of the First International Satellite Land Surface Climatology Project (ISLSCP) Field Experiment (FIFE). *J. Geophys. Res.*, 97, 18345–18371.
SELLERS, P. J., HEISER, M. D. and HALL, F. G., 1992b. Relations between surface conductance and Spectral Vegetation Indices at intermediate (100 m^2 to 15 km^2) length scales. *J. Geophys. Res.*, 97: 19033–19059.
SHUKLA, J. and MINTZ, Y., 1982. Influence of land-surface evapotranspiration on the Earth's climate. *Science*, 215: 1498–1501.
SIMPSON, J., ADLER, R. F. and NORTH, G. R., 1988. A proposed Tropical Rainfall Measuring Mission (TRMM) satellite. *Bull. Am. Meteorol. Soc.*, 69: 279–295.
SKOLE, D. and TUCKER, C., 1993. Tropical deforestation and habitat fragmentation in the Amazon: satellite data from 1978 to 1988. *Nature*, 260: 1849–2024.
SLINGO, A. and SLINGO, J. M., 1991. Response of the National Center for Atmospheric Research Community Climate Model to improvements in the representation of clouds. *J. Geophys. Res.*, 96: 15341–15358.
SMITH, L. D. and VONDER HAAR, T. H., 1990. Cloud-radiation interactions in a general circulation model: impact upon the planetary radiation balance. *J. Geophys. Res.*, 96: 893–914.
SMITH, E. A., COOPER, H. J., CROSSON, W. L. and HENG-YI, W., 1993. Estimation of surface heat and moisture fluxes over a prairie grassland. 3. Design of a hybrid physical/remote sensing biosphere model. *J. Geophys. Res.*, 98: 4951–4978.
SODEN, B. J. and BRETHERTON, F. P., 1994. Evaluation of water vapor distribution in general circulation models using satellite observations. *J. Geophys. Res.*, 99: 1187–1210.
SOMERVILLE, R. C. J. and REMER, L. A., 1984. Cloud optical thickness feedbacks in the CO_2 climate problem. *J. Geophys. Res.*, 89: 9668–9672.
SPENCER, R. W. and CHRISTY, J. R., 1992. Precision and radiosonde validation of satellite gridpoint temperature anomalies. Part I: MSU channel 2. *J. Climate*, 5: 847–857.
STARKS, P. J., NORMAN, J. M., BLAD, B. L., WALTER-SHEA, E. A. and WALTHALL, C. L., 1992. Estimation of shortwave hemispherical reflectance from bidirectionally reflected radiance data. *Remote Sensing Environ.*, 33: 123–134.
STEPHENS, G. L. and GREENWALD, T. J., 1991. The Earth's radiation budget and its relation to atmospheric hydrology: 2. Observations of cloud effects. *J. Geophys. Res.*, 96: 15325–15340.

SUD, Y. C., CHAO, W. C. and WALKER, G. K., 1993. Dependence of rainfall on vegetation: theoretical considerations, simulation experiments, observations and inferences from simulated atmospheric soundings. *J. Arid Environ.*, 25: 5–18.

SUSSKIND, J., 1993. Water vapor and temperature. In: R. J. GURNEY, J. L. FOSTER and C. L. PARKINSON (Editors), *Atlas of Satellite Observations Related to Global Change.* Cambridge University Press, Cambridge, pp. 89–128.

SUSSKIND, J., ROSENFIELD, J. and REUTER, D., 1984. Remote sensing of weather and climate parameters from HIRS2/MSU on TIROS-N. *J. Geophys. Res.*, 89: 4677–4697.

THOMAS, R. H., 1993. Ice sheets. In: R. J. GURNEY, J. L. FOSTER and C. L. PARKINSON (Editors), *Atlas of Satellite Observations Related to Global Change.* Cambridge University Press, Cambridge, pp. 385–400.

TJEMKES, S. A., STEPHENS, G. L. and JACKSON, D. L., 1991. Spaceborne observation of columnar water vapour: SSMI observations and algorithm. *J. Geophys. Res.*, 96: 10941–10954.

TRENBERTH, K. E., 1991. Climate diagnostics from global analyses: conservation of mass in ECMWF analyses. *J. Climate*, 4: 707–722.

TRENBERTH, K. E., BRANSTATOR, G. W. and ARKIN, P. A., 1988. Origins of the 1988 North American drought. *Science*, 242: 1640–1645.

TSELIOUDIS, G., ROSSOW, W. B. and RIND, D., 1992. Global patterns of cloud optical thickness variation with temperature. *J. Climate*, 5: 1484–1495.

WENTZ, F., 1992. Measurement of oceanic wind vector using satellite microwave radiometers. *IEEE Trans. Geosci. Remote Sensing*, 30: 960–972.

WHITLOCK, C. H., CHARLOCK, T. P., STAYLOR, W. F., PINKER, R. T., LASZLO, I., DIPASQUALE, R. C. and RITCHEY, N. A., 1993. *WCRP Surface Radiation Budget Shortwave Data Product Description - Version 1.1.* NASA Technical Memorandum 107747. National Aeronautics and Space Administration, Langley Research Center, Hampton, VA.

WILHEIT, T. J., CHANG, A. T. C. and CHIU, L. C., 1991. Retrieval of monthly rainfall indices from microwave radiometric measurements using probability distribution functions. *J. Atmos. Ocean. Tech.*, 8: 118–137.

WILLSON, R. C., 1993. Solar irradiance. In: R. J. GURNEY, J. L. FOSTER and C. L. PARKINSON (Editors), *Atlas of Satellite Observations Related to Global Change.* Cambridge University Press, Cambridge, pp. 5–18.

WILLSON, R. C. and HUDSON, H. S., 1991. The Sun's luminosity over a complete solar cycle. *Nature*, 361: 42–44.

WU, M. L. C. and CHANG, L. A., 1991. Differences in global data sets of atmospheric and surface parameters and their impact on outgoing longwave radiation and surface downward flux calculations. *J. Geophys. Res.*, 96: 9227–9262.

where the horizontal energy flux vector $\boldsymbol{H}$ is given by

$$\boldsymbol{H} = \frac{1}{g}\int_0^{p_0} \overline{\beta(c_p T + \Phi + Lq)\boldsymbol{V}}\,\mathrm{d}p \tag{24}$$

and the source sink term S is

$$S = R_\mathrm{T} - R_\mathrm{S} + H_\mathrm{S} + LE \tag{25}$$

Thus S represents the net radiative flux of energy into the atmosphere at the top and out of the atmosphere at the surface, and the sensible and latent energy fluxes into the atmosphere at the surface.

The energy flux vector $\boldsymbol{H}$ can be further decomposed into its mean and transient parts as (as in equation (12)) where

$$\boldsymbol{H}_\mathrm{M} = \frac{1}{g}\int_0^{p_0} \overline{\beta}(c_p \overline{T} + \overline{\Phi} + L\overline{q})\overline{\boldsymbol{V}}\, dp \tag{26}$$

and

$$\boldsymbol{H}_\mathrm{T} = \frac{1}{g}\int_0^{p_0} \beta(c_p T' + \Phi' + Lq')\boldsymbol{V}'\,\mathrm{d}p \tag{27}$$

Further decomposition of the flux vector into its rotational and divergent components then provides $\boldsymbol{H} = \boldsymbol{H}_\psi + \boldsymbol{H}_\chi$ (as in equation (13)) where the divergent component $\boldsymbol{H}_\chi$ is related directly to the distribution of sources and sinks.

Figure 3, taken from TRENBERTH and SOLOMON (1994), shows the divergent component of the net energy flux vector and its associated potential function for the month of January. The flow of energy between sources and sinks is clearly shown. Energy source regions are asso-

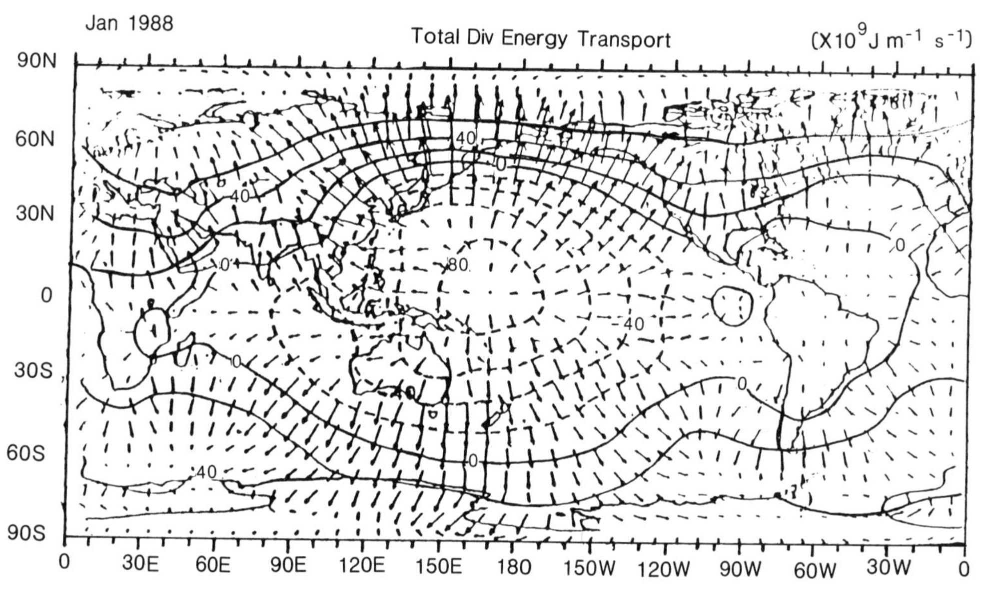

Fig. 3. The total energy flux and potential function for the month of January 1988 (from TRENBERTH and SOLOMON, 1994).

ciated with the tropical oceans and with the summer hemisphere land masses in general whereas sink regions are associated with cold land and ocean regions particularly in the winter hemisphere. This distribution of sources and sinks is compatible with the expression for the source sink term above, where source regions should correspond to regions of positive surface heat flux and evaporation and sink regions should correspond to negative radiative flux contributions. The primary region of energy divergence is located in the heat-source region of the western Pacific. Energy divergence is also apparent over Brazil, equatorial Africa and along the ITCZ of the Indian Ocean. Centres of energy divergence are also evident just off the east coast of Asia and North America. Divergence in these regions is associated with the strong sensible and latent heat fluxes occurring in cyclogenetic areas over the relatively warm ocean currents. These findings are consistent with those of JOHNSON (1989) for the FGGE period.

The divergent energy flux can be decomposed separately into its mean and transient parts. Figure 4 (from TRENBERTH and SOLOMON, 1994) shows the transient part of the divergent energy flux and the associated potential functions. The transient component of the divergent energy flux is an important part of this net energy flux. The transient mode of energy divergence shows a local dominance in extratropical latitudes whereas the stationary component (not shown) is more global in character.

In the Northern Hemisphere, transient energy convergence tends to occur in the region of troughs in the mass stream function field (not shown) whereas transient energy divergence occurs over the ridges. In the Southern Hemisphere, the transient energy transport is generally convergent in a meridional belt extending from 45 to 70°S whereas the total transport is divergent over this region. Equatorward of this transient energy convergence belt is another nearly continuous belt of transient energy divergence which stretches from South America through the South Atlantic, across the Indian Ocean, Australia and the South Pacific. Throughout this systematic distribution of energy exchange in the baroclinic region of the Southern Hemisphere, the component of the transient energy transport is directed pole-

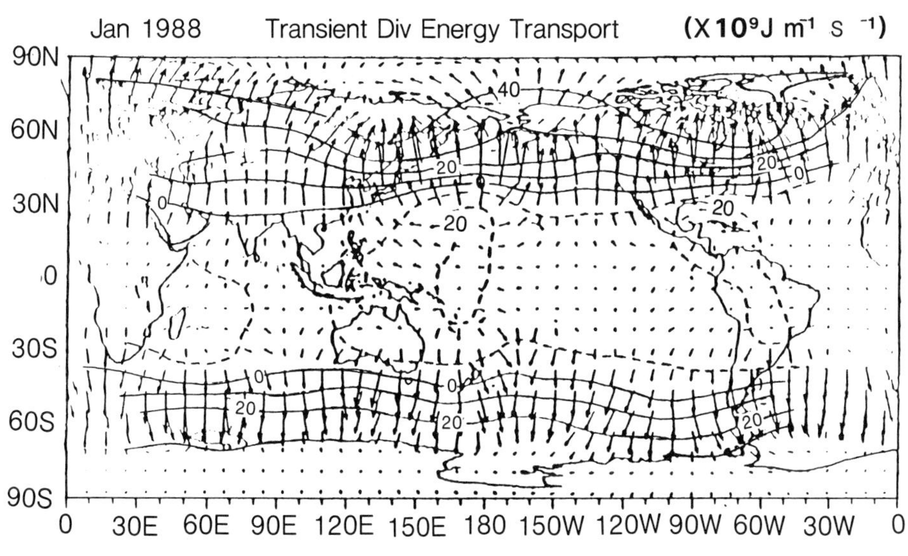

Fig. 4. The transient component of the energy flux and potential function for the month of January 1988 (from TRENBERTH and SOLOMON, 1994).

ward whereas the stationary component (not shown) is directed equatorward. In the Northern Hemisphere between 30 and 60°N, the poleward and equatorward meridional transports by the two components have the same sense. However, there is substantial zonal variation of the pattern of energy exchange in association with the longwave structure in this winter hemisphere.

JOHNSON (1989) has provided an explanation for the inverse relationship between the transient and stationary components of the energy divergence. Divergence of the total energy transport by geostrophic motion is primarily restricted to the advection of enthalpy. Consequently, for the middle latitudes of a hydrostatic atmosphere, to maintain the ratio of geopotential to internal energy, any increase in temperature in the lower troposphere through convergence of enthalpy transport requires a corresponding increase of geopotential energy in and above the level of transient enthalpy convergence. In turn, the adjustment to hydrostatic balance drives a divergent component of geopotential energy transport directed out of the region. This divergence of energy transport is reflected as a stationary component of the energy transport because it occurs through the mass circulation. In regions of transient energy divergence in isobaric layers the stationary component of geopotential energy at upper levels is convergent.

Water vapour

From the generalised time-averaged prototype budget equation of BOER (1982) (equation (6)) discussed in the section on Climate analysis, we can write the vertically integrated moisture budget equation as

$$\frac{1}{g}\frac{\partial}{\partial t}\int_0^{p_0}\overline{\beta q\,\mathrm{d}p}+\frac{1}{g}\nabla\cdot\int_0^{p_0}\overline{\beta q\boldsymbol{V}}\,\mathrm{d}p=\overline{E-P} \tag{28}$$

The vertically integrated flux vector for moisture is often termed the "aerial run-off" and is conventionally written as

$$\boldsymbol{Q}=\frac{1}{g}\int_0^{p_0}\overline{\beta q\boldsymbol{V}}\,dp$$

The total amount of water vapour is conventionally written as

$$W=\frac{1}{g}\int_0^{p_0}\overline{\beta q}\,\mathrm{d}p$$

so that the more usual form of the moisture budget equation emerges

$$\frac{\partial \boldsymbol{W}}{\partial t}+\nabla\cdot\boldsymbol{Q}=\overline{E-P} \tag{29}$$

Decomposition of modes of transport of the aerial run-off into mean, eddy, rotational and divergent components follows naturally using the formalism presented in the section on Climate analysis.

Analyses of the aerial run-off in order to gain a better understanding of the mechanisms involved in the maintenance of the local atmospheric water vapour content have been limited. Because of severe data limitations, PEIXÓTO and OORT (1983, 1992) concentrated on zon-

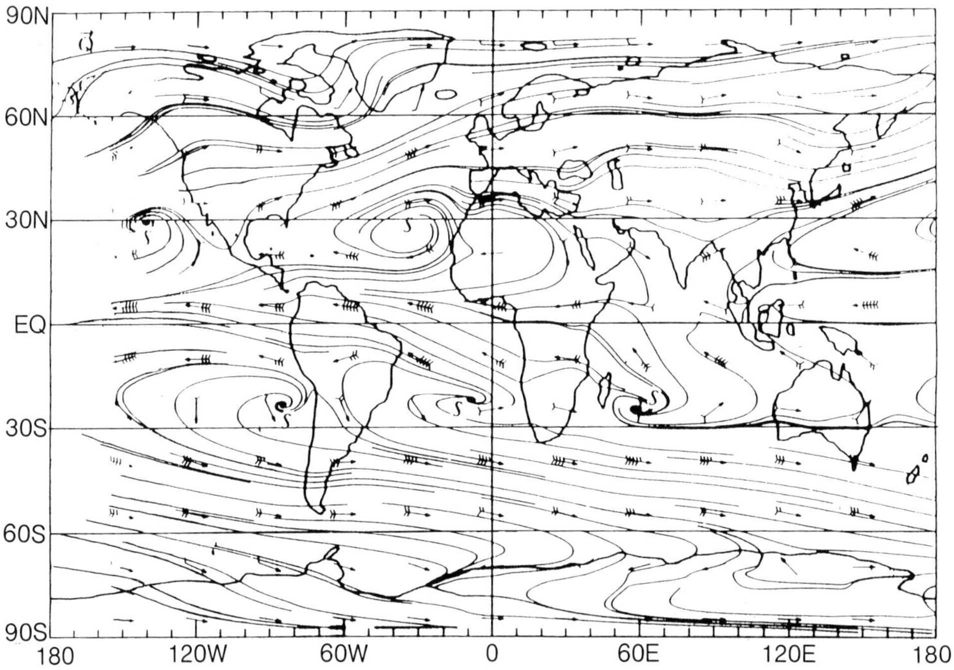

Fig. 5. The global distribution of the annual mean aerial run-off determined by PEIXOTO and OORT (1983) together with corresponding stream lines. Each barb represents 2 g kg^{-1} m s^{-1}.

ally averaged distributions of water vapour transport although as shown in Fig. 5, they were able to deduce global distributions of the mean aerial run-off which support reasonable expectations regarding the sources and sinks of moisture. Using global data assimilation analyses for a single year (FGGE), CHEN (1985) found that the stationary non-divergent mode shown in Fig. 6(a) describes most of the total water vapour transport. The stationary divergent mode composed of a combination of local Hadley circulation and the local Walker circulation, is responsible for the maintenance of high regional values of the water vapour content in equatorial regions. In mid-latitudes, the transient divergent mode shown in Fig. 6(b) is an important component of the poleward transient water vapour transport in the storm tracks. Problems associated with difficulties in moisture analysis, especially over the Southern Oceans, suggest some caution is necessary in assessing the reliability of these computations of water vapour transport. Further analysis is needed over a longer time period. It is interesting, however, to note that many of the models used in AMIP show quite large interannual variations in the stationary divergent mode of water vapour transport in the tropics which is strongly linked to the El Niño-Southern Oscillation (ENSO). An example from the BMRC model is shown in Fig. 7.

Large scale interactions

Teleconnections across the globe are a fundamental component of the climate system. Canonical examples include the ENSO phenomenon (PHILANDER, 1983, 1990); the Pacific-North American (PNA) pattern (HOREL and WALLACE, 1981; HOSKINS and KAROLY, 1981; WALLACE and GUTZLER, 1981) and the Madden Julian Oscillation (MADDEN and JULIAN,

Storm tracks

The synoptic climatology of storms, cyclone and anticyclone tracks, fronts and "storm tracks" (i.e. regions of maximum variance in geopotential at synoptic time scales less than a week) are important components of climate, especially because of their social and economic impacts on affected populations. The interesting "weather" experienced in mid-latitudes is primarily associated with "storminess" due to the passage of cyclones, anticyclones and their accompanying fronts. The usual interpretation of such systems is that they are a manifestation of the conversion of the potential energy associated with the north-south temperature gradient into the energy of atmospheric eddies through the process of baroclinic instability.

The more traditional synoptic approach of "storm track" analysis usually only considers the low centres when individual "storms" are tracked (e.g. MURRAY and SIMMONDS, 1991; KÖNIG et al., 1993). As TRENBERTH (1991a) argues, it is probably more meaningful to include both highs and lows since "weather" events are associated not only with the approach of low pressure systems but also with the withdrawal of high pressure systems. WALLACE et al. (1988) present a good summary of the relationships between traditional indicators of storm activity (tracks of cyclones and anticyclones) and the distribution of time-filtered geopotential height. Figure 11, from TRENBERTH (1991a), illustrates the link between the storm track and the mean flow through the polar jet streams; Figure 11(a) shows the zonal mean of the standard deviation of the geopotential height at 300 hPa as a direct measure of the storm track activity over the annual cycle, Fig. 11(b) shows the zonal average of the zonal wind at 300 hPa to illustrate the annual cycle of the subtropical and polar jet streams and Fig. 11(c) shows the meridional gradient of temperature at 700 hPa as a direct measure of baroclinicity in the lower troposphere. One of the most striking things about storm track activity is the persistent storminess in the Southern Hemisphere throughout the year compared to the Northern Hemisphere where the storminess shows a considerable seasonal variation in both intensity and location. The local meridional temperature gradient is a reasonable indicator of the location and strength of the storm track. There is also a strong relationship between the storm track and the major tropospheric jet streams. In the Southern Hemisphere, there is a noticeable semiannual variation with a slight shift of the polar jet and the storm tracks towards the pole during the transition seasons and is most probably related to the existence of the strong heat sink of Antarctica.

TRENBERTH (1991a) has demonstrated that the observed relationships amongst bandpass filtered eddy quantities can be understood in terms of geostrophic theory and perturbation analysis applied to baroclinic systems. A detailed study of the interannual variability of storm tracks and relations with the mean flow must await the emergence of global data from data assimilation "re-analysis projects". However, since large interannual variations in the mean flow are well documented and these variations have a profound influence on high frequency storms, it is to be expected that there will be significant interannual variations in storm tracks. TRENBERTH (1991a) has also investigated the question of the impact of the storm track eddies on the mean flow itself using a localised Eliassen-Palm flux. He finds that baroclinic effects from the poleward heat flux dominate in winter and the eddies act so as to decelerate the mean westerlies in the upper troposphere. The barotropic effect due to the flux convergence of meridional momentum contributes to the maintenance of the mean

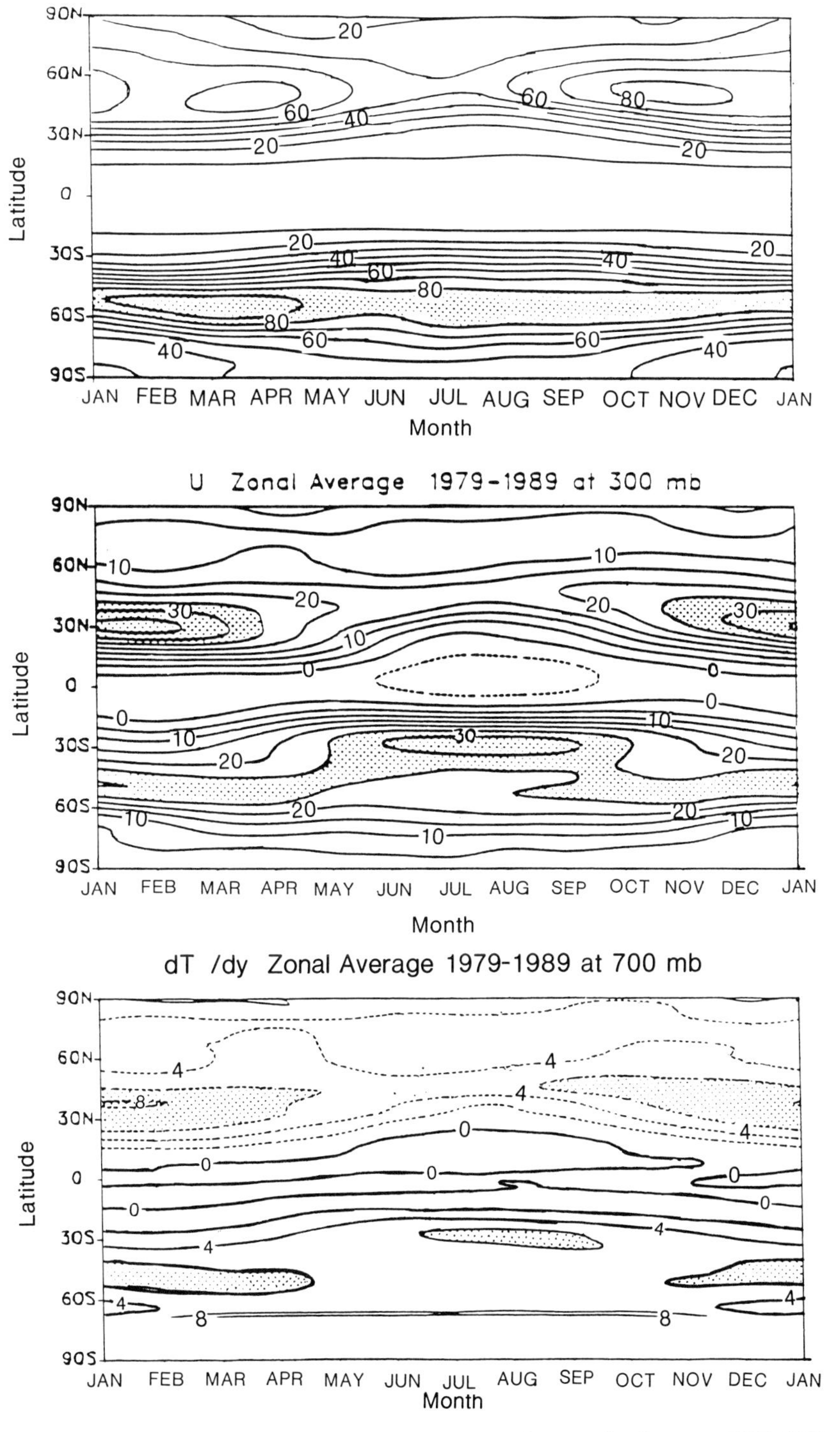

Fig. 11. The seasonal variation in monthly mean zonal means for the years 1979–1989 from ECMWF analyses. (a) Time filtered zonal average of the standard deviation in 300 hPa geopotential, contour interval 10 m, values greater than 80 m are stippled; (b) zonal mean of the zonal component of the wind at 300 hPa, contour interval 5 m s^{-1}, values greater than 25 m s^{-1} are stippled and negative values (easterlies) are dashed; (c) meridional temperature gradient at 700 hPa, contour interval 2×10^{6} K m^{-1}, magnitudes grater than 6×10^{-6} K m^{-1} are stippled and negative values are dashed (from TRENBERTH, 1991).

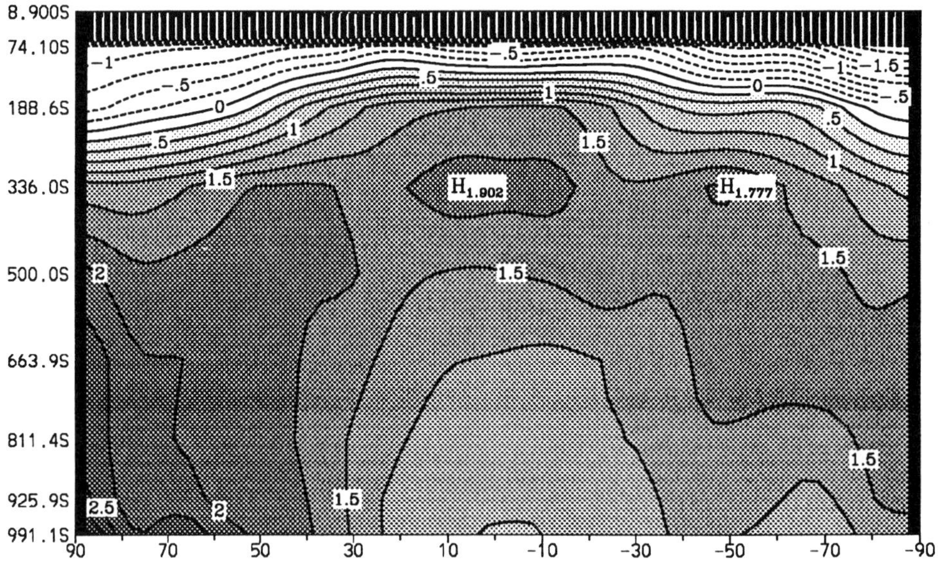

Fig. 14. The ensemble average of the zonal average annual change (transient – control2) in air temperature (K) for an transient increase in CO_2 of 1% per year in an experiment conducted by the Australian Bureau of Meteorology Research Centre using a coupled ocean-atmosphere model without flux correction. The ensemble average was for 20 years centred around the time of doubling of CO_2. The contour interval is 0.25 K and negative contours are dashed. Areas of warming are shaded.

troposphere is still warmed. This slightly modified behaviour is illustrated in Fig. 14 which shows the annual mean temperature response at the time of doubling in the transient CO_2 increase experiment conducted by BMRC (POWER et al., 1993). This transient experiment, and other similar experiments conducted by other modelling groups, indicates that the response in the baroclinic eddies may be different between the two hemispheres in enhanced greenhouse conditions.

HELD (1993) suggests, on the basis of linear instability theory, that the temperature gradient in the lower troposphere should dominate the response under greenhouse warming and hence eddies in the Northern Hemisphere should transport less energy while those in the Southern Hemisphere should transport more. However, as HELD (1993) points out, no definitive statement can be made since, even if the eddy transport of heat responds more to changes in the temperature gradient in the lower troposphere, it may be that the eddy transport of momentum is also sensitive to the gradient in the upper troposphere resulting in eddies with a larger fraction of their total energy in the middle and upper troposphere. Further complications arise due to consideration of the role of moist processes in conjunction with the growth and decay of the baroclinic eddies. Any warming of the atmosphere will tend to increase the moisture holding capacity in the eddies. The intensity of the resulting storms will depend on the balance between the competing effects of direct enhancement of the energy of eddies by latent heat release versus the determination of the size and strength of the eddies through the balance between the poleward transport of energy by the eddies and the gradient in the heating (HELD, 1993). Much further work is required before a more complete understanding of what controls the size and strength of the eddies in mid-latitudes. A further issue that requires further study is the relationship between the cloud field that is associated with regions of high eddy activity and the temperature gradient which is responsible for their formation.

Consideration of potential changes in the distribution of the rainfall associated with a movement of mid-latitude depressions is also of great importance since in most regions much of the rain falls from the equatorward edge of these depressions. The way in which the latitudinal distribution of eddy activity is affected by changes in the horizontal temperature gradient is not clear. Preferred regions for the growth and decay of depressions exist through zonal asymmetries of the large scale temperature and flow fields. These asymmetries arise in turn from major asymmetries in the lower boundary either through topographic features or contrasts between the heating over land compared to an adjacent ocean. Linear stability analyses of realistic planetary scale flows (e.g. FREDERIKSEN, 1985) and sensitivity studies with AGCMs are gradually leading to a better understanding of the factors which control the location of the tracks of depressions. However, because of the fundamental non-linearities involved, it is very difficult to predict how the dominant planetary-wave pattern and associated storm tracks might change in response to a given change in forcing.

The potential changes in the circulation of the Atlantic Ocean due to changes in freshwater input at high latitudes are discussed in Chapter 14 by PENG. The stability of North Atlantic deep water formation and the possibility of different modes of oceanic circulation have been the subject of much recent research using ocean only models. In most of these studies, feedback mechanisms involving salinity operate. The robustness of distinct climatic ocean states remains to be fully explored with three-dimensional coupled ocean-atmosphere models. However, in the context of global warming, it is possible that, with higher temperatures, a modest freshening of high latitude waters is possible due to increased moisture convergence producing enhanced excess of precipitation over evaporation (a requirement of moisture balance). Any freshening of polar water would weaken vertical overturning in the ocean and thereby increase the time that the surface waters were exposed to the freshening, reducing the overturning still further. Such a weakening in the overturning is found in many coupled ocean-atmosphere experiments. Whether an irreversible transition to a state where there is no deep-water formation will occur remains an open question.

Extensive public discussion on climatic changes to severe weather systems, such as hail storms, mid-latitude wind storms and tropical cyclones have engendered much debate because of the potentially catastrophic consequences to the global insurance industry and economies of individual countries. Much of the discussion is conjectural as it is difficult to consider regional-scale changes let alone changes in very rare severe weather events, for which there is often not even suitable records available for current climate (HOLLAND, 1981; NICHOLLS, 1992). We therefore examine the potential for changes associated with tropical cyclones, as an example of the problems and solutions to be applied.

Some studies have shown skill in climate model predictions of systems that have some of the large-scale characteristics of tropical cyclones (BENGTSSEN et al., 1994). However, we have shown that tropical cyclone development is fundamentally linked to complex scale-interaction processes at the regional scale. Since these scale interactions cannot be confidently predicted in future climate scenarios, and since climate models cannot resolve the cyclone core and actual intensity, we need to utilise some other measure for estimating climate changes. Whilst there are some creative attempts at applying climatic indices to model fields (e.g. RYAN et al., 1992), most methods have relied on application of thermodynamic approximations, such as by EMANUEL (1988) to tropical cyclones or to extrapolating current SST relationships (GRAY, 1968) to changes in ocean temperatures.

These methods were investigated in a review paper by LIGHTHILL et al. (1994), who concluded that there were no consistent statistical relationships between cyclone numbers and SST changes. For example the study of El Niño effects by HOLLAND et al. (1988) showed that the direct effects of SST changes were negligible and that it was the subtle circulation changes, and scale interactions that were responsible for any observed relationship. A careful study by LANDSEA and GRAY (1994) also found no relationship between the proportion of intense cyclones and surface temperature anomalies over the entire northern hemisphere.

The thermodynamic model predictions of maximum potential intensity (EMANUEL, 1987) indicate quite strong sensitivity to small ocean changes at the very intense cyclone stage. The observational evidence for such changes is equivocal, with some conclusions of no change (EVANS, 1993) and other evidence of some change (DEMARIA and KAPLAN, 1994), albeit with less amplitude than predicted from purely thermodynamic considerations. Other theoretical studies of the impact of spray and ocean energy transfers (FAIRALL et al., 1994) indicates that there may be counteracting processes occurring for extreme cyclones.

The best conclusion for future climates is that the regional characteristics can be expected to change in manners that are difficult to predict with current knowledge and techniques. We support the conclusions by LIGHTHILL et al. (1994), that SST changes of a few degrees will have negligible direct effect on tropical cyclones. Any small changes in cyclones are likely to be difficult to ascertain above the high natural variablity that occurs. However, substantial indirect changes in tropical cyclones at a regional level can be expected, and have been observed over recent decades. These changes arise from readjustments in the regional response to the global climate system.

References

BENGTSSON, L. and SHULKA, J., 1988. Integration of space and in situ observations to study climate change. *Bull. Am. Meteorol. Soc.,* 69: 1130–1143.

BENGTSSON, L., BOTZET, M. and ESCH, M., 1994. Hurricane-type vortices in a general circulation model, Part I. *Rep. No. 123*, Max-Planck-Institut für Meteorologie, Hamburg, Germany, ISSN 0937–1060, 42 pp.

BOER, G. J., 1982. Diagnostic equations in isobaric coordinates. *Mon. Weather Rev.,* 110: 1801–1820.

BOER, G. J., 1986. A comparison of mass and energy budgets from two FGGE datasets and a GCM. *Mon. Weather Rev.,* 114: 885–902.

BOER, G. J., 1990. Earth-atmosphere exchange of angular momentum simulated in a general circulation model and implication for the length of day. *J. Geophys. Res.*, 95: 5511–5531.

BOER, G. J. and SARGENT, N. E., 1985. Vertically integrated budgets of mass end energy for the globe. *J. Atmos. Sci.*, 42: 1592–1613.

BRIEGEL, L. M., 1993. An observational study of tropical cyclogenesis in the western Pacific Ocean. *Proc. 20th Conf. on Hurricanes and Tropical Meteorology, San Antonio, May*. American Meteorological Society, Boston, MA, pp. 397–400.

CHANG, H.-R. and WEBSTER, P. J., 1990. Energy accumulation and emanation at low latitudes. Part II: nonlinear response to strong episodic equatorial forcing. *J. Atmos. Sci.*, 47: 2624–2644.

CHARNEY, J. G., 1947. The dynamics of long waves in a baroclinic westerly current. *J. Meteorol.*, 4: 135–162.

CHEN, T.-C., 1985. Global water vapour flux and maintenance during FGGE. *Mon. Weather Rev.*, 113: 1801–1819.

DEMARIA, M. and KAPLAN, J., 1994. Sea surface temperature and the maximum intensity of Atlantic tropical cyclones. *J. Climate,* 7: 1324–1334.

EMANUEL, K. A., 1987. The dependence of hurricane intensity on climate. *Nature,* 326: 483–485.

EMANUEL, K. A., 1988. The maximum intensity of hurricanes. *J. Atmos. Sci.,* 45: 1143–1155.

EVANS, J. L., 1993. Sensitivity of tropical cyclone intensity to sea surface temperature. *J. Climate,* 6: 1133–1140.

FAIRALL, C. W., KEPERT, J. D. and HOLLAND, G. J., 1994. The effect of sea spray on surface energy transports over the ocean. *The Atmosphere Ocean System*, in press.

FARRELL, B. and WATTERSON, I., 1985. Rossby waves in opposing currents. *J. Atmos. Sci.*, 42: 428–449.

FIORINO, M. J. and ELSBERRY, R. L., 1989. The role of vortex structure in tropical cyclone motion. *J. Atmos. Sci.*, 46: 975–990.

FORTELIUS, C. and HOLOPAINEN, E., 1990. Comparison of energy source estimates derived from atmospheric circulation data with satellite measurements of net radiation. *J. Climate,* 3: 646–660.

FRASER, J. R., 1994. Vorticity and momentum budget analysis of ENSO anomalies in a GCM. *M. Sc. Thesis,* Monash University, Melbourne.

FREDERIKSEN, J. S., 1985. The geographic locations of Southern Hemisphere storm tracks: linear theory. *J. Atmos. Sci.,* 42: 710–723.

FUJIWHARA, S., 1921. The natural tendency towards symmetry of motion and its application as a principle of motion. *Q. J. R. Meteorol. Soc.*, 47: 287–293.

GILL, A. E., 1982. *Atmosphere-Ocean Dynamics.* International Geophysics Series, Vol. 30, Academic Press, New York, ISBN 0-12-283520-4, 662 pp.

GRAY, W. M., 1968. Global view of the origin of tropical disturbances and storms. *Mon. Weather Rev.,* 96: 669–700.

HALL, N. M. J., HOSKINS, B. J., VALDES, P. J. and SENIOR, C. A., 1994. Storm tracks in a high resolution GCM with doubled CO_2. *Q. J. R. Meteorol. Soc.,* 120: 1209–1230.

HALTINER, G. J. and WILLIAMS, R. T., 1979. *Numerical Prediction and Dynamic Meteorology*, 2nd Edition. Wiley, New York, ISBN 0-471-05971-4, 477 pp. .

HELD, I. M., 1993. Large-scale dynamics and global warming. *Bull. Am. Meteorol. Soc.,* 74: 228–241.

HIDE, R., DICKEY, J. O., MARCUS, S. L., ROSEN, R. D. and SALSTEIN, D. A., 1994. Diagnostic Sub-Project (No 15) for atmospheric angular momentum of the Atmospheric Model Intercomparison Project (AMIP). In: G. J. BOER (Editor), *Research Activities in Atmospheric and Oceanic Modelling.* WMO/TD 592, Geneva.

HOLLAND, G. J., 1981. On quality of the Australian tropical cyclone data base. *Aust. Meteorol. Mag.*, 29: 169–181.

HOLLAND, G. J., 1994. Scale interaction in the western Pacific monsoon. *Meteorol. Atmos. Phys.*, in press.

HOLLAND, G. J. and DIETACHMAYER, G. S., 1993. On the interaction of tropical-cyclone scale vortices. II: Interacting vortex patches. *Q. J. R. Meteorol. Soc.*, 119: 1363–1379.

HOLLAND, G. J. and LANDER, M., 1993. The meandering nature of tropical cyclone tracks. *J. Atmos. Sci.,* 50: 1254–1266.

HOLLAND, G. J., MCBRIDE, J. L. and NICHOLLS, N. N., 1988. Australian tropical cyclones and the Greenhouse effect. In: *Greenhouse. Planning for Climate Change.* E. J. Brill, Leiden, The Netherlands, pp. 438–455.

HOLTON, J. R., 1992. *An Introduction to Dynamic Meteorology*, 3rd Edition. Academic Press, New York.

HOREL, J. D. and WALLACE, J. M., 1981. Planetary-scale atmospheric phenomena associated with the southern oscillation. *Mon. Weather Rev.*, 109: 813–829.

HOSKINS, B. J., 1983. Dynamical processes in the atmosphere and the use of models. *Q. J. R. Meteorol. Soc.,* 109: 1–21.

HOSKINS, B. J. and KAROLY, D., 1981. The steady linear response of spherical atmosphere to thermal and orographic forcing. *J. Atmos. Sci.*, 38: 1179–1196.

HOSKINS, B. J., MCINTYRE, M. E. and ROBERTSON, A. W., 1985. On the use and significance of isentropic potential vorticity maps. *Q. J. R. Meteorol. Soc.*, 111: 877–946.

There have been many calculations of the radiative forcing due to observed increases of greenhouse gases since the pre-industrial era and its potential increase during the next few decades. The radiative forcing values for the individual gases have been used not only to rank the relative contribution of the individual gases to past climate change but also to provide guidelines for policy considerations such as the decision to phase out the CFCs from both climate warming and O_3 depletion viewpoints.

HANSEN and LACIS (1990) have calculated the radiative forcing due to the added greenhouse gases CO_2, CH_4, N_2O and CFCs for the periods 1850–1957 and 1958–1989. For the first period, the net forcing is calculated to be 0.87 W m^{-2} with contributions mainly from CO_2 (0.58 W m^{-2}) and CH_4 (0.24 W m^{-2}). However, for the second period, the calculated contributions are CO_2 (0.66 W m^{-2}), CFCs (0.25 W m^{-2}) and CH_4 (0.24 W m^{-2}) with a total net forcing of 1.17 W m^{-2}. The total radiative forcing of ~2 W m^{-2} since the industrial era is consistent with several other calculations documented in HOUGHTON et al. (1992). Estimates of the future radiative forcing for the period 1990–2100 were conducted by WIGLEY and RAPER (1992) using the IPCC scenarios IS92a–e. The calculated range of the forcing values is between 3.38 and 8.48 W m^{-2} with a value of 6.68 W m^{-2} for IS92a.

In recent years, GCMs have been used to evaluate the global climate change and its regional distribution due to the greenhouse effect. In general, there are two types of GCM experiments: equilibrium simulations using twice the pre-industrial CO_2 concentration and transient simulations with a gradual increase of CO_2, usually a 1% increase per year (see HOUGHTON et al., 1992). The equilibrium simulations allow scientists to study the climate feedbacks which represent an important source of uncertainty in climate prediction. For example, a warming climate directly caused by CO_2 increase can increase the atmospheric capacity of holding more water vapour which itself is a greenhouse gas. Model results indicate that the increase in water vapour can further enhance the CO_2-induced warming by 50%. Changes in temperature and moisture are likely to change cloud amount and cloud altitude. Changes in cloud cover would also play an important role in affecting the initial warming. However, the feedback is very complex since an increase in low clouds tends to reflect solar radiation and thus to cool the surface, but an increase in high clouds will warm the surface by trapping more infrared radiation. Model results suggest that a warming climate tends to increase the amounts of high clouds, but to decrease the low cloud, both contributing to an additional warming effect. However, increases in atmospheric moisture will tend to increase the amount of water in clouds and allow less solar radiation to pass through, thereby reducing the surface warming. Thus the magnitude of climatic change due to the greenhouse effect will depend significantly on the climate feedbacks.

The transient experiments are needed to take account of the response of ocean and a gradual increase of the greenhouse gases' concentration. The ocean plays a predominant role in regulating the climate change. The ocean circulation and its change associated with changes in the atmosphere through atmosphere-ocean coupling are the fundamental issues related to the use of models for prediction of the magnitude and timing of climatic change due to the greenhouse effect in the future.

Table III summaries several transient CO_2 experiments conducted using AOGCM. The globally averaged annual mean increase of surface air temperature at the time of effective CO_2 doubling is between 1.3 and 2.3°C. These values are about 60% of the individual models' equilibrium warming with doubled CO_2 calculated with a mixed-layer ocean. The

TABLE III

CONTROL AND TRANSIENT CO_2 SIMULATIONS WITH COUPLED OCEAN-ATMOSPHERE GENERAL CIRCULATION MODELS (AFTER GATES ET AL., 1993)

	GFDL	MPI	NCAR	UKMO
AGCM	R15 L9	T21 L19	R15 L9	2.5 × 3.75° L11
OGCM	4.5 × 3.75° L12	4 × 4° L11	5° L4	2.5 × 3.75° L17
CO_2 (ppmv)	300	390[a]	330	323
CO_2 (time)	1%/year (compound)	SA90 (~1.3%) (compound)	1%/year (linear)	1%/year (compound)
Simulation length (year)	100	100	70	75
Time for CO_2 doubling (year)	70	60	100	70
Warming (°C) at 2× CO_2	2.3	1.3	~2.3	1.7
2× CO_2 sensitivity (°C)	4.0	2.6	4.0	~2.7

L, number of vertical levels; R, number of spectral waves (rhomboidal truncation); T, no spectral waves (triangular truncation).
[a]Equivalent CO_2 value for trace gases.

lower values of the surface warming reflect mainly the large thermal inertia of the oceans which have not reached equilibrium at the time of effective CO_2 doubling. Comparisons between the equilibrium and transient experiments show that the patterns of the transient response are similar to those in an equilibrium responses, except over the high-latitude southern ocean and northern North Atlantic ocean where the delayed warming is caused by the oceanic thermohaline circulation.

Here, we discuss in more detail the changes in large-scale climate using the equilibrium experiments of WANG et al. (1992) and the transient experiments of MANABE et al. (1991, 1992) and GREGORY (1993).

Equilibrium experiment

WANG et al. (1992) describe the only GCM experiments to treat all the individual greenhouse gases using the IPCC SA90 (see Table I for concentration values) in the simulations while most of the other GCM studies use CO_2 as a surrogate (see HOUGHTON et al., 1992). The main difference between CO_2 and the trace gases CH_4, N_2O, $CFCl_3$ and CF_2Cl_2 lies in their difference in distribution of the longwave radiative forcing perturbations which may lead to different climate response (WANG et al., 1991).

Fig. 3 shows the latitude-altitude distribution of the longwave radiative heating rates between 1990 and 2050 for two cases: CO_2 alone and all the gases included. During this period, the concentration changes will provide a tendency for the stratosphere to cool and the troposphere to warm. In the stratosphere, CO_2 uniformly cools the temperature while the other gases further enhance the stratospheric cooling. However, these gases induce a warming effect in the lower stratosphere and upper troposphere, which is a unique characteristic distinguishing these gases from CO_2 despite the fact that all these gases are also uniformly

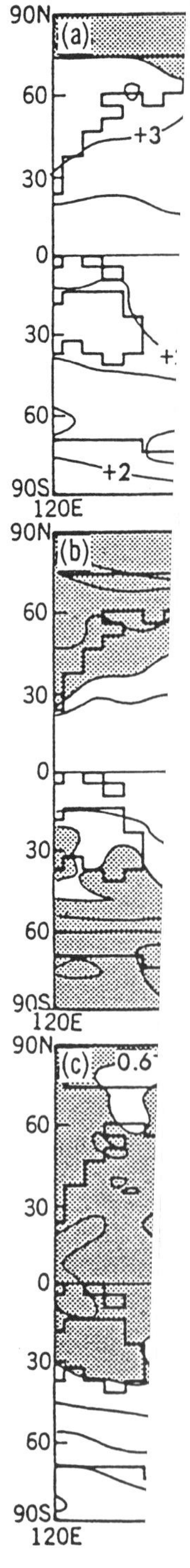

Fig. 7. (a) Th
eral circulati
surface air te
responses (af

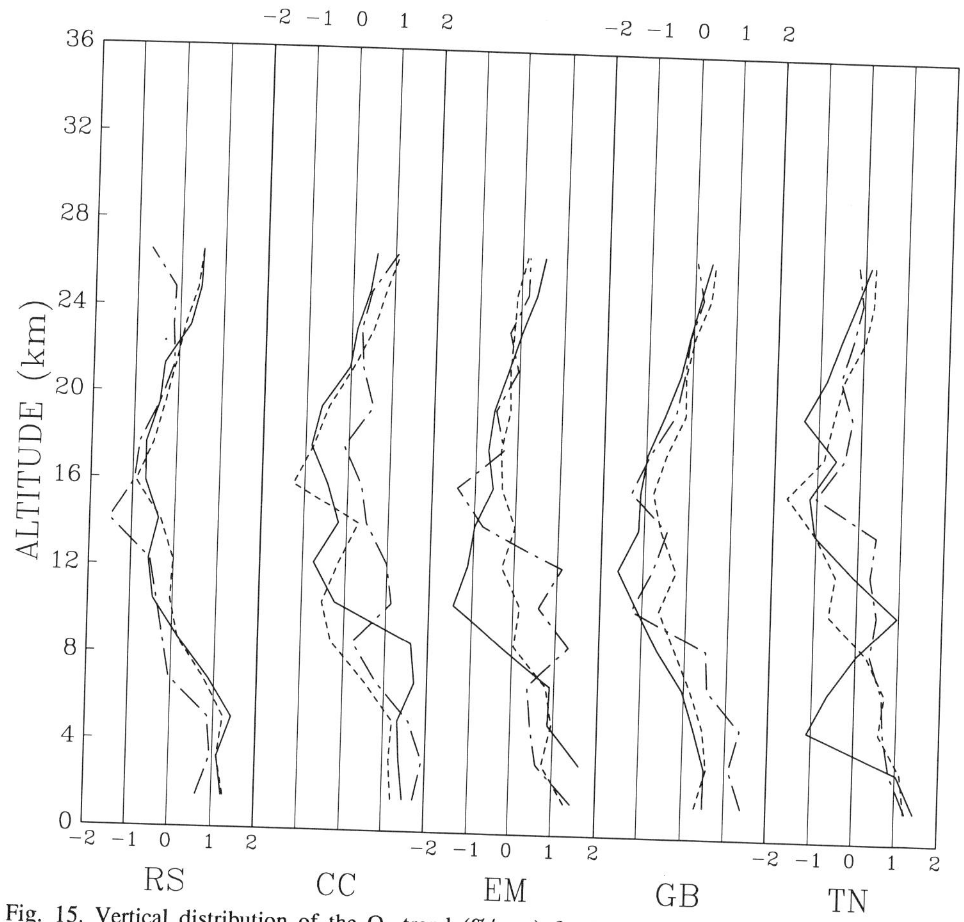

Fig. 15. Vertical distribution of the O_3 trend (%/year) for December–March (dashed line), May–September (solid line), and October–November (dash-dot-dash line) calculated for Resolute (RS), Churchill (CC), Edmonton (EM), Goose Bay (GB) and Tateno (TN).

These results clearly suggest that the observed changes in vertical O_3 distribution, in particular in the lower stratosphere and upper troposphere, can have significant climatic implications. While the lower stratospheric O_3 depletion can be attributed to increases of chlorofluorocarbons (WMO, 1991), the causes for the large increase in tropospheric O_3 in middle and high latitudes of the Northern Hemisphere is less clear. However, a recent study (THOMPSON, 1992) suggests that the increases may be caused by increasing amounts of O_3 precursor trace gases (CO, NO_x and hydrocarbons). Nevertheless, in light of the large effect on the radiative forcing, it will be important to quantify the sources for the increases during the last two decades so that the future trend can be better assessed. The large sensitivity of the O_3 radiative forcing to lower stratospheric temperatures points out the need to study the climatic implications of O_3 changes using GCMs in which the interactions between radiation and dynamics are properly incorporated.

Future climates

It is quite clear that emissions resulting from human activities are substantially increasing

TABLE VIII

COMPARISON OF THE CHANGES IN RADIATIVE FORCING (W M^{-2}) BETWEEN O_3 CHANGES AND THE COMBINED O_3 CHANGES AND INCREASES OF GREENHOUSE GASES CO_2, CH_4, $CFCL_3$, CF_2CL_2 AND N_2O (AFTER WANG ET AL., 1993)

	O_3 Changes				Combined O_3 and other gases	
	Total		Troposphere			
	FT	FD	FT	FD	FT	FD
Hohenpeissenberg						
January	0.410	0.327	0.222	0.136	0.794	0.710
July	0.570	0.548	0.490	0.467	1.050	1.027
Payerne						
January	0.356	0.190	0.242	0.074	0.797	0.630
July	0.397	0.136	0.097	−0.167	0.879	0.617
Resolute						
Winter	−0.038	−0.085	0.012	−0.036	0.230	0.183
Summer	0.194	−0.009	0.052	−0.153	0.620	0.415
Goose Bay						
Winter	0.025	−0.066	0.013	0.012	0.379	0.288
Summer	0.043	−0.218	−0.153	−0.415	0.497	0.235

The calculated changes in radiative forcing are between 1971–1980 and 1981–1990. Concentration increases for other gases also correspond to the indicated periods. FT and FD refer to fixed-temperature and fixed-dynamics treatments in the radiative forcing calculations.

the atmospheric concentrations of the greenhouse gases. Since pre-industrial times the radiative forcing of these greenhouse gases has increased by about 2 W m^{-2}, which is consistent with observed surface warming. The trend of increased emissions will continue for at least the next few decades, which will enhance the greenhouse effect. It is calculated that the rate of increases in radiative forcing is about 1 W m^{-2} in each 20-year period. The continued increase in the radiative forcing will likely further warm the global climate. As summarized in Fig. 16, the few available coupled atmosphere-ocean general circulation models all predict a warming of the surface of about 2.5°C at year 70 when CO_2 doubling is occurring. Presently, the issue of great concern is the magnitude and timing of the regional climate changes in a globally warming environment and the subsequent effect and impact of such climatic changes.

On the other hand, increasing tropospheric sulphate aerosols at mid-latitudes in the Northern Hemisphere (CHARLSON et al., 1992; KIEHL and BRIEGLEB, 1993; Chapter 9 by ANDREAE) may induce a cooling effect which certainly needs to be addressed within the context of establishing the total effect on radiative forcing due to increases of the greenhouse gases.

Observations have clearly indicated that stratospheric O_3 is decreasing around the globe and scientific evidence suggests that the phenomenon is caused by the use of chlorofluorocarbons. In addition, from limited ozonesonde measurements, there is an increasing trend for tropospheric O_3 at the middle and high latitudes of the Northern Hemisphere. Although the causes for such an increase remain to be understood, there are reasons to believe that human activities in the future will continue affecting the O_3 in similar fashion as observed in the last few decades. In addition, extensive aircraft fleet may be in operation in the lower stratosphere and upper troposphere in the coming decades. The emissions from aircraft in these viously highly dependent on assumptions about the further implementation of emission controls and about the rate of technological development in the developing world. Assuming a rapid increase of coal consumption in Asia, especially China, and a modest level of industrialization elsewhere in the developing world, GALLOWAY (1989) expects sulphur emissions to reach some 240 Tg by the year 2020. In the 1992 supplement to the IPCC Report, HOUGHTON et al. (1992) present a more optimistic view, suggesting that anthropogenic sulphur emissions will reach about 120 Tg/year by 2025 and will level out at about 150 Tg/year around the year 2050 (Fig. 3).

Aerosol production from precursor gases involves oxidation of DMS and other reduced sulphur gases to SO_2 and other intermediate species, among them methane sulphonic acid (MSA), dimethyl sulphoxide (DMSO), and dimethyl sulphone ($DMSO_2$). SO_2 and some of the other species can then be further oxidized photochemically to sulphuric acid which then forms sulphate aerosol. These usually incorporate gaseous ammonia, in the remote atmosphere typically at a mole ratio of about 1:1. Methane sulphonate appears to become associated with pre-existing aerosols rather than forming new aerosol particles, while for sulphate both pathways are used. LANGNER et al. (1992) have modelled the atmospheric transformations of biogenic DMS and anthropogenic SO_2, and estimate that about 50% of DMS is converted to sulphate. In the case of anthropogenic SO_2, about 50% is thought to be removed by wet and dry deposition before it can be oxidized to sulphate. For DMS, deposition losses of the intermediate, SO_2, probably play a lesser role, since DMS has a lifetime of about 2 days in the remote troposphere and can therefore be mixed away from the Earth's surface. The SO_2 formed from its oxidation is thus less likely to be immediately redeposited in the vicinity of the source than anthropogenic SO_2, which is injected close to the ground surface. On the other hand, in the case of DMS there are additional sinks, especially the wet and dry deposition of intermediates like MSA, DMSO, and $DMSO_2$, which lead to a loss of biogenic sulphur before it can become incorporated into aerosol. In Table I, we therefore use a yield of 50% for the transformation of both biogenic DMS and anthropogenic SO_2 to sulphate (or MSA) aerosol. In both cases, it is assumed that the final product has an ammonium-to-sulphate ratio of 1:1.

The total global aerosol source flux is about 3.5 Pg/year (1 Pg = 10^{15} g) (Table II). This is near the high end of the range of estimates by other authors, which typically fall between 2 and 4 Pg/year (JAENICKE, 1988; WARNECK, 1988; PREINING, 1991). The uncertainty in the total mass flux of aerosols is dominated by the uncertainties in the dust and sea-salt sources, both of which are difficult to assess and to define. Furthermore, as we will show below, these uncertainties do not influence our assessment of the role of anthropogenic aerosols in climate forcing very strongly, because they relate to the fraction of the aerosol which is optically the least efficient.

Climatic effects

As mentioned above, aerosols interact with the Earth's radiation budget both directly by scattering and absorbing radiation, and indirectly by modifying the extent and radiative properties of clouds. Unfortunately, neither effect is related in a straightforward, linear fashion to a single variable, such as the total mass burden of atmospheric aerosol. Instead, the effects are dependent on the size spectrum of the aerosol particles, on their chemical composition, on their spatial and temporal distribution etc. This makes an assessment of their climatic effects much more difficult than that of the long-lived and relatively evenly distributed greenhouse gases CO_2, CH_4, and N_2O. The following paragraphs discuss the results of recent efforts to estimate the effects of changing atmospheric aerosol levels on climate.

The discussion focuses on climate forcing, defined as an imposed change that modifies the planetary radiation balance and thus acts as the driving force of climate change. Climate forcing is expressed in terms of a change in energy input into the climate system (in W m^{-2}). Using climate forcing rather than climate change makes the discussion of aerosol effects more straightforward, since it eliminates the numerous feedback processes, including changes in clouds, water vapour, ice and snow cover etc., which introduce a high degree of uncertainty into the discussion on climate effects of anthropogenic emissions. Thus, while the forcing due to a doubling of CO_2 can be predicted quite precisely to be about 4–4.5 W m^{-2}, the resulting global warming may lie somewhere in the range of 1.5–5.5°C, depending on the parameterization of the various feedback mechanisms. We will therefore use the estimate of present-day greenhouse gas forcing (about +2.0–2.5 W m^{-2}; HANSEN and LACIS, 1990; HOUGHTON et al., 1990, 1992) as a yardstick against which we can measure aerosol climate forcings without having to address the complications of climate feedbacks.

Direct radiative effects

The net effect of atmospheric aerosols is the sum total of the scattering and absorption of incoming solar (shortwave) radiation and outgoing (longwave, IR) radiation. In principle, only

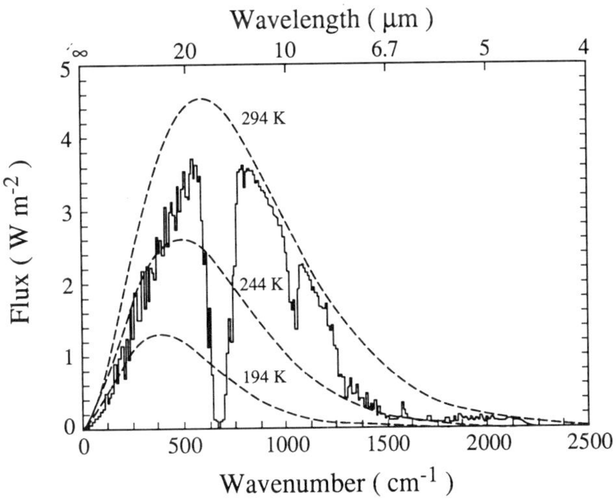

Fig. 4. Infrared radiation emission from the Earth. The dashed lines show the emission from a black body (W m^{-2} per 10 cm^{-1} spectral interval) across the thermal infrared for temperatures of 294 K, 244 K and 194 K. The solid line shows the net flux at the tropopause (W m^{-2}) in each 10 cm^{-1} interval, using a standard narrow-band radiation scheme and a clear-sky mid-latitude summer atmosphere with a surface temperature of 294 K. In general, the closer this line is to the dashed line for 294 K, the more transparent the atmosphere (from HOUGHTON et al., 1990).

the back-scattering of incoming radiation has a cooling effect, while the absorption of incoming radiation and the trapping of outgoing radiation all act in the direction of warming. The latter effect is, however, usually relatively minor, since particles interact most strongly with light which has a wavelength close to their physical size. Since most of the outgoing IR radiation is at wavelengths above 7 μm (Fig. 4), only the sea-salt aerosol and some desert dust aerosols have the potential to absorb significant amounts of outgoing radiation. Furthermore, most of the aerosol, especially the coarse particle aerosol, is present in the lowest kilometres of the troposphere, where its temperature (and therefore its radiational spectrum) is close to that of the Earth's surface. We can therefore ignore this term for most environmental situations. This view is supported by the model calculations of COAKLEY et al. (1983) who find that the longwave effect (warming) is less than 2% of the shortwave radiative effect (cooling). However, some model calculations of the radiation balance over the Sahara desert do suggest that longwave absorption may play an important role in this environment (GRASSL, 1988 and references therein).

The effect of particle size

The efficiency of scattering and absorption of visible light by aerosol particles is highly dependent on their size. Fig. 5 (from COVERT et al., 1980) shows the dependence of the mass extinction, scattering and absorption cross sections as a function of aerosol particle diameter for light with a wavelength of 550 nm and a particle complex refractive index, $m = (1.5-0.02i)$, i.e. a moderately scattering and fairly weakly absorbing aerosol. Comparison between this figure and the typical size distribution of atmospheric aerosols (Fig. 6) shows that, by a remarkable coincidence, there is a maximum in the aerosol/mass size distribution at about 0.4 μm diameter, very close to the maximum of scattering efficiency (SHAW, 1987). This maximum is caused by the fact that aerosols with a diameter below 0.1 μm are effectively removed by diffusion and coagulation, while aerosols with diameters above 1 μm are rapidly deposited by impaction and precipitation. As a result, atmospheric turbidity is, under most circumstances, dominated by the effect of the particles in the range of 0.1–1 μm diameter (SHAW, 1980). This is clearly evident in Fig. 7 (from COVERT et al., 1980) which

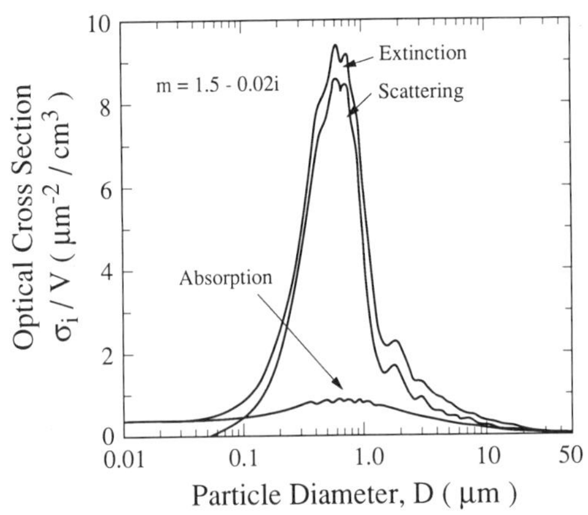

Fig. 5. Particle extinction, scattering and absorption cross-sections per unit of aerosol volume, as calculated from Mie theory (from COVERT et al., 1980).

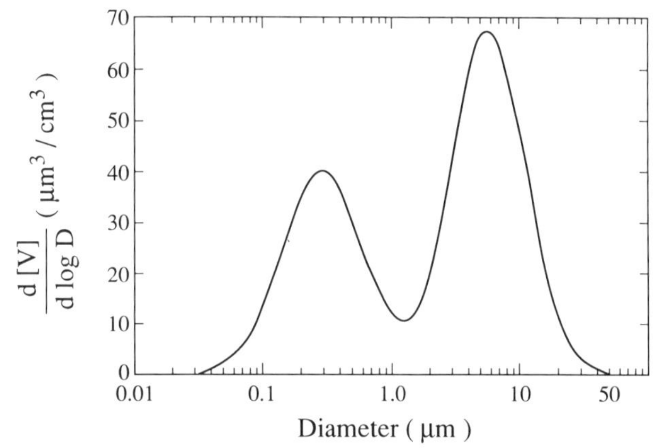

Fig. 6. Typical bimodal atmospheric aerosol volume-size distribution (from COVERT et al., 1980).

shows the extinction and scattering resulting from a typical particle size population. Ninety percent of the extinction is due to particles with diameters <2 μm.

Global-mean radiative forcing perturbation by direct effect

As a first step in the analysis of the climatic effect of changing atmospheric aerosol levels, we appraise their effect on global-mean radiative forcing. It must be emphasized here that this is a very crude way of looking at the problem, since the distribution of aerosols in the atmosphere is highly uneven due to their short lifetime and therefore their effects on radiative forcing are even more variable in space and time than the effects of the relatively evenly distributed greenhouse gases. The fact that even the radiative forcing due to the greenhouse gases is highly variable regionally is often little appreciated. This variability is due to the

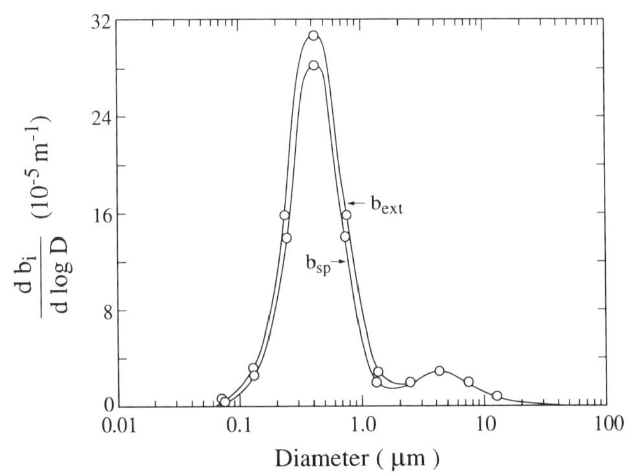

Fig. 7. Particle extinction and scattering-size distributions as calculated from cross-sections and size distributions in Figs. 5 and 6. The integral of $db_{ext} = 1.9 \times 10^{-4}$. The integral of $db_{sP} = 1.7 \times 10^{-4}$ m^{-1}. Measured $b_{sP} = 1.7 \times 10^{-4}$ m^{-1} on the occasion that the above distribution was made. In the calculation of b_{ext}, it is assumed that the absorption coefficient of the coarse mode is zero. It can be seen that about 90% of the extinction is due to particles <2 μm in diameter (from COVERT et al., 1980).

small, since tropospheric clouds are nearly opaque to infrared radiation already, and a further increase in absorbance has little additional effect. A fractional increase in low clouds caused by increased CCN concentrations would result in a proportional increase in long-wave absorption, but for low clouds with temperatures relatively close to that of the Earth's surface, the cooling component dominates.

Estimating indirect radiative forcing

The effect of changing CCN concentrations on global climate was assessed quantitatively by CHARLSON et al. (1987) in connection with the discussion of a hypothesis which linked the production of biogenic sulphur gases by marine phytoplankton to the concentration of sulphate aerosol in the atmosphere, and thus to climate (Fig. 11). These authors used the relationships between droplet concentration, effective droplet radius, and cloud thickness given in equations (5) to (8) above to predict the changes in cloud albedo resulting from changes in CCN, and showed that under the conditions of the clean marine atmosphere, modest changes in CCN concentrations would significantly influence cloud albedo. This is summarized in Fig. 12, which shows the change of cloud albedo as a function of the original albedo of the reference cloud and the change in CCN number density. These results indicate that cloud-top albedo is most sensitive to change in cloud droplet number (N) when the original albedo of the cloud is in the range 0.2–0.7, i.e. for clouds of average and below-average al-

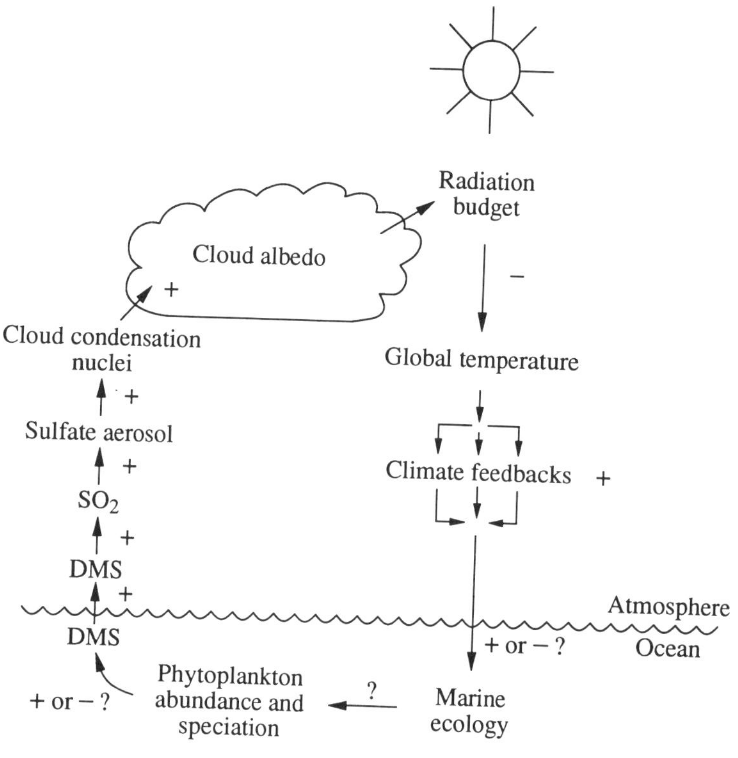

Fig. 11. Climatic feedback loop linking oceanic phytoplankton, emission of DMS, atmospheric sulphur chemistry, cloud albedo and climate. The plus and minus signs indicate the sign of the feedback.

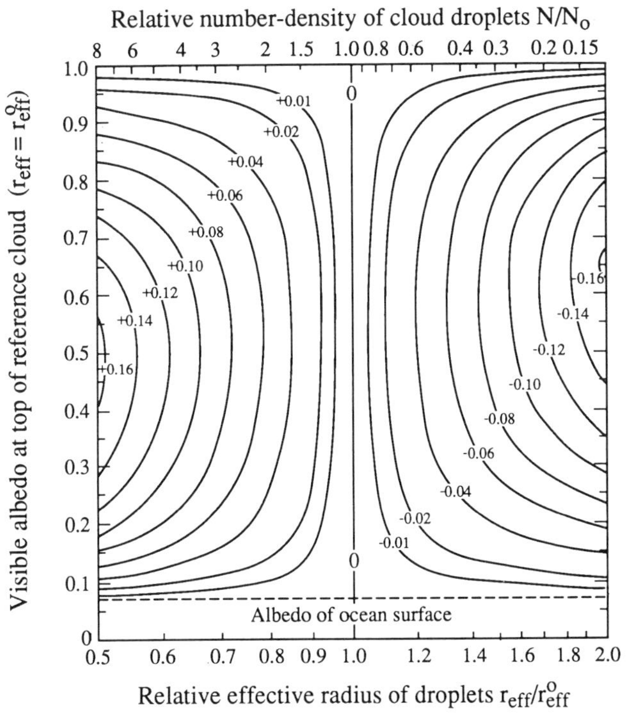

Fig. 12. Change (Δa) of visible albedo (0.6-μm wavelength) at cloud top caused by changing droplet number-density N while holding vertically integrated liquid water content (liquid water path, LWP) constant. Δa is plotted as a function of the albedo of the reference cloud and the effective radius (r_{eff}, the surface-area-weighted mean radius) of the dropsize distribution relative to that of the reference cloud (r_{eff}^0) (from CHARLSON et al., 1987).

bedo, such as stratus clouds. For such clouds, an increase in N by 1% would increase cloud-top albedo by about 0.08%.

CLAW estimated that a change in CCN concentration over the oceans by +30% would result in a decrease in the effective droplet radius by 10%. This would change the albedo of low stratus and stratocumulus clouds (the type of cloud most sensitive to this effect) by +0.020 (2%) at the top of the cloud and for the wavelength region 500–700 nm. Adjusting for the absorption in the 0.6 μm band of ozone in the layer of air between the top of the cloud and the top of the atmosphere reduces this albedo change to +0.018 (WISCOMBE et al., 1984). Finally, when the entire solar spectrum is considered, the increase in top-of-atmosphere (TOA) albedo over the cloud is reduced to 0.016, since cloud albedo is lower in the near-infrared than in the visible part of the spectrum (SHINE et al., 1984; WISCOMBE et al., 1984). On the basis of a more accurate method of resolution of the radiative transfer equation than the simplified approach used by CLAW, FOUQUART and ISAKA (1992), come to essentially the same estimate of the effect of a 10% change in effective radius on TOA albedo. When the area covered by the type of cloud modelled here is taken into account, an average albedo change for the total surface area of the Earth of +0.005 is calculated.

As discussed above for the case of the direct effect of aerosols, there is a linear relationship between a change in global mean albedo and the resulting change in radiative forcing:

$$\delta F_C = -\tfrac{1}{4} F_T \delta A_{GM} \qquad (9)$$

Consequently, an albedo change of +0.005 would lead to a negative radiative forcing of −1.7 W m^{-2}, or change in global-average surface temperature of −1.3 K. This assumes, of course, that a change in CCN and N does not alter the average area covered by clouds, i.e. effects on cloud lifetime are ignored here.

Following the publication of the CLAW hypothesis, SCHWARTZ (1988) argued that, if this hypothesis were correct, the production of sulphate aerosol from anthropogenically released SO_2 would have already led to a pronounced increase in CCN concentrations in the Northern Hemisphere. Based on an assessment of the rather uncertain database presently available on the concentrations of sulphate aerosol over the world oceans, and assuming that number concentration would be proportional to mass concentration, he postulated that CCN in the Northern Hemisphere were greater by 30% than in the relatively unpolluted Southern Hemisphere, and that therefore there should be a radiative forcing of −2 W m^{-2} due to the indirect effect from anthropogenic sulphate aerosols in the Northern Hemisphere. Schwartz's assumption that the enhancement in CCN number concentration is the same as the enhancement in mass concentration is probably not justified, since the particle number density decreases faster during transport than the mass concentration. CHARLSON et al. (1992) assume a 15% global mean enhancement in *N*, and obtain a global mean forcing of −1 W m^{-2} from cloud effects. They emphasize that there is a great amount of uncertainty (at least a factor of 2; KAUFMAN et al., 1991) related to the enhancement of CCN number concentrations and to the resulting enhancement of cloud-droplet number density.

Most authors have considered only the anthropogenic sulphate aerosol as a source of increasing CCN concentrations. However, other aerosol sources, especially biomass burning, also release large amounts of aerosols which may act as CCN (WARNER and TWOMEY, 1967; HOBBS and RADKE, 1969; DÉSALMAND, 1987; RADKE, 1989; CRUTZEN and ANDREAE, 1990; HUDSON et al., 1991; ROGERS et al., 1991; PENNER et al., 1992). Smoke particles from biomass burning appear to be effective CCN in spite of the fact that they consist predominantly of organic material (HALLET et al., 1989; HUDSON et al., 1991). This is because soluble materials, like nitric, ammonia, and short-chain organic acids, are rapidly incorporated from the gas phase of the smoke into the aerosol and thus render it hygroscopic (ANDREAE et al., 1988, TALBOT et al., 1988). RADKE et al. (1991) estimated the global production rate of CCN from biomass burning to be in the range of 10^{20}–10^{21} particles per second, which corresponds to a global average pyrogenic CCN concentration of 10^2–10^3 cm^{-3}, assuming a CCN lifetime of 5×10^5 s and a scale height of 1 km. In laboratory combustion experiments with Guinean savanna vegetation, PHAM-VAN-DINH et al. (1994) found CCN (at 0.1% supersaturation) emission factors around 2×10^{11} g^{-1} dry matter, with the highest emission factors observed during the smoldering stage. This corresponds to a CCN production rate of (3–6) $\times 10^{19}$ s^{-1} for biomass burning in the African savannas. Based on airborne measurements over savanna fires in southern Africa, the global production of CCN from biomass fires is estimated to be about 2–4 $\times 10^{20}$ s^{-1} (M. O. ANDREAE, unpublished data).

Most of these CCN will be present over the continents, and a considerable concentration decrease must be expected during transport over the remote oceans, where they would have the strongest effect on climate. However, maps of the distribution of aerosol optical depth over the oceans clearly show plumes extending over thousands of kilometres from the biomass burning regions (HUSAR and STOWE, 1994). Therefore, PENNER et al. (1992) have estimated that the cooling effect of smoke aerosols from biomass burning via the enhancement of cloud albedo is about 1 W m^{-2}, if a mean pyrogenic CCN concentration of 10 cm^{-3} prevails over the remote oceans. This forcing due to biomass burning is of a magnitude similar to the cooling due to sulphate CCN.

The soot content of smoke and industrial aerosols adds a light-absorbing component to cloud water, which in principle could reduce the albedo of clouds and counteract the whitening effect of larger droplet numbers (TWOMEY, 1977). The resulting increased absorption would have the strongest effect in the case of optically thick clouds, which have a high initial albedo and are not appreciably brightened by increasing droplet number. This prediction is confirmed by satellite observations made on Amazonian cumulus clouds, which showed that in spite of a reduction of droplet size from 15 to 9 μm due to dense smoke from biomass burning, cloud reflectance decreased from 0.71 to 0.68 (KAUFMAN and NAKAJIMA, 1993). On the other hand, the albedo of warm, relatively thin clouds with an initial albedo of 0.3–0.4 increased by 0.08–0.15 in the presence of smoke aerosols (Y. J. KAUFMAN, 1993, personal communication).

On a global scale, the effect of increasing albedo is thought to dominate over that of increased absorption (TWOMEY et al., 1984). This is supported by the observation of increased albedo in marine stratus clouds which have incorporated stack effluents from ships at sea (COAKLEY et al., 1987; KING et al., 1993). Radiative transfer calculations based on the measurement of light absorbing substances in cloud droplet nuclei in marine clouds impacted by polluted continental air also suggest that light absorption from soot does not significantly influence the cloud optical properties under normal circumstances (TWOHY et al., 1989). Direct evidence for a net cloud-brightening effect of aerosols is also found in satellite observations, which show higher cloud reflectance in areas with elevated aerosol loadings (DURKEE, 1994). However, in a satellite study of low-level cloud albedo over the North Atlantic, FALKOWSKI et al. (1992) was able to observe increased cloud albedo due to anthropogenic sulphate only over the area immediately adjacent to the North American coast. Marine sulphur emissions appeared to govern the variability in cloud albedo over most of the North Atlantic.

As in the case of the direct aerosol effects, the cloud effect is very unevenly distributed in space and time. Obviously, it is limited to cloudy regions, while the aerosol effect is significant only in cloud-free areas. However, an assessment of the regional distribution of the impact of the enhancement of cloud albedo is made extremely difficult by the very strong dependence of the albedo increase on the absolute CCN concentration. Figure 13 shows the dependence of albedo susceptibility (dA/dN, the increase in albedo per unit increase in CCN concentration) on CCN concentration and cloud albedo. This figure shows that the strongest effect would result from an even relatively slight increase in CCN in the cleanest and most remote parts of the world. Since the aerosol or its precursors can travel over considerable distances (thousands of kilometres) from their sources, especially in the free troposphere, the effects could be on a hemispheric scale. While the mass concentration of anthropogenic aerosols at such large distances might be low, an addition of even some 10–20 CCN cm^{-3} into a clean marine air mass with some 50 CCN cm^{-3} would have a considerable effect on cloud albedo. At the present time, it is usually suggested that the impact of increasing CCN is felt predominantly in the Northern Hemisphere, although the effect of smoke from biomass burning as well as SO_2 emissions from increasingly industrializing regions in the developing world may already significantly impact the Southern Hemisphere. Smoke from chemistry in remote regions confirm the influence of anthropogenic emissions in areas far from continental sources (LAWSON and WINCHESTER, 1979; ANDREAE et al., 1984; WARNECK, 1988; SAVOIE and PROSPERO, 1989; CHURCH et al., 1991; DUCE et al., 1991). Based on aircraft measurements, ANDREAE et al. (1988) showed that mid-tropospheric sulphate concentrations over the North Pacific off the west coast of the United States were dominated by long-range transport from Asia. This was established by air-mass trajectory analyses and supported by radon measurements. Analyses of the interrelationship between ozone, nitrate and sulphate concentrations in North Atlantic air masses sampled at Barbados also showed pervasive influence of anthropogenic sources (SAVOIE et al., 1989, 1992b). Shipboard measurements of aerosol and precipitation chemistry over the Atlantic document widespread enhancement of sulphate levels above the biogenic background (CHURCH et al., 1991). Long-term ground-level measurements of non-sea-salt sulphate concentrations at island stations also show clearly elevated concentrations over the Northern Hemisphere oceans relative to those in the Southern Hemisphere at corresponding latitudes (Table III). This hemispheric difference is also clearly seen in the relatively few airborne measurements available in remote areas (Table IV).

The increase in CCN due to anthropogenic emissions is well documented in industrialized regions (WARNER and TWOMEY, 1967; HOBBS et al., 1970; BRAHAM, 1974; SCHMIDT, 1974; TWOMEY et al., 1978). That elevated sulphate levels over the North Atlantic cause an increase in CCN is suggested by the data of HOPPEL (1979), who finds that CCN concentra-

TABLE III

ANNUAL AVERAGE SURFACE CONCENTRATION OF NSS-SO_4^{2-} IN AIR (PMOL/MOL AIR) (MODIFIED FROM LANGNER AND RODHE, 1991)

Location	Latitude	Longitude	nss-SO_4^{2-}	References
Northern Hemisphere				
Canadian Arctic[a]	76°N	87°W	210	BARRIE and HOFF (1985)
Spitsbergen	79°N	20°E	168	HEINTZENBERG and LARSSEN (1983)
Shemya	52°N	174°E	155	PROSPERO et al. (1985)
Bermuda	32°N	65°W	455	WOLFF et al. (1986)
Midway	30°N	175°W	133	SAVOIE and PROSPERO (1989)
Oahu	20°N	158°W	119	SAVOIE and PROSPERO (1989)
Barbados	13°N	60°W	175	SAVOIE et al. (1989)
Guam	12°N	144°E	119	SAVOIE and PROSPERO (1989)
Belau	8°N	133°E	147	SAVOIE and PROSPERO (1989)
Fanning	3°N	160°W	154	SAVOIE and PROSPERO (1989)
Southern Hemisphere				
Am. Samoa	15°S	170°W	84	SAVOIE and PROSPERO (1989)
New Caledonia	22°S	166°E	98	SAVOIE and PROSPERO (1989)
Norfolk Island	30°S	168°E	56	SAVOIE and PROSPERO (1989)
Cape Grim	40°S	144°E	21[b]	AYERS et al. (1991)
Cape Grim			26[c]	ANDREAE et al. (1991)
Mawson	68°S	61°E	21	SAVOIE et al. (1992a)
South Pole			20	TUNCEL et al. (1989)

[a]Mean for stations Alert, Mould Bay and Iglooik.
[b]Submicrometre fraction.
[c]In fraction <2.0 mm diameter. Coarser particles contain an additional 39 pmol mol^{-1} nss-SO_4^{2-}.

TABLE IV

OBSERVATIONS OF SO_4^{2-} IN THE TROPOSPHERE (PMOL/MOL AIR) (MODIFIED FROM LANGNER AND RODHE, 1991)

Location	Median	Range	No. of samples	References
Boundary layer (<2 km)				
Northern Hemisphere				
NW Europe	2,238[a]	–	597	OEDC (1977)
W Atlantic	290	<47–1211	18	WHELPDALE et al. (1987)
North Atlantic	424[a]	89–963	26	CHURCH et al. (1991)
North Atlantic	–	150–2600	23	BÜRGERMEISTER and GEORGII (1991)
North Atlantic				GALLOWAY et al. (1990)
Tropical Air masses	213	128–280	16	
Air masses from N Am	939	181–2500	6	
North Atlantic				BERRESHEIM et al.(1991)
Marine air masses	327[a]	±316	35	
Continental Influence	905	±437	30	
North Pacific	352[a]	43–717	26	QUINN et al. (1990)
Pacific, West of USA	95[a]	–	13	GILLETTE and BLIFFORD (1971)
Pacific (Eq.-37°N)	150	60–906	13	HUEBERT and LAZRUS (1980)
North Pacific W of USA	96	45–130	13	ANDREAE et al. (1988)
W North America	120	60–1148	16	HUEBERT and LAZRUS (1980)
Western USA	182[a]	–	23	GILLETTE and BLIFFORD (1971)
Central USA	480[a]	–	24	BOATMAN et al. (1989)
Southern Hemisphere				
South Atlantic	–	110–376	12	BÜRGERMEISTER and GEORGII (1991)
Pacific (57° S-Eq.)	150	60–906	10	HUEBERT and LAZRUS (1980)
Equatorial Pacific	90	40–193	18	QUINN et al. (1990)
South Pacific	–	18–358	19	BATES et al. (1992)
Tasmania	30	19–54	12	BERRESHEIM et al. (1990)
Southern Ocean	7	5	9	BERRESHEIM et al. (1989)
Amazonia				BINGEMER et al. (1990)
Dry season	125a	±50	41	
Wet season	72a	±56	45	
Middle troposphere (2–7 km)				
Northern Hemisphere				
Greenland	50	24–155	12	FLYGER et al. (1976)
North Atlantic	66	20–450	21	FLYGER et al. (1976)
North Atlantic				GALLOWAY et al. (1990)
Tropical air masses	67	38–78	3	
Air masses from N Am	1,470	396–1560	3	
Western USA	67a	–	22	GILLETTE and BLIFFORD (1971)
W North America	<45b	<45–332	28	HUEBERT and LAZRUS (1980)
Central USA	251a	–	24	BOATMAN et al. (1989)
Pacific, W of USA	31a	–	11	GILLETTE and BLIFFORD (1971)
North Pacific, W of USA	180	130–200	9	ANDREAE et al. (1988)
Pacific (Eq.-37° N)	<45b	<46–230	29	HUEBERT and LAZRUS (1980)
Southern Hemisphere				
Tasmania	33	15–45	11	BERRESHEIM et al. (1990)
Pacific (57° S-Eq.)	<45b	–	16	HUEBERT and LAZRUS (1980)
Amazonia	14a	7c	32	BINGEMER et al. (1990)

cleating particles. It may therefore be argued that the production of CCN is not limited by the mass of pollutants injected into the atmosphere. Yet some field studies have shown pronounced enhancements of CCN number in air that has passed over urban areas and received inputs of aerosols from anthropogenic sources. For example, BRAHAM (1974) found an increase of 500 cm^{-3} in air downwind of St. Louis, MO, which resulted in an increase in the droplet number concentration in low stratus clouds by about the same number. HUDSON'S (1983) observations in stratus clouds off the California coast and PUESCHEL et al.'s (1986) measurements at Whiteface Mountain also support a close connection between CCN and droplet numbers.

It is therefore indisputable that the increase in overall pollutant emissions, which accompanies the expected continued increase in greenhouse gas emissions, will also lead to growing CCN number concentrations, albeit not necessarily in linear proportion. We can therefore predict with some confidence that clouds in the future will contain more droplets per unit volume and, other things being equal, these droplets must become smaller. Besides the increase in albedo, other cloud microphysical effects will result from such a change. The most important could be a change in precipitation efficiency which has a cooling effect if it leads to higher cloud amounts and longer cloud lifetime, but may also have a warming component if the result is an overall increase of the amount of water vapour present in the atmosphere. It is clear that the many interacting and non-linear effects in this system can only be satisfactorily considered by using rather complex climate models which are still in the process of being assembled at this time.

In addition to increased emissions of substances which lead to the formation of submicrometre aerosols, human activities and the resulting climate change may strongly influence the amount of mineral dust present in the atmosphere. Desertification in Africa, "dust bowl conditions" in China, increasing erodibility of land due to deforestation in the Amazon Basin and equatorial Africa, and increasing storm intensities resulting from rising atmospheric temperature gradients, all may lead to an enhanced mineral dust load in the atmosphere, and a resulting albedo increase. These effects may be regionally very pronounced; for example, African dust outbreaks over the Atlantic can produce optical depths of about 1.0 over areas of 10^6 km^2 (CARLSON and CAVERLY, 1977; CARLSON and BENJAMIN, 1980; PROSPERO and NEES, 1986). Evidence for a dramatic increase in the dust flux from Africa in the last two decades is presented by PROSPERO and NEES (1986).

Another potential feedback mechanism is the influence of global change on marine DMS emissions. As we have discussed above, biogenic DMS is the dominant source of sulphate aerosol, and consequently CCN, in the natural atmosphere. A modest change in this source, which could be effected by shifts in phytoplankton speciation, food-web structure etc., would have considerable influence on the albedo of clouds over large ocean areas, with obvious climatic consequences. Such ecological changes are likely to occur as a result of a change in ocean temperatures or an increase in the flux of UV radiation to the ocean surface, but their consequences on the biogenic sulphur emissions cannot be predicted at this time due to unsatisfactory knowledge about the linkages between plankton ecology and DMS emission.

In this chapter, the effects of possible increases in stratospheric aerosols on climate have not been considered. Early studies, performed during the first phase of concern about the impact of a stratospheric aircraft fleet, have suggested that significant cooling at the Earth's surface may result, especially at higher latitudes (HERMAN et al., 1976). Due to light absorption by these aerosols, which would certainly contain soot carbon, stratospheric warming would result. More significant may be the indirect effects on climate resulting from the reduction of stratospheric ozone catalysed by increasing stratospheric aerosols. Lower ozone concentrations would cause a reduction in the amount of radiative energy absorbed by stratospheric ozone, and thus a stratospheric cooling (see Chapter 9 by WANG et al.). Numerous feedbacks exist in this system, which will have to be analysed by comprehensive models. This appears to be especially pertinent in view of the observation that stratospheric aerosol levels appear to have been increasing over the past decade (HOFMANN, 1990).

Some words of caution are necessary at the end of this chapter. The effect of non-linearities in the aerosol effects on direct and cloud-amplified forcing have already been pointed out above. These effects are superimposed on the climate system, which itself is described by a system of coupled non-linear differential equations. The chaotic behaviour characteristic of such systems results in unpredictable fluctuations at all time-scales and a tendency to jump between highly disparate states of the system, e.g. ice-ages and interglacials, or different modes of ocean circulation (BROECKER et al., 1985; BROECKER, 1987; Chapter 14 by PENG, 1994). The tendency for the Earth's climate system to undergo fast jumps at time-scales of decades to centuries has been recently documented in ice cores obtained from Greenland (GRIP (GREENLAND ICE-CORE PROJECT) MEMBERS, 1993). This study suggests that the stable climate of the present interglacial is a rather unusual case, and that previous interglacials were characterized by frequent, short-term oscillations between considerably warmer and colder states. Temperature changes of up to 10°C in a couple of decades appear to be possible in interglacials, making human and ecological adaptation to climate change highly unlikely.

The internal noise of the system makes the real-time observation of externally induced changes quite difficult, viz. the discussions in recent years about the question "are we seeing the effects of global warming?". Our present "global experiment", which we conduct by imposing various forcings on the climate system and seeing if we can discern their effect above the intrinsic fluctuations, is therefore a rather hazardous one. We may only obtain satisfactory answers, if ever, at the point when the system jumps into another state, quite different from the present one, and quite unpredictable in its consequences for our and the planet's well-being.

References

ALBRECHT, B. A., 1989. Aerosols, cloud microphysics, and fractional cloudiness. *Science*, 245: 1227–1230.

ANDERSON, T. L., WOLFE, G. V. and WARREN, S. G., 1992. Biological sulphur, clouds and climate. In: W. A. NIERENBERG (Editor), *Encyclopedia of Earth System Science.* Academic Press, New York, pp. 363-376.

ANDREAE, M. O., 1983. Soot carbon and excess fine potassium: long-range transport of combustion-derived aerosols. *Science*, 220: 1148–1151.

ANDREAE, M. O., 1986. The ocean as a source of atmospheric sulphur compounds. In: P. BUAT-MÉNARD (Editor), *The Role of Air-Sea Exchange in Geochemical Cycling*. Reidel, Dordrecht, pp. 331–362.

ANDREAE, M. O., 1990. Ocean-atmosphere interactions in the global biogeochemical sulphur cycle. *Mar. Chem.*, 30: 1–29.

ANDREAE, M. O., 1993. The influence of tropical biomass burning on climate and the atmospheric environment. In: R. S. OREMLAND (Editor), *Biogeochemistry of Global Change: Radiatively Active Trace Gases*. Chapman and Hall, New York, pp. 113–150.

ANDREAE, M. O. and JAESCHKE, W. A., 1992. Exchange of sulphur between biosphere and atmosphere over temperate and tropical regions. In: R. W. HOWARTH, J. W. B. STEWART and M. V. IVANOV (Editors), *Sulphur cycling on the Continents: Wetlands, Terrestrial Ecosystems, and Associated Water Bodies*. SCOPE 48, Wiley, Chichester, UK, pp. 27–61.

ANDREAE, M. O., ANDREAE, T. W., FEREK, R. J. and RAEMDONCK, H., 1984. Long-range transport of soot carbon in the marine atmosphere. *Sci. Total Environ.*, 36: 73–80.

ANDREAE, M. O., BERRESHEIM, H., ANDREAE, T. W., KRITZ, M. A., BATES, T. S. and MERRILL, J. T., 1988. Vertical distribution of dimethylsulfide, sulphur dioxide, aerosol ions, and radon over the northeast Pacific Ocean. *J. Atmos. Chem.*, 6: 149–173.

ANDREAE, M. O., BERRESHEIM, H., BINGEMER, H., JACOB, D. J., LEWIS, B. L., LI, S. and TALBOT, R. W., 1990. The atmospheric sulfur cycle over the Amazon Basin, 2. Wet Season. *J. Geophys. Res.*, 95: 16813–16824.

ANDREAE, M. O., ELBERT, W. and ANDREAE, T. W., 1991. Non-seasalt sulfate concentrations at Cape Grim, 1988–1990. In: S. R. WILSON and J. L. GRAS (Editors), *Baseline 89*. Impress Studios, Newcastle, Australia, pp. 59–60.

ANDREAE, M. O., DE MORA, S. J. and ELBERT, W., 1995. Biogenic sulfur emissions and aerosols over the tropical South Atlantic, 3. Atmospheric dimethylsulfide, aerosols and cloud condensation nuclei. *J. Geophys. Res*, in press.

ANDREAE, T. W., ANDREAE, M. O. and SCHEBESKE, G., 1994. Biogenic sulfur emission and aerosol particles over the tropical South Atlantic. 1. Dimethylsulfide in seawater and in the atmospheric boundary layer. *J. Geophys. Res.*, 99: 22819–22829.

ARTAXO, P., MAENHAUT, W., STORMS, H. and VAN GRIEKEN, R., 1990. Aerosol characteristics and sources for the Amazon Basin during the wet season. *J. Geophys. Res.*, 95: 16971–16986.

AYERS, G. P., IVEY, J. P. and GILLETT, R. W., 1991. Coherence between seasonal cycles of dimethyl sulphide, methanesulphonate and sulphate in marine air. *Nature*, 349: 404–406.

BAKER, M. B. and CHARLSON, R. J., 1990. Bistability of CCN concentrations and thermodynamics in the cloud-topped boundary layer. *Nature*, 345: 142–145.

BARRIE, L. A. and HOFF, R. M., 1985. Five years of air chemistry observations in the Canadian Arctic. *Atmos. Environ.*, 19: 1995–2010.

BATES, T. S., CALHOUN, J. A. and QUINN, P. K., 1992. Variations in the methanesulfonate to sulfate molar ratio in submicrometer marine aerosol particles over the South Pacific Ocean. *J. Geophys. Res.*, 97: 9859–9865.

BATES, T. S., CLINE, J. D., GAMMON, R. H. and KELLY-HANSEN, S. R., 1987. Regional and seasonal variations in the flux of oceanic dimethylsulfide to the atmosphere. *J. Geophys. Res.*, 92: 2930–2938.

BERRESHEIM, H., ANDREAE, M. O., AYERS, G. P. and GILLETT, R. W., 1989. Distribution of biogenic sulphur compounds in the remote southern hemisphere. In: E. S. SALTZMAN and W. J. COOPER (Editors), *Biogenic Sulfur in the Environment*. ACS Symposium Series, Washington, DC, pp. 352–366.

BERRESHEIM, H., ANDREAE, M. O., AYERS, G. P., GILLETT, R. W., MERRILL, J. T., DAVIS, V. J. and CHAMEIDES, W. L., 1990. Airborne measurements of dimethylsulfide, sulphur dioxide and aerosol ions over the Southern Ocean south of Australia. *J. Atmos. Chem.*, 10: 341–370.

BERRESHEIM, H., ANDREAE, M. O., IVERSON, R. L. and LI, S.-M., 1991. Seasonal variations of dimethylsulfide emissions and atmospheric sulphur and nitrogen species over the western North Atlantic Ocean. *Tellus*, 43B: 353–372.

BETZER, P. R., CARDER, K. L., DUCE, R. A., MERRILL, J. T., TINDALE, N. W., UEMATSU, M., COSTELLO, D. K., YOUNG, R. W., FEELY, R. A., BRELAND, J. A., BERNSTEIN, R. E. and GRECO, A. M., 1988. Long-range transport of giant mineral aerosol particles. *Nature*, 336: 568–571.

BIGG, E. K., 1986. Discrepancy between observation and prediction of concentrations of cloud condensation nuclei. *Atmos. Res.*, 20: 81–86.

BINGEMER, H. G., ANDREAE, M. O., ANDREAE, T. W., ARTAXO, P., HELAS, G., JACOB, D. J., MIHALOPOULOS, N. and NGUYEN, B. C., 1992. Sulfur gases and aerosols in and above the equatorial African rainforest. *J. Geophys. Res.*, 97: 6207–6217.

BLANCHARD, D. C., 1983. The production, distribution, and bacterial enrichment of the sea-salt aerosol. In: P. S. Liss and W. G. N. Slinn (Editors), *Air-Sea Exchange of Gases and Particles*. Reidel, Boston, MA, pp. 407–454.

Bluth, G. J. S., Schnetzler, C. C., Krueger, A. J. and Walter, L. S., 1993. The contribution of explosive volcanism to global atmospheric sulphur dioxide concentrations. *Nature*, 366: 327–329.

Boatman, J. F., Wellman, D. L., Van Valin, C. C., Gunter, R. L., Ray, J. D., Sievering, H., Kim, Y., Wilkinson, S. K. and Luria, M., 1989. Airborne sampling of selected trace chemicals above the central United States. *J. Geophys. Res.*, 94: 5081–5093.

Bolin, B. and Charlson, R. J., 1976. On the role of the tropospheric sulphur cycle in the shortwave radiative climate of the Earth. *Ambio*, 5: 47–54.

Braham Jr., R. R., 1974. Cloud physics of urban weather modification-a preliminary report. *Bull. Am. Meteorol. Soc.*, 55: 100–106.

Broecker, W. S., 1987. Unpleasant surprises in the greenhouse? *Nature*, 328: 123–126.

Broecker, W. S., Peteet, D. M. and Rind, D., 1985. Does the ocean-atmosphere system have more than one stable mode of operation? *Nature*, 315: 21–26.

Bryson, R. A., 1974. A perspective on climate change. *Science*, 184: 753–760.

Bürgermeister, S. and Georgii, H.-W., 1991. Distribution of methanesulfonate, NSS sulfate and dimethylsulfide over the Atlantic and the North Sea. *Atmos. Environ.*, 25A: 587–595.

Cachier, H., Brémond, M.-P. and Buat-Ménard, P., 1989. Carbonaceous aerosols from different tropical biomass burning sources. *Nature*, 340: 371–373.

Carlson, T. N. and Benjamin, S. G., 1980. Radiative heating rates for sarahan dust. *J. Atmos. Sci.*, 37: 193–213.

Carlson, T. N. and Caverly, T. S., 1977. Radiative characteristics of Sahara dust at solar wavelengths. *J. Geophys. Res.*, 82: 3141–3152.

Charlson, R. J., 1988. Have the concentrations of tropospheric aerosol particles changed? In: F. S. Rowland and I. S. A. Isaksen (Editors), *The Changing Atmosphere*. Wiley, Chichester, UK, pp. 79-90.

Charlson, R. J. and Pilat, M. J., 1969. Climate: the influence of aerosols. *J. Appl. Meteorol.*, 10: 1001–1002.

Charlson, R. J., Covert, D. S. and Larson, T. V., 1984. In: L. H. Ruhnke and A. Deepak (Editors), *Hygroscopic Aerosols*. Deepak, Hampton, VA, pp. 35–44.

Charlson, R. J., Lovelock, J. E., Andreae, M. O. and Warren, S. G., 1987. Oceanic phytoplankton, atmospheric sulphur, cloud albedo and climate. *Nature*, 326: 655–661.

Charlson, R. J., Lovelock, J. E., Andreae, M. O. and Warren, S. G., 1989. Sulphate aerosols and climate (a reply). *Science*, 340: 437–438.

Charlson, R. J., Langner, J., Rodhe, H., Leovy, C. B. and Warren, S. G., 1991. Perturbation of the northern hemisphere radiative balance by backscattering from anthropogenic sulfate aerosols. *Tellus*, 43AB: 152–163.

Charlson, R. J., Schwartz, S. E., Hales, J. M., Cess, R. D., Coakley Jr., J. A., Hansen, J. E. and Hofmann, D. J., 1992. Climate forcing by anthropogenic aerosols. *Science*, 255: 423–430.

Church, T. M., Tramontano, J. M., Whelpdale, D. M., Andreae, M. O., Galloway, J. N., Keene, W. C., Knap, A. H. and Tokos, J., 1991. Atmospheric and precipitation chemistry over the North Atlantic Ocean: shipboard results from April-May 1984. *J. Geophys. Res.*, 96: 18705–18725.

Cipriano, R. J., Monahan, E. C., Bowyer, P. A. and Woolf, D. K., 1987. Marine condensation nucleus generation inferred from whitecap simulation tank results. *J. Geophys. Res.*, 92: 6569–6576.

Clarke, A. D., 1989. Aerosol light absorption by soot in remote environments. *Aerosol Sci. Tech.*, 10: 161–171.

Clarke, A. D. and Charlson, R. J., 1985. Radiative properties of the background aerosol: absorption component of extinction. *Science*, 229: 263–265.

Coakley Jr., J. A. and Chylek, P., 1975. The two-stream approximation in radiative transfer: including the angle of the incident radiation. *J. Atmos. Sci.*, 32: 409.

Coakley Jr., J. A. and Davies, R., 1986. The effect of cloud sides on reflected solar radiation as deduced from satellite observations. *J. Atmos. Sci.*, 43: 1025–1035.

Coakley Jr., J. A., Cess, R. D. and Yurevich, F. B., 1983. The effect of tropospheric aerosols on the Earth's radiation budget: a parameterization for climate models. *J. Atmos. Sci.*, 40: 116–138.

Coakley Jr., J. A., Bernstein, R. L. and Durkee, P. A., 1987. Effect of ship-track effluents on cloud reflectivity. *Science*, 237: 1020–1022.

$$NO_2 + h\nu \rightarrow NO + O \quad (14a)$$

$$NO + HO_2 \rightarrow NO_2 + OH \quad (14b)$$

$$O + O_2 + M \rightarrow O_3 + M \quad (14c)$$

The rate-limiting process is reaction (14b) which involves the hydroperoxy radical (HO_2). Similar reactions with organic peroxy radicals (RO_2 where R is CH_3, C_2H_5, C_3H_7 etc.) are also contributing to the conversion of NO to NO_2 (without consumption of O_3) and hence to the formation of ozone. The presence of peroxy radicals results from the oxidation of methane, non-methane hydrocarbons and carbon monoxide. These latter compounds, as well as the nitrogen oxides are released in the atmosphere as a result of human activities (e.g. industrial combustion, transportation, biomass burning etc.). The control of nitrogen oxides and of hydrocarbons and CO are key measures to reduce unacceptable levels of ozone observed, for example, in industrialized regions during summer and under stable meteorological conditions. The type of strategy to be implemented (NO_x versus hydrocarbon limitations) varies with background conditions and with the relative importance of natural versus anthropogenic emissions of the ozone precursors.

The photochemical destruction of tropospheric ozone is provided by the reaction

$$O_3 + HO_2 \rightarrow OH + 2O_2 \quad (15)$$

as well as by the photolysis of ozone and subsequent reaction between the electronically excited oxygen atom and a water molecule.

Ozone in the polar regions

The development of an ozone hole in Antarctica since the late 1970s (FARMAN et al., 1985) has perhaps been the most spectacular geophysical event of the 20th century. Observations made by space-borne and ground-based instrumentation have revealed that nearly 90% of the ozone is destroyed during August and October ("polar dawn") in a region extending from 15 to 22 km altitude (Fig. 4) and covering an area nearly as large as the Antarctic continent. In November, as the polar vortex breaks down and ozone is transported towards the pole, ozone concentrations are restored to near their pre-hole values.

The commonly accepted explanation for the formation of the ozone hole is that chlorine reservoirs (HCl, $ClONO_2$), which sequester a large fraction of inorganic chlorine in the lower stratosphere, are converted into reactive chlorine on the surface of ice particles in polar stratospheric clouds (PSCs) observed in cold air masses. When the temperature decreases below a thermodynamic threshold of approximately 193 K, small solid particles containing a mixture of nitric acid and water are formed and produce the so-called type I PSCs. At temperatures below 187 K, pure ice crystals are produced and the so-called type II PSCs are formed. Typical diameters for type I and type II PSC particles are 0.5 and 10 μm, respectively. The heterogeneous conversion mechanisms which have been identified in the laboratory are the following (see e.g. WMO, 1991):

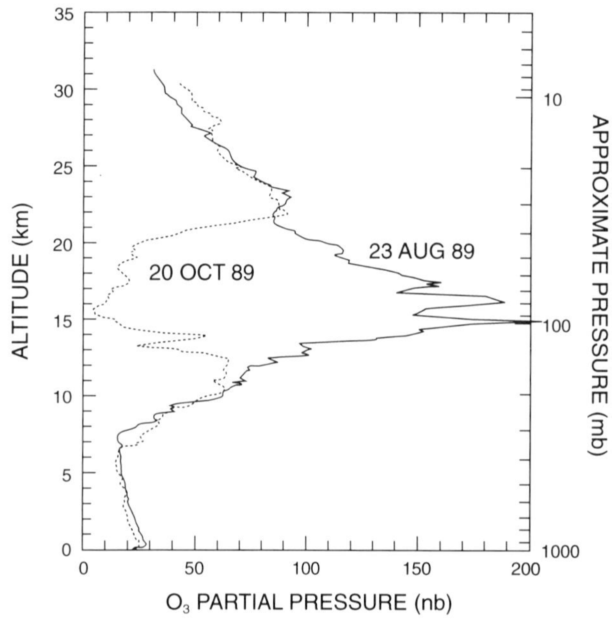

Fig. 4. The formation of the Antarctic ozone hole: vertical distributions of the ozone partial pressure as a function of altitude on 23 August 1989 and 20 October 1989. From DESHLER et al. (1990).

$ClONO_2$ (g) + H_2O (s) → HOCl (g) + HNO_3 (s)

N_2O_5 (g) + H_2O (s) → 2 HNO_3 (s)

$ClONO_2$ (g) + HCl (s) → Cl_2 (g) + HNO_3 (s)

HOCl (g) + HCl (s) → Cl_2 (g) + H_2O (s)

N_2O_5 (g) + HCl (s) → $ClNO_2$ (g) + HNO_3 (s)

where (g) and (s) refer to the gas phase and the solid solutions, respectively. The gravitational sedimentation of the largest particles (e.g. type II crystals) can produce a substantial denitrification and dehydration of the lower Antarctic stratosphere during winter and early spring. As a consequence, the concentration of NO_x (= NO + NO_2) is low and the ClO radical produced after photolysis of Cl_2, HOCl and $ClNO_2$ does not recombine with NO_2 to reproduce $ClONO_2$, but accumulates to reach concentrations close to 1 ppbv. ClO radicals remain therefore available for destroying large quantities of ozone and for producing the Antarctic ozone hole.

Similar chemical processes are observed in the Arctic in winter. For example, the Microwave Limb Scanner (MLS) instrument on board the Upper Atmosphere Research Satellite detected in January 1992 ClO mixing ratios close to 2 ppbv (WATERS et al., 1993). Observations made during several airborne and ground-based Arctic expeditions have provided evidence for limited denitrification in air masses processed by polar stratospheric clouds. There are, however, several differences which explain why no ozone hole is formed during model (KAYE and JACKMAN, 1993). The solid line corresponds to a case in which the productions of CFCs and halons remain constant at their 1986 level. The effect of the 50% and 95% cutbacks in the production of CFCs and halons is also shown by the dashed and dash-dotted lines, respectively. The ozone decrease in year 2060 reaches 8% when the first scenario is adopted but is reduced to 4 and 2%, respectively, when the second and third scenarios are adopted. Note that, because the atmospheric lifetime of the CFCs and halons is of the order of decades to more than a century, the effects of the decrease in the CFCs and halons emissions will only be noticeable after a relatively long period of time.

Ozone changes and the climate system

It is now well established that the observed changes in the concentration of CO_2 and other "greenhouse" gases have produced perturbations in the radiative forcing of the climate system. Figure 13 shows, for example, the magnitude of this radiative perturbation at the tropopauseas a function of latitude for the period 1850–1990, as estimated by HAUGLUSTAINE et al. (1993). Although the contribution of CO_2 is the largest (1.5 W m^{-2} on the global average), the radiative perturbations associated with the change since 1850 in the abundance of methane, nitrous oxide, and the CFCs (0.7, 0.2 and 0.35 W m^{-2}, respectively) are also substantial. The increase in the radiative forcing resulting from the increasing concentrations of tropospheric ozone is estimated to be 0.55 W m^{-2} on the global average, with a maximum of 0.9 W m^{-2} at mid-latitudes in the Northern Hemisphere. The relative contributions of the various gases to the global enhancement (since pre-industrial times) in the radiative forcing at the tropopause are, according to the model of HAUGLUSTAINE et al. (1993), 45% for CO_2,

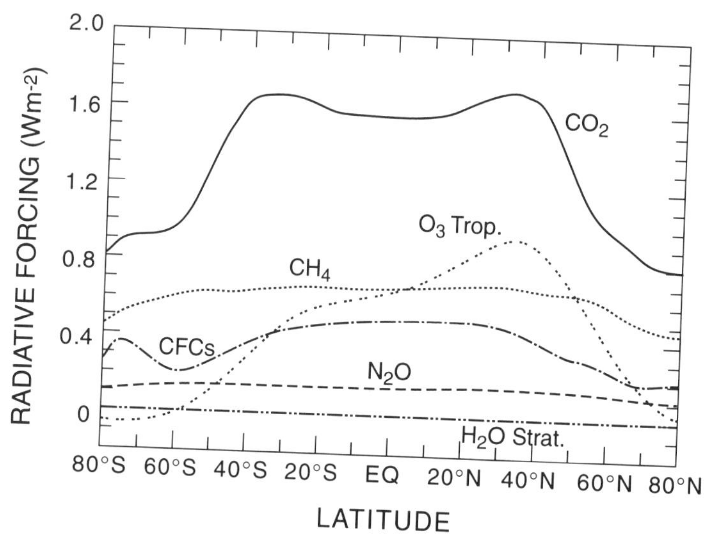

Fig. 13. Change in the radiative forcing (W m^{-2}) calculated as a function of latitude by HAUGLUSTAINE et al. (1993) for the period 1850–1990. The contribution of several radiative gases is shown. Chemical feedbacks are taken into account.

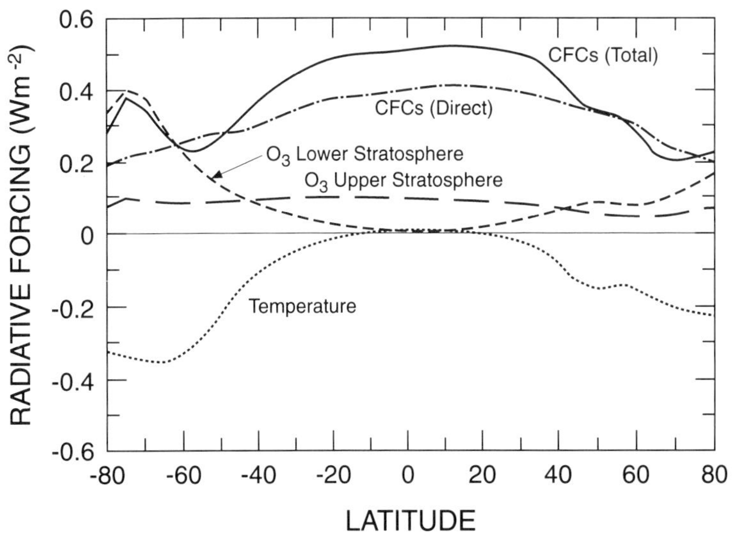

Fig. 14. Latitudinal dependence of changes in the radiative forcing (W m^{-2}) resulting from changes in stratospheric ozone induced by increases in chlorofluorocarbon abundances since pre-industrial times.

20% for CH_4, 17% for tropospheric ozone, 13% for the CFCs, 4.9% for N_2O, and 0.2% for stratospheric water vapour.

As pointed out by RAMASWAMY et al. (1992), the depletion of stratospheric ozone resulting from elevated concentrations of CFCs in the stratosphere produces significant perturbations in the radiative budget (ultraviolet and infrared) of the atmosphere. This chemically induced perturbation of ozone has to be compared to the direct radiative forcing of the CFCs. The ozone depletion produced in the upper stratosphere, near 45 km altitude, reduces the absorption of solar ultraviolet, so that more solar energy is provided to the troposphere-surface system (warming). As suggested by Fig. 14, the latitudinal dependence of this radiative effect is weak. Similarly, the ozone depletion in the lower stratosphere contributes to the warming of the troposphere-surface system by solar radiation, but in this case, the effect becomes more pronounced with increasing latitude, since most of the ozone is chemically destroyed in the polar regions. Note, however, that the local radiative heating, and hence the temperature of the lower stratosphere, is reduced near the tropopause as a result of the ozone depletion in this region of the atmosphere. As a consequence, the emission of longwave radiation towards the troposphere is reduced (RAMASWAMY et al., 1992) and an additional cooling mechanism for the troposphere-surface system (curve labelled "temperature" in Fig. 14) needs to be considered.

In the model of HAUGLUSTAINE et al. (1993), the net effect of ozone changes (associated with the chemical action of the CFCs) is to slightly amplify the primary forcing of the CFCs (see Fig. 14) even when including the temperature effect. In the tropics, the indirect effect is produced by the ozone depletion in the upper stratosphere, while at high latitudes all the different effects included in the analysis have to be taken into account. As noted recently by SCHWARTZKOPF and RAMASWAMY (1993), the magnitude and the sign of the ozone forcing are critically dependent on the vertical profile of the ozone perturbation near the tropopause.

Increases in ultraviolet radiation due to ozone depletion

An expected consequence of stratospheric ozone reductions is the increase in solar ultraviolet (UV) radiation reaching the lower atmosphere and the surface of the Earth (WMO, 1991; UNEP, 1991). Oxygen and ozone molecules combine to remove all UV-C (200–280 nm) and shorter radiation, but some UV-B (280–320 nm) does penetrate, the amount depending sensitively on the ozone column amount (UV-A, 320–400 nm, is only weakly affected by ozone). For example, a 10% ozone depletion increases the radiation by 3% at 320 nm, and by 11% at 310 nm (solar zenith angle of 45°, 300 DU).

The UV-A and UV-B energy ranges have the potential to induce significant amounts of photo-chemical activity, variously expressed as changes in tropospheric chemistry, as perturbations to terrestrial and aquatic ecosystems, and as potential health effects on humans and animals (UNEP, 1991). Many of these effects have been studied in detail, and in some cases quantitative sensitivity studies of the effect can be made (MADRONICH et al., 1991).

The effectiveness of UV radiation with regard to inducing specific photo-chemical processes is shown in Fig. 15. Such plots are commonly dubbed monochromatic action spectra, and can be exceedingly difficult to determine. Most are found to decrease with increasing wavelength, and the steepness of this fall-off largely determines the sensitivity to ozone depletion. Atmospheric transmission, on the other hand, increases with wavelength as shown in Fig. 16, so that the active radiation peaks at some intermediate wavelength. Combining

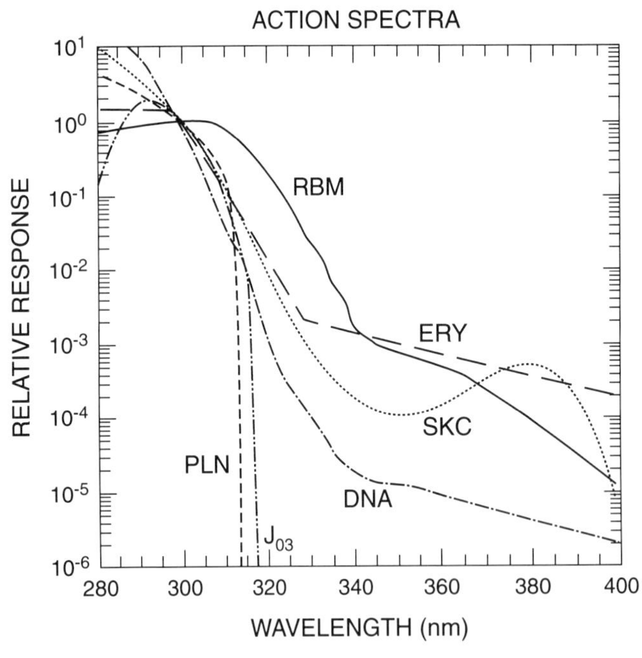

Fig. 15. Relative effect of different wavelengths on some biological and chemical processes (action spectra). J_{O3} = product of the ozone absorption cross section (MOLINA and MOLINA, 1986) and quantum yield (DE MORE et al., 1990) at 256K, divided by 3.38×10^{-19} cm^2; RBM, response function for Robertson-Berger meter (URBACH et al., 1974); ERY, standard spectrum for erythema induction (skin reddening), based on an extensive review of exposure studies on fair-skinned humans (MCKINLEY and DIFFEY, 1987); SKC, action spectrum for UV-induced skin cancer in SKH:HR1 mice (DE GRUIJL et al., 1993); DNA, action spectrum for in vitro DNA damage (SETLOW, 1976); PLN, generalized plant damage (CALDWELL et al., 1986).

the spectral effectiveness and the atmospheric transmission at each wavelength leads to the following definitions:

$$\text{spectral dose rate} = F(\lambda)\, B(\lambda)$$

$$\text{dose rate} = \int F(\lambda)\, B(\lambda)\, d\lambda$$

$$\text{dose} = \iint F(\lambda)\, B(\lambda)\, d\lambda\, dt$$

where $F(\lambda)$ is the spectral irradiance at wavelength λ, $B(\lambda)$ is an action spectrum, and t is any integration time (MADRONICH, 1993a,b).

Daily doses for DNA damage caused by UV-B radiation are illustrated in Fig. 17, for clear sky conditions, and recent average ozone column as measured by the Total Ozone Mapping Spectrometer (TOMS) from the Nimbus 7 satellite (STOLARSKI et al., 1992). Trends in the daily doses are shown in Figs. 18 and 19, and are based on TOMS total ozone values between January 1979 and December 1989. No significant change is seen in the tropics, but relative trends (Fig. 18) generally increase towards the polar regions while the energy increments (Fig. 19) are shifted in the direction of the summer and the sub-solar tropics. Comparable increases in UV doses can be computed for many different action spectra. Table I summarizes the results for three different spectra, and for the incidence of non-melanoma skin cancer.

The relative sensitivity to a 1% ozone depletion is shown for a number of biological and chemical processes in Table II. The large variation of these factors and possibly large uncer-

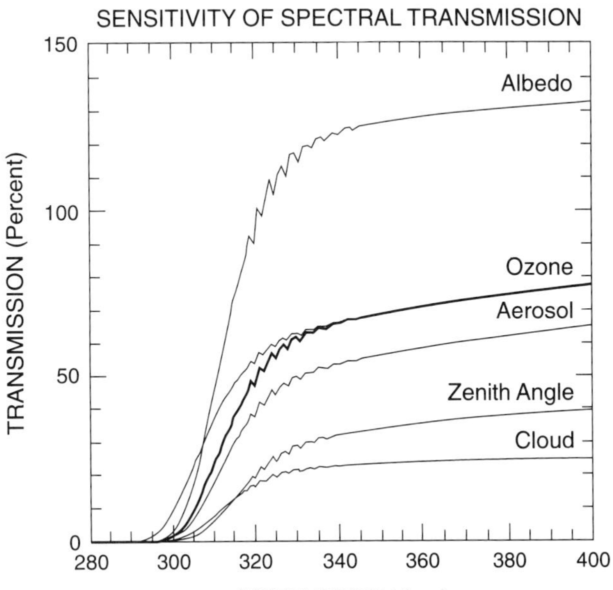

Fig. 16. Spectral transmission of the atmosphere. Thick line gives results of a delta-Eddington radiative transfer model for clear sky (no aerosol or clouds), ozone column = 300 DU, solar zenith angle = 30° and surface albedo = 0.05. Effect of changing these parameters individually (i.e. one at a time) is shown by thin solid lines. New values are cloud optical depth = 32, solar zenith angle = 60°, aerosol visible range = 10 km, ozone column = 150 DU and surface albedo = 0.60 (from MADRONICH,

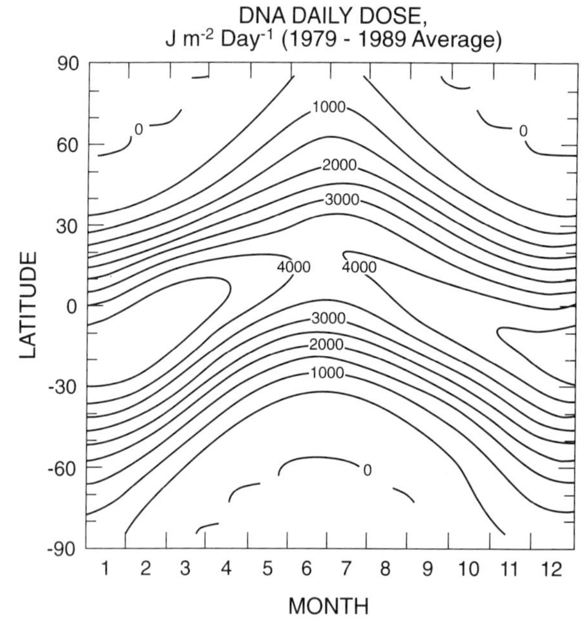

Fig. 17. Latitudinal and seasonal dependence of the daily DNA-damaging UV-B doses, calculated for clear skies and low surface albedo (0.05) using measured TOMS ozone column averages over 1979–1989. Units of contours are J m^{-2}, with action spectra normalized to unity at 300 nm (from MADRONICH, 1993a).

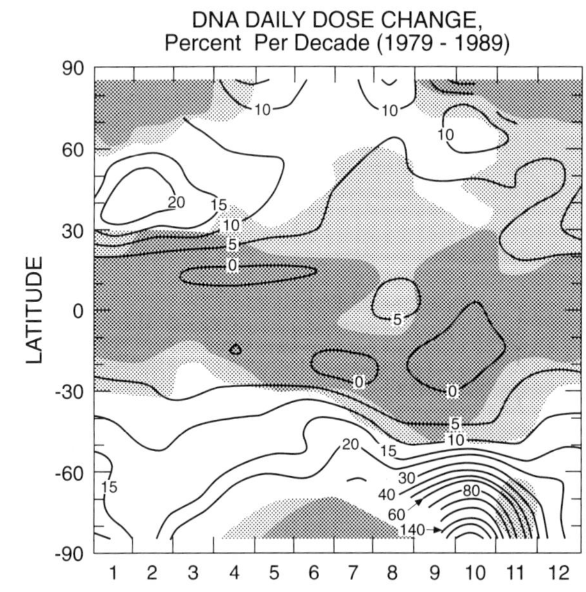

Fig. 18. Trends in daily effective UV dose for DNA damage, computed using TOMS ozone column data over the 11-year period 1979–1989. Dark shading indicates regions where the trends differ from zero by less than one standard deviations (1σ), light shading by less than 2σ. Trend in percent per decade (from MADRONICH, 1992).

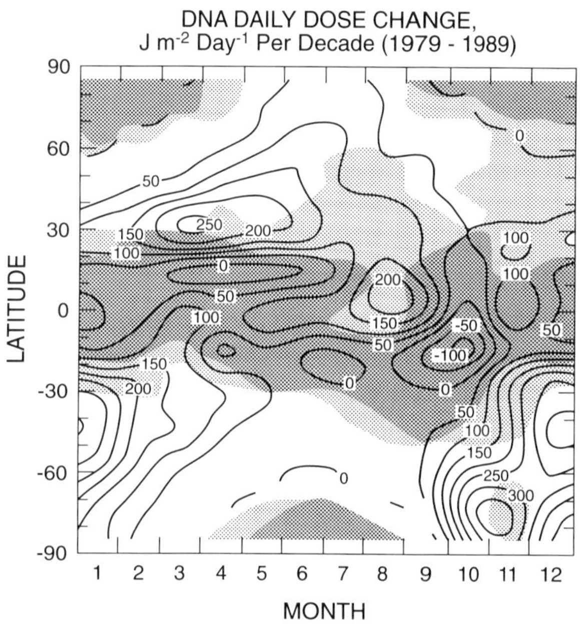

Fig. 19. Same as Fig. 3, but with trend in absolute units of J m^{-2} per decade, with normalization to unity photon effectiveness at 300 nm (from MADRONICH, 1992).

TABLE I

EXPECTED INCREASES (PERCENT)[a] IN ANNUAL UV DOSES AND SKIN CANCER INCIDENCE DUE TO STRATOSPHERIC OZONE DEPLETION FROM 1979 TO 1992

Latitude	Total ozone	Erythema induction dose	DNA damage dose	Skin cancer dose	Basal cell carcinoma incidence[b]	Squamous cell carcinoma incidence[b]
85 N	−8.8 ± 3.2	7.1 ± 1.7	14.8 ± 3.6	10.6 ± 2.5	15.1 ± 5.6	28.5 ± 11.2
75 N	−9.0 ± 2.9	7.6 ± 1.7	14.9 ± 3.3	10.8 ± 2.4	15.4 ± 5.8	29.1 ± 11.5
65 N	−7.4 ± 1.7	7.6 ± 1.7	14.1 ± 3.2	10.3 ± 2.3	14.7 ± 5.6	27.7 ± 11.0
55 N	−7.4 ± 1.3	7.2 ± 1.7	12.9 ± 3.2	9.5 ± 2.3	13.5 ± 5.3	25.4 ± 10.3
45 N	−6.6 ± 1.2	6.5 ± 1.7	10.9 ± 3.0	8.1 ± 2.2	11.6 ± 4.7	21.6 ± 9.0
35 N	−4.8 ± 1.4	5.0 ± 1.8	8.2 ± 3.0	6.1 ± 2.2	8.6 ± 4.0	16.0 ± 7.6
25 N	−2.7 ± 1.5	3.0 ± 1.9	4.8 ± 3.1	3.5 ± 2.2	5.0 ± 3.5	9.0 ± 6.4
15 N	−1.5 ± 1.1	1.7 ± 1.4	2.6 ± 2.3	1.9 ± 1.6	2.7 ± 2.4	4.8 ± 4.4
5 N	−0.6 ± 1.6	0.7 ± 1.9	1.2 ± 3.0	0.8 ± 2.2	1.2 ± 3.1	2.1 ± 5.5
5 S	−1.1 ± 1.4	1.2 ± 1.7	2.0 ± 3.0	1.4 ± 2.0	2.0 ± 2.8	3.6 ± 5.2
15 S	−1.9 ± 1.3	2.2 ± 1.5	3.5 ± 2.4	2.5 ± 1.7	3.6 ± 2.6	6.5 ± 4.8
25 S	−2.6 ± 1.6	3.1 ± 1.7	5.0 ± 2.4	3.6 ± 1.9	5.1 ± 3.1	9.2 ± 5.8
35 S	−4.0 ± 1.6	4.8 ± 1.6	7.9 ± 2.6	5.7 ± 1.9	8.1 ± 3.6	14.9 ± 6.8
45 S	−5.6 ± 1.4	7.2 ± 1.6	12.5 ± 2.7	8.9 ± 1.9	12.7 ± 4.8	23.9 ± 9.2
55 S	−9.0 ± 1.5	10.9 ± 2.0	19.7 ± 3.6	14.2 ± 2.6	20.4 ± 7.4	39.3 ± 15.1
65 S	−15.0 ± 2.0	16.3 ± 3.0	30.5 ± 5.8	21.9 ± 4.1	31.9 ± 12.2	64.0 ± 26.6
75 S	−19.5 ± 2.6	24.1 ± 5.4	49.8 ± 12.0	34.0 ± 7.7	50.6 ± 21.4	107.7 ± 52.0
85 S	−21.1 ± 3.0	31.0 ± 6.8	72.0 ± 17.6	46.5 ± 10.5	70.6 ± 31.2	159.6 ± 83.6

[a]Evaluated over the 14-year data record, expressed as percent relative to the 1979 intercept. Uncertainties are one standard deviation.

[b]Based on skin cancer dose increase and relevant biological amplification factors.

Plains (Riebsame, 1990) did not and that on Nigeria (Udo et al., 1990) described primarily the impact of the Sahelian drought.
Moreover, small-scale climatic changes cannot impact other regions unless their effect is "transmitted" by the atmosphere or oceans. To date, such remote impacts have only been shown to occur for relatively large-scale land-use changes (e.g. Bonan et al., 1992). This chapter therefore focusses on possible large-scale climatic impacts and on the physical mechanisms underpinning the response to likely human-induced land-use change.
Firstly, a review of all types of land-use change (see next section) and of the mechanisms by which these changes can affect climate (section on Land-cover change influences on climate) is given. Then the possible impacts of tropical land-use changes, especially deforestation and desertification, are considered in the section on Regional-scale impacts of land-cover change on climate. This acts as a prelude to an assessment of future climates in the final section.

Human-induced land-use change

The major types of human-induced land-use change are deforestation (and reforestation), desertification (which often includes overgrazing and excessive exploitation of vegetation), agricultural expansion and soil erosion. These changes are widespread and prompt other responses such as soil degradation and salinization (Fig. 2). Dam-building, reclamation and erosion of coastal zones, irrigation of arid land, urbanization and industrialization are often very important locally but seem unlikely to cause anything other than local climatic changes (e.g. Changnon, 1992). Their secondary importance is partly because of their smaller spatial extent which also exacerbates the other reason: their smaller feedbacks into the climatic system.

Agriculture and engineering

Food demand will increase as populations increase and diet patterns change. In the remaining years of this century, about 1.3 billion people will be added to the global population (see Introduction and Table I); rising incomes, however, may account for 30–40% of the increased demand for food in developing countries and about 10% in industrial nations. Thus over the next few decades, the global food system must be managed in order to increase food production by 3–4% every year. Table IIa compares the extent of land areas converted to agricultural crops in two periods: 1860–1919 and 1920–1978. Despite the increasing population, Richard's (1990) estimates of conversion rates do not differ much between the two periods, in agreement with the estimates listed in Table IIb which shows a decrease in the per capita gross cropped area but a considerable increase in fertilizer use on agricultural land.
Increases in cropped areas in recent decades have often extended cultivation on to marginal lands prone to degradation and erosion (e.g. Myers, 1984; Glantz, 1994). For example, expansion of agricultural areas encompasses increased use of irrigation and hence the possibility of salinization. In 1984, Henderson-Sellers and Gornitz estimated that the extent of irrigation in 21 arid countries increased by 37 million ha over the preceding 30 years

TABLE IIa

LAND AREAS CONVERTED INTO REGULAR CROPPING (10^6 HA) FROM 1860 TO 1978 (AFTER RICHARDS, 1990; REPRINTED WITH THE PERMISSION OF CAMBRIDGE UNIVERSITY PRESS)

World region	First period (1860–1919)		Second period (1920–1978)	
	To crops	From crops	To crops	From crops
Africa	15.9	–	90.5	–
North America	163.7	2.5	27.9	29.4
Central America and the Caribbean	4.5	–	18.8	0.4
South America	35.4	–	65.0	–
Middle East	8.0	–	31.1	–
South Asia	49.9	–	66.7	–
Southeast Asia	18.2	–	39.0	–
East Asia	15.6	0.2	14.5	8.4
Europe (excluding Soviet Europe)	26.6	6.0	13.8	12.7
Former Soviet Union	88.0	–	62.9	–
Australia/New Zealand	15.1	–	40.0	–
TOTAL	440.9	8.7	470.2	50.9
Net area converted to crops		432.2		419.3

(FRAMJI and MAHAJAN, 1969; FAO Production Yearbooks). However, salinization of irrigated fields destroys up to 125,000 ha year^{-1} (UNCOD, 1977), or 3.75 million ha in 30 years. By the late 1970s, soil erosion was estimated to exceed soil formation on about a third of US cropland primarily in the midwestern agricultural heartland (BROWN, 1987). In

TABLE IIb

AGRICULTURAL DEVELOPMENT FROM 1964 TO 1984 (AFTER BRUNDTLAND, 1987)

Region	Per capita gross cropped area (ha)		Per hectare fertilizer use (kg)	
	1964	1984	1964	1984
World	0.44	0.31	29.3	85.3
North America	1.05	0.90	47.3	93.2
Western Europe	0.31	0.25	124.4	224.3
Eastern Europe and former Soviet Union	0.84	0.71	30.4	122.1
Africa	0.74	0.35	1.8	9.7
Near East[a]	0.53	0.35	6.9	53.6
Far East[b]	0.30	0.20	6.4	45.8
Latin America	0.49	0.45	11.6	32.4
CPEs of Asia[c]	0.17	0.10	15.8	170.3

Source: Based on FAO data and modified from BRUNDTLAND, G. H. (1987) *Our Common Future*, by permission of Oxford University Press.

[a]An FAO grouping that includes West Asia plus Egypt, Libya and Sudan.

[b]An FAO grouping that covers South and South-East Asia excluding the centrally planned economies of Asia.

[c]An FAO grouping of centrally planned economies of Asia that covers China, Kampuchea, North Korea, Mongolia and Vietnam.

increase itself. In particular, it is not clear whether surface temperatures will increase or decrease.

Roughness length

The roughness decrease imposed in deforestation experiments is usually an alteration from a forest roughness length of around 2 m to one more appropriate to a grassland of, say, 0.05 m. For desertification, the decrease would be somewhat less. Imposing a decrease in

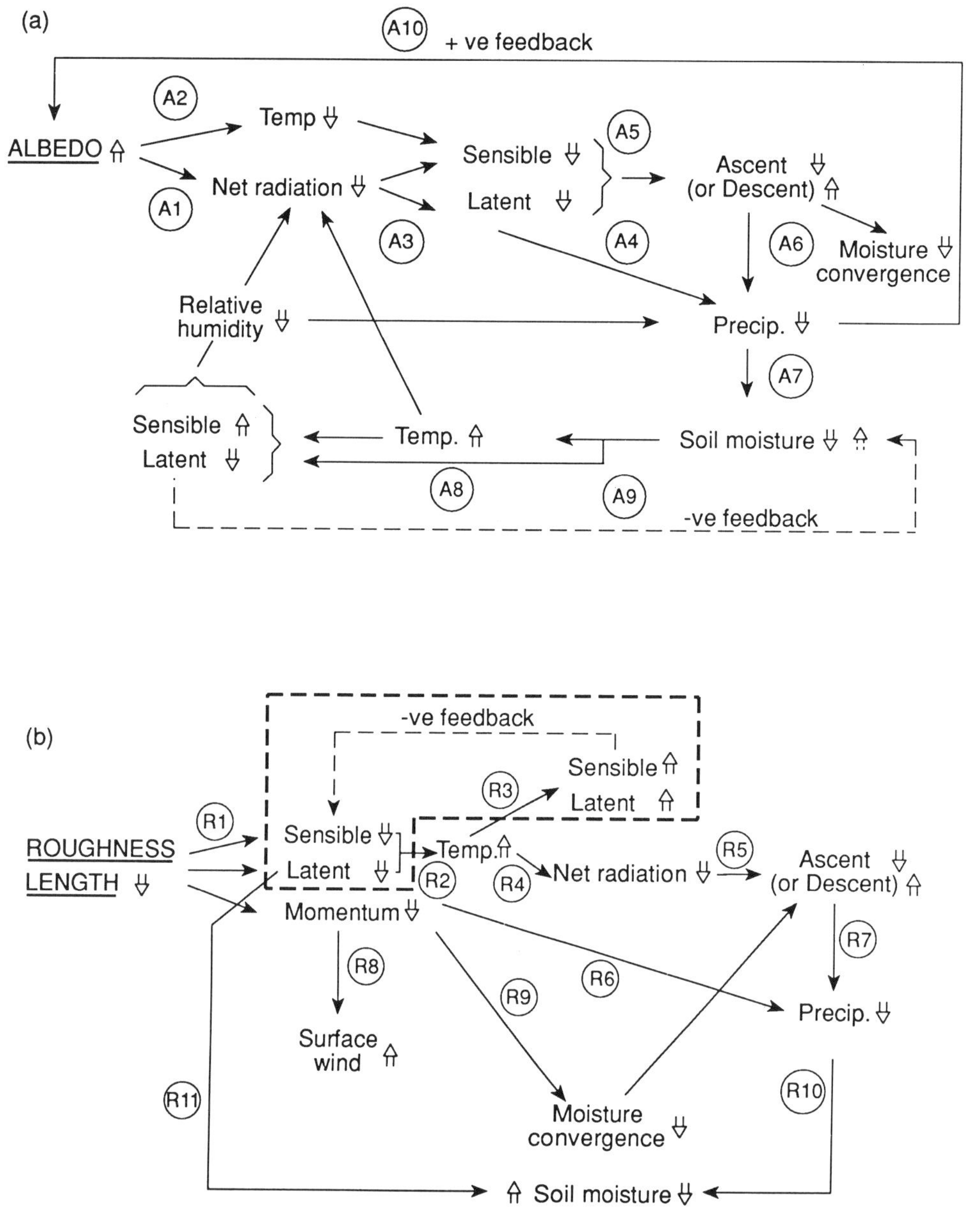

Fig. 7. Schematic representation of some of the climatic responses to prescribed land-surface changes: (a) following an albedo increase; (b) following a roughness length decrease. The numbered processes are described in the text. In desertification experiments, those processes in part (a) might be anticipated. In the case of deforestation, processes in both (a) and (b) might occur. (Part (a) was modified from Mylne and Rowntree, 1992; reprinted by permission of Kluwer Academic Publishers).

roughness (Fig. 7b) acts to reduce the turbulent fluxes of sensible and latent heat and of momentum between the continental surface and the atmosphere (process R1). This reduction in the fluxes tends to cause an increase in the surface temperature (process R2). The temperature increase prompts a negative feedback (R3) since it will tend to cause an increase in sensible flux and also in latent heat. The effect of an increase in surface temperature (process R2) and thus increased longwave radiation from the surface is decreased net radiation (process R4) prompting decreased ascent (process R5). In addition to the reduction in evaporation due to the reduced roughness length (process R1), there is an additional reduction due to the reduction in the transpiration from the (prescribed) reduction in biomass. This combined reduction in evaporative flux acts to reduce precipitation (process R6) as will the reduction in ascent (or increased descent) (process R7).

The decrease in surface roughness decreases the turbulent momentum flux which decreases the surface drag and increases the wind strength in the planetary boundary layer (process R8). Decreased drag results in decreased cross-isobaric flow and hence decreased moisture convergence (process R9). This will contribute further to the reduction in ascent and hence the decrease in precipitation (reinforcing process R7).

The decrease in precipitation will tend to cause a decrease in soil moisture (process R10) although this will be balanced by the tendency to increase soil moisture amounts caused by the decrease in evaporation (process R11). Not only is the outcome of these two effects uncertain but there is also the potential for a positive feedback on evaporation caused by a decrease in soil moisture (this would follow the arrow of process R11 but in the other direction).

As in the case of the prescribed albedo increase, there are ambiguities over the direction of change in some climatic parameters as a result of feedbacks and differing magnitudes of changes. In addition, it is not clear how to relate these local impacts to regional changes such as convergence which is also controlled by larger-scale circulation characteristics (cf. SUD et al., 1988; ROWNTREE, 1991).

In tropical deforestation, of course, both major disturbances in Fig. 7 occur: albedo is increased and roughness length decreases. As can be surmised from Fig. 7, it is very difficult to anticipate either the resulting direction or the magnitude of changes in many climatic parameters. The imposition of an albedo increase might cool the surface and thus prompt descent in the atmospheric column. The imposition of a reduced roughness length is likely to reduce evaporation, and hence warm the surface, but also to prompt descent or at least decreased ascent in the air column.

Global climate model experiments designed to assess the effects of land-use and land-cover change might then be expected to reveal local surface climatic changes and, perhaps, more widespread dynamical responses including disturbances in the regional-scale moisture convergence (e.g. SUD et al., 1988; NOBRE et al., 1991). The possibility of regional-scale atmospheric responses to land-cover change has been noted by, for example, HENDERSON-SELLERS and GORNITZ (1984) and SHUKLA et al. (1990) with reference to tropical deforestation and by CHARNEY et al. (1977) and LAVAL and PICON (1986) with reference to desertification. Major atmospheric features worthy of detailed investigation are the Hadley and Walker circulations and the tropical jets. These are included in the assessment of the climate model sensitivity studies in the next section of this chapter.

Technical Note, NCAR/TN-387+STR. National Center for Atmospheric Research, Boulder, CO, 80 pp.

DIRMEYER, P. A., 1992. GCM studies of the influence of vegetation on the general circulation: the role of albedo in modulating climate change. Ph. D. Thesis, University of Maryland, MD, unpublished, 226 pp.

DOUGLAS, I., 1990. Sediment transfer and siltation. In: B. L. TURNER II, W. C. CLARK, R. W. KATES, J. F. RICHARDS, J. T. MATHEWS and W. B. MEYER (Editors), *The Earth as Transformed by Human Action: Global and Regional Changes in the Biosphere over the Past 300 Years*. Cambridge University Press, Cambridge, pp. 215–234.

DREGNE, H. E., 1977. Desertification of arid lands. *Econ. Geog.*, 53: 322–331.

DREGNE, H. E., 1985. Aridity and land degradation. *Environment*, 27: 16–20; 28–33.

DZIETARA, S. and JANICOT, S., 1980. Etude du champ de vent au-dessus de l'Afrique Occidentale. *Rapport de stage de 3e année*. Juin, Ecole de la Météorologie, France.

ECKHOLM, E. and BROWN, L. R., 1977. Spreading deserts – the hand of Man. *Worldwatch Paper Nr 13*. Washington, DC, 40 pp.

EL-BADRY, M. A., 1992. World population change: a long-range perspective. *Ambio*, XXI: 18–23.

EMANUEL, W. R., SHUGART, H. H. and STEVENSON, M. P., 1985. Climatic change and the broad-scale distribution of terrestrial ecosystem complexes. *Clim. Change*, 7: 29–44; and Response to Rowntree. *Clim. Change*, 7: 457–460.

ETHERIDGE, D. M., PEARMAN, G. I. and DE DILVA, F., 1988. Atmospheric trace-gas variations as revealed by air in an ice core from Law Dome, Antarctica. *Ann. Glaciol.*, 10: 28–33.

EUROPEAN TIMBER TRENDS and PROSPECTS, 1950–2000. Econ. Comm. Europe. *Timber Bull. for Europe, 24, Suppl. 3*. FAO, Geneva, 1976, 308 pp.

FAO Production Yearbooks, 1951, 1960, 1970, 1979, 1980.

FAO, 1984. *Land, Food and People*. Food and Agriculture Organization, Rome.

FAO, 1985. *Production Yearbook*, Vol. 37. United Nations, Rome.

FAO, 1990. *Production Yearbook*, Vol. 43. FAO, Rome.

FAO, 1991. Forest Resource Assessment 1990 Project. *FAO/UN Second Interim Report on the State of Tropical Forests*. Paper presented at the 10th World Forestry Congress, Sept., Paris.

FAO/UNEP, 1981. *Tropical Forest Resources Assessment Project*. FAO, Rome.

FELS, E. and KELLER, R., 1973. Manmade lakes: their problems and environmental effects. In: W. C. ACKERMAN, G. F. WHITE and E. B. WORTHINGTON (Editors), *Am. Geophys. Union Mono.*, 17: 43–49.

FOLLAND, C. K., KARL, T. and VINNIKOV, K. YA., 1990. Observed climate variations and change. In: J. T. HOUGHTON, G. J. JENKINS and J. J. EPHRAUMS (Editors), *Climate Change. The IPCC Scientific Assessment*. Cambridge University Press, Cambridge, pp. 195–238.

FRAMJI, K. K. and MAHAJAN, I. K., 1969. *Irrigation and Drainage in the World*, 2nd edition. Int. Comm. on Irrigation and Drainage, New Delhi, 1344 pp.

FRANKE, R. and CHASIN, B., 1980. *Seeds of Famine: Ecological Destruction and the Development Dilemma in the West African Sahel*. Allenheld and Osmun, Montclair, NJ.

GLANTZ, M. H., 1994. *Drought Follows the Plow: Cultivating Marginal Areas*. Cambridge University Press, Cambridge.

GORNITZ, V., 1985. A survey of anthropogenic vegetation changes in West Africa during the last century - climatic implications. *Clim. Change*, 7: 285–325.

GORSE, J., 1985. Desertification in the Sahelian and Sudanian zones of West Africa. *Unasylva*, 37: 2–18.

GRAEDEL, T. E. and CRUTZEN, P. J., 1990. Atmospheric trace constituents. In: B. L. TURNER II, W. C. CLARK, R. W. KATES, J. F. RICHARDS, J. T. MATHEWS and W. B. MEYER (Editors), *The Earth as Transformed by Human Action: Global and Regional Changes in the Biosphere over the Past 300 Years*. Cambridge University Press, Cambridge, pp. 295–312.

GRAETZ, D., FISHER, R. and WILSON, M., 1992. *Looking Back: The Changing Face of The Australian Continent, 1972–1992*. CSIRO Office of Space Science and Applications, Canberra, 159 pp.

HANSEN, J., JOHNSON, D., LACIS, A., LEBEDEFF, S., LEE, P., RIND, D. and RUSSELL, G., 1981. Climate impact of increasing atmospheric carbon dioxide. *Science*, 213: 957–966.

HANSEN, J., LACIS, A., RIND, D., RUSSELL, G., STONE, P., FUNG, I., RUEDY, R. and LERNER, J., 1984. Climate sensitivity: analysis of feedback mechanisms. In: J. E. HANSEN and T. TAKAHASHI (Editors), *Climate Processes and Climate Sensitivity. Geophys. Monogr.*, 295: 130–163.

HAO, W. M., LIU, M.-H. and CRUTZEN, P. J., 1990. Estimates of annual and regional releases of CO_2 and other trace gases to the atmosphere from fires in the tropics, based on FAO statistics for the period 1975–1980. In: J. G. GOLDAMMER (Editor), *Fire in the Tropical Biota, Ecosystem Processes and Global Challenges.* Springer-Verlag, New York, pp. 440–462.

HART, T. L., 1990. Estimates of horizontal divergence over northern Australia, and implications for biogeophysical feedback. *Aust. Meteorol. Mag.*, 38: 107–122.

HARVEY, L. D. D., 1991. A commentary on: tropical deforestation and atmospheric carbon dioxide. *Clim. Change*, 19: 119–121.

HEATHCOTE, R. L., 1991. Managing droughts? Perception of resource management in the face of the drought hazard in Australia. In: A. HENDERSON-SELLERS and A. J. PITMAN (Editors), *Vegetation and Climate Interactions in Semi-Arid Regions.* Kluwer, Dordrecht, pp. 219–230.

HELLDEN, U., 1984. *Drought Impact Monitoring.* Lund University Naturgeografiska Institute, Lund, Sweden.

HELLDEN, U., 1991. Desertification – time for an assessment, *Ambio*, 20(8): 372–383.

HENDERSON-SELLERS, A., 1987. Effects of change in land use on climate in the humid tropics. In: R. E. DICKINSON (Editor), *Geophysiology of Amazonia. Vegetation and Climate Interactions.* Wiley, New York, pp. 463–493.

HENDERSON-SELLERS, A., 1990a. Getting GARP to grow grass: land surfaces in global climate models. *Aust. Meteorol. Mag.*, 38: 245–254.

HENDERSON-SELLERS, A., 1990b. Predicting generalized ecotype groups with the NCAR CCM: first steps towards an interactive biosphere. *J. Climate*, 3: 917–940.

HENDERSON-SELLERS, A., 1992. Continental vegetation as a dynamic component of a global climate model: a preliminary assessment. *Clim. Change*, 23: 337–377.

HENDERSON-SELLERS, A., 1993. A factorial assessment of the sensitivity of the BATS land-surface parameterization scheme. *J. Climate*, 6(2): 227–247.

HENDERSON-SELLERS, A. and GORNITZ, V., 1984. Possible climatic impacts of land cover transformations, with particular emphasis on tropical deforestation. *Clim. Change*, 6: 231–258.

HENDERSON-SELLERS, A. and MCGUFFIE, K., 1994. Landsurface characterization in greenhouse climate simulations. *Int. J. Climatol.*, 14: 1065–1094.

HENDERSON-SELLERS, A. and PITMAN, A. J. (Editors), 1991. *Vegetation and Climate Interactions in Semi-Arid Regions.* Kluwer, Dordrecht, 238 pp.

HENDERSON-SELLERS, A. and WILSON, M. F., 1983. Surface albedo data for climatic modelling. *Rev. Geophys. Space Phys.*, 21: 1743–1778.

HENDERSON-SELLERS, A., DICKINSON, R. E., DURBIDGE, T. B., KENNEDY, P. J., MCGUFFIE, K. and PITMAN, A. J., 1993. Tropical deforestation: modelling local to regional-scale climate change. *J. Geophys. Res.*, 98(D4): 7289–7315.

HOLDRIDGE, L. R., 1947. Determination of world plant formations from simple climatic data. *Science*, 105: 367–368.

HOLDRIDGE, L. R., 1964. *Life Zone Ecology.* Tropical Science Center, San Jose, Costa Rica.

HOUGHTON, J. T., JENKINS, G. J. and EPHRAUMS, J. J. (Editors), 1990. *Climate Change. The IPCC Scientific Assessment.* Cambridge University Press, Cambridge, 365 pp.

HOUGHTON, J. T., CALLANDER, B. and VARNEY, S. K. (Editors), 1992. *Climate Change 1992. Supplementary Report to The IPCC Scientific Assessment.* Cambridge University Press, 200 pp.

HOUGHTON, R. A., 1990a. The global effects of tropical deforestation. *Environ. Sci. Technol.*, 24: 414–432.

HOUGHTON, R. A., 1990b. The future role of tropical forests in affecting the carbon dioxide concentration of the atmosphere. *Ambio*, 19: 204–209.

HOUGHTON, R. A., 1991. Tropical deforestation and atmospheric carbon dioxide. *Clim. Change*, 19: 99–118.

HOUGHTON, R. A. and SKOLE, D. L., 1990. Carbon. In: B. L. TURNER II, W. C. CLARK, R. W. KATES, J. F. RICHARDS, J. T. MATHEWS and W. B. MEYER (Editors), *The Earth as Transformed by Human Action: Global and Regional Changes in the Biosphere over the Past 300 Years.* Cambridge University Press, Cambridge, pp. 393–408.

HOUGHTON, R. A., SCHLESINGER, W. H., BROWN, S. and RICHARDS, J. F., 1985. Carbon dioxide exchange between the atmosphere and terrestrial ecosystems. In: J. R. TRABALKA (Editor), *Atmospheric Carbon Dioxide and the Global Carbon Cycle.* Department of Energy, DOE/ER-0239, Washington, DC, pp. 113–140.

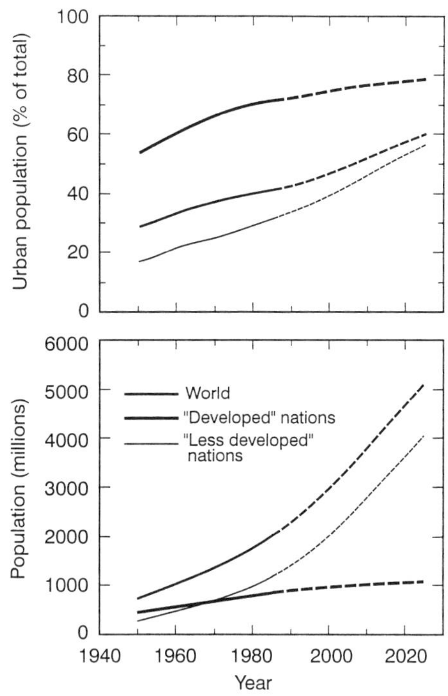

Fig. 1. Urban population growth from 1900 to 2025, expressed both as a percentage of the total population and in total numbers. *Note:* Figures are actual from 1900 to 1985, and projected from 1985 to 2025 (dashed lines). Based on data from OKE (1986). "Developed" nations are those nations in North America, Europe, the former USSR, Australia plus Japan; "Less developed" nations are those in Africa, Latin America and Asia (excluding Japan, Melanesia, Micronesia and Polynesia). These are the groupings used in the United Nations report.

In this way cities indirectly modify those atmospheric variables, such as temperature, humidity, airflow and rainfall, whose ensemble averages define the climate (OKE, 1980). This happens at local (<1 km horizontally; <100 m vertically) and meso (<100 km horizontally; <~1 km vertically) spatial scales. The magnitude and extent of these weather and climate effects cannot be generalised for all cities because they depend on city-specific factors such as the pre-urban land use and the regional climate.

Interest in urban climates began in 1833 when Luke Howard (HOWARD, 1833) first documented that cities appeared to be warmer than the surrounding countryside. Many studies in the 160 years since have well established this urban heat island effect; a phenomenon linked to differences in urban/rural cooling rates and best seen at night under clear skies and light winds. Longer term studies show that cities modify climates, at least at the local spatial scale, e.g. KARL et al. (1988) found urban effects on air temperatures in towns with only 10,000 inhabitants. Much research in the last 20 years has been directed towards not only describing the nature of urban climates, but understanding the processes that lead to the observed urban climate. In extrapolating this current knowledge to predict future urban climates, the following questions must be considered:

- What will be the nature of future cities? Urban climates in the future will be determined by many factors, including: height and spacing of buildings; nature of construction materials; location of city; type and pattern of energy use; transport mode; type and location of industries; amount of greenspace; city layout etc.

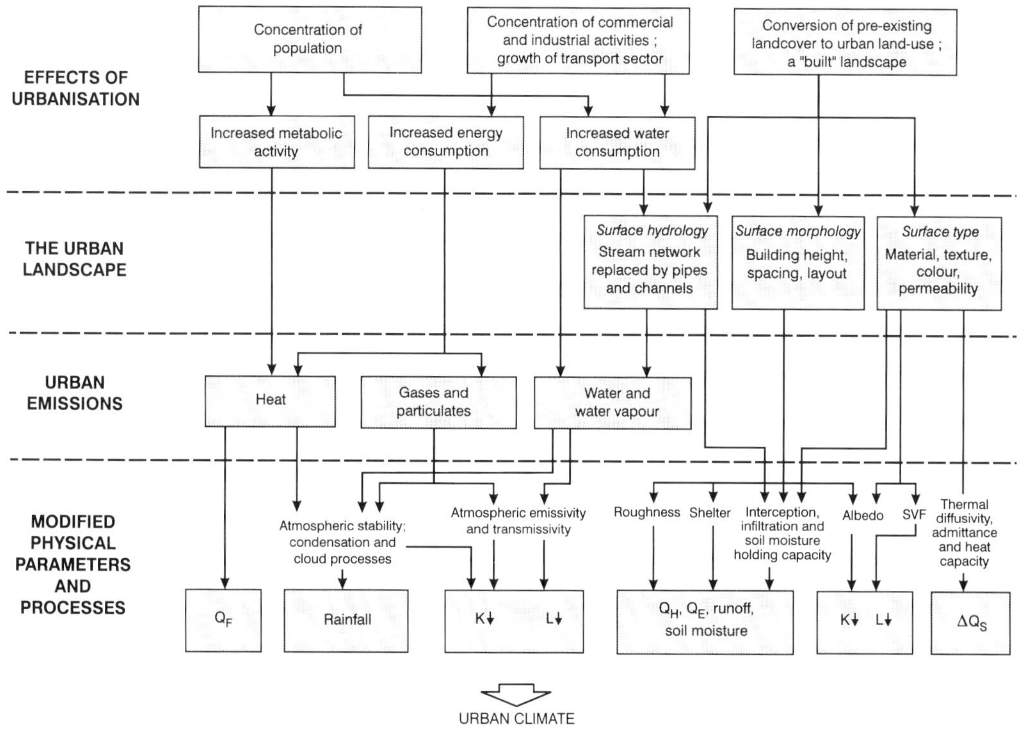

Fig. 2. Conceptual flowchart of the effects of urbanisation on the radiation, energy balance and local climate (modified from YAMASHITA and SEKINE, 1990/91). Note that feedbacks between the modified physical processes are not included.

- Will there be an urban influence on the regional and global climate, and how do we separate this effect from that due to global climate changes, such as result from the greenhouse effect?.
- What will be the nature of the future global climate? Urban effects are superimposed on the regional and global climate, thus the climate of cities in the future has to be considered in light of global climate changes.

The last, and to a lesser extent the second, question can really only be answered using GCMs with an ability to include local scale urban effects, a task beyond the reach of current GCMs. Recent analyses of the US temperature record (e.g. KARL et al., 1988; KARL and JONES, 1989; JONES et al., 1989) have provided observational evidence of the contribution of urban warming to the continental and global temperature signal. The variation between estimates from different studies illustrates that they are by no means definitive.

Numerical models potentially have an important role in answering questions about urban effects, comprehensive reviews of urban climate models can be found in OKE (1974, 1979a) and BORNSTEIN (1986, 1989). Many mesoscale numerical models that have been specifically applied to urban meteorology are limited by their crude land surface schemes (ROSS and OKE, 1988). Two such examples are the modelling studies of HJELMFELT (1982) and SEAMAN et al. (1989) which are of great interest because they separate urban and topographic effects on mesoscale airflow and thermal features using a series of model "experiments". However, the results are limited by the use of a simple moisture availability factor to partition sensible and latent heat fluxes. Thus far models have helped to elucidate urban processes but at present they are of limited use in predicting future climate change in was made between suburban and "dry" rural, or suburban and "wet" rural. Their study is a good reminder that "rural" land use is often just as likely to be influenced by human activities as urban land use and is not an appropriate surrogate for pre-urban land use and climates. CLEUGH and OKE (1986) avoided this dilemma by using energy balances measured simultaneously at a suburban site and a "control" site. This control site was flat with short grass which eliminated the complexity introduced by buildings and surface types that characterise suburban land use and enabled urban effects on atmospheric processes to be clarified. Unless such measurements are conducted over a long term, this methodology does not allow for an assessment of urban effects on climate, only on atmospheric processes. That this approach is not a substitute for Lowry's methodology illustrates the difficulty in proving urban climate effects – a point that should be borne in mind throughout this chapter. This methodological difficulty also illustrates the important role that physically based models have in understanding and detecting urban effects. There is an urgent need for development and validation of appropriate urban surface parameterisation schemes that can be linked to current generation mesoscale models and embedded into GCMs.

Scales

A key to interpreting and/or modelling urban effects on atmospheric processes is to develop a methodology or conceptual framework that accounts for the heterogeneity of the urban "surface". As with any natural landscape, this heterogeneity is scale-dependent and OKE (1976) developed such a framework that took advantage of the natural range of scales present in urban morphology. Table I shows these scales and their spatial dimensions. Oke divided the urban surface and atmosphere into urban canopy and urban boundary-layers as shown in Fig. 3. The urban canopy layer (UCL), which lies below the mean roof level, is made up of an array of vertical, horizontal, sloping and multi-level surface elements. These surface elements each have varying thermal, moisture-holding, radiative and aerodynamic properties which, in turn, lead to a range of microclimates within the UCL. The urban

TABLE I

SPATIAL SCALES IN THE URBAN ENVIRONMENT

Urban units	Urban features	Spatial dimensions (m)		Scale	Atmospheric layer
		Width	Length		
Building	Single building, tree or garden	10	10	Micro	UCL including roughness or wake layer
Canyon	Urban street and bordering buildings	30	300		
Block	City block, park, factory complex	500	500	Local	Surface, or constant flux, layer
Land use zone	Residential, commercial, industrial etc.	5000	5000		UBL including surface layer
City	Urban area	25,000+	25,000+	Meso	

Adapted from OKE (1984).

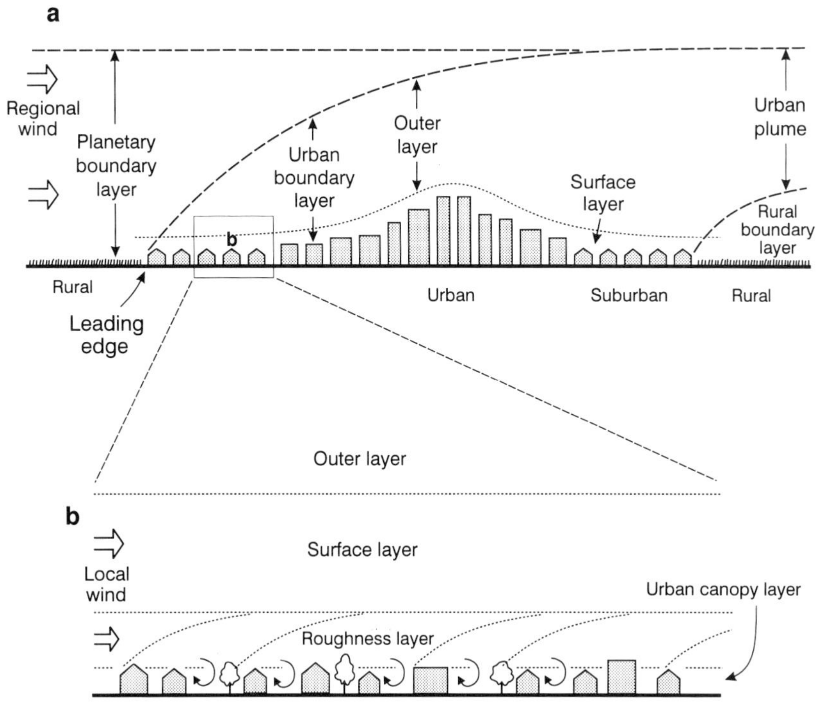

Fig. 3. Atmospheric layers in an urban environment (after OKE, 1976).

boundary-layer (UBL) extends from mean roof level upwards. The UBL is derived from the concept of an internal boundary-layer growing with distance downwind from the leading edge, which in this case is the surface transition from upwind "rural" to urban land use (Fig. 3). The air flowing over the city becomes modified by, and adjusted to, the new surface, starting with the lowest air layers. If the horizontal extent of the city is sufficiently large, this urban-influenced boundary-layer (the UBL) will occupy the entire PBL. The constant flux layer (also referred to as the surface layer) lies at the base of the UBL and above the roughness layer (also known as the wake layer) where flow is influenced by individual buildings. Areally averaged turbulent fluxes from the underlying canopy are assumed to be constant with height in the surface layer, whose depth may reach ca. 100 m by day.

Nature of the urban surface

Although we can identify many individual urban surface elements, Oke's approach treats these as simply facets of an integrated urban surface, which is actually a volume. Fig. 4 is a conceptualisation of this urban canopy volume, containing roughness elements such as trees and buildings; horizontal surfaces such as roads, pavements and rooftops; and the air contained within the canopy layer. Turbulent exchanges between the urban surface and atmosphere are represented by areally averaged fluxes flowing across the plane ABCD in Fig. 4a, into/out of the urban canopy volume. Provided our perspective is from some location at a distance above the top of the UCL, the urban canopy volume can be conceptualised as a two-dimensional "surface" representative of a larger spatial scale, e.g. a land use zone (Table I). We can then begin to note some characteristic features of this urban surface.

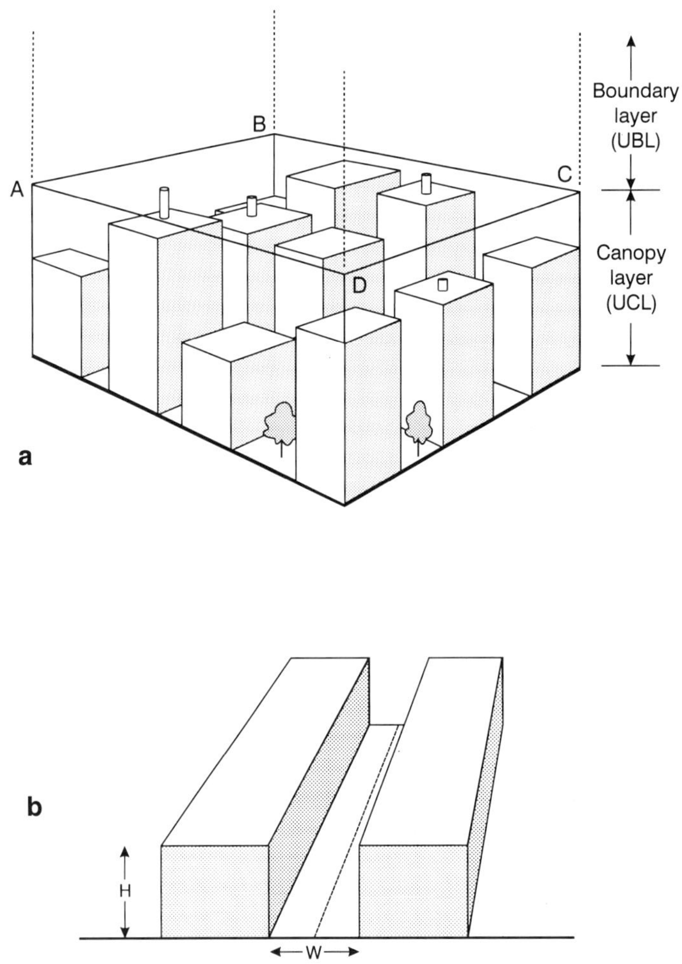

Fig. 4. (a) The urban canopy volume (after OKE, 1987). (b) The urban canyon with height (*H*) and width (*W*).

In suburban land use the dominant roughness elements are buildings (mostly houses and commercial buildings) and trees whose heights are typically of order 10 m. These roughness elements are sparsely distributed, spacings of ca. 20–50 m were identified for a suburban, North American city (Vancouver, SCHMID, 1988; SCHMID and OKE, 1992). This roughness element spacing is strongly influenced by the street pattern and hence can be anisotropic. The surface is also aerodynamically rough, buildings act as bluff bodies which extract momentum from airflow via skin friction and form drag. Roughness parameters vary with roughness element height and spacing. Suburban land use has typical roughness lengths of 0.5-1 m while urban roughness lengths may be as large as 4 m (OKE, 1980). The open canopy structure and the "patchy" array of horizontal surfaces with contrasting properties (e.g. dry roads and wet lawns) leads to active horizontal exchanges of heat and mass via microscale advection.

It is important to realise that the atmospheric feedbacks and interactions that occur within the UCL mean that we cannot simply sum individual surface elements to determine the effect of the entire urban surface on atmospheric processes and climates. Likewise, horizontal pavements or lawns cannot be used as surrogates for the urban surface. Rather, we have to understand the way that the entire UCL functions. Thus, we recognise the urban canyon (Fig. 4b) as a fundamental unit of the built landscape that is replicated throughout the UCL. The urban canyon is simply two building walls, with height H, separated by a road of width, W. The aspect ratio (H/W) describes building spacing and is an important descriptor of the UCL. Radiative exchanges and airflow characteristics depend on the size of the aspect ratio.

Urban climate processes

Fig. 2 shows the connections between urban activities and climatic processes, this section explains these connections and so addresses the question of "how do urban areas modify these physical processes?" The magnitude and direction of these urban-induced changes are quantified by reviewing the current state of knowledge, most of which has derived from a combination of observational and modelling studies.

Radiative effects

The incoming shortwave radiation ($K\downarrow$) exerts a first order forcing on the climate of any location, as has been shown in earlier chapters. When discussing the radiation balance for the UCL, $K\downarrow$ refers to the radiation impinging on a plane extending across the top of the UCL (plane ABCD in Fig. 4a). Shortwave radiation comprises both direct beam and diffuse components, the latter refers to radiation reflected and scattered by atmospheric particulates and gases. The amount of shortwave radiation absorbed within the UCL volume depends on the latter's albedo (α). Cities also affect the longwave radiation budget; downwelling longwave radiation from the urban atmosphere impinging on the plane ABCD in Fig. 4a is defined as $L\downarrow$. We begin by discussing atmospheric effects on downward shortwave and longwave radiation and then discuss the influence of the urban canopy on shortwave and longwave absorption. Readers are referred to Chapter 1 by HENDERSON-SELLERS which describes the pathways for radiative transfer in the atmosphere.

A depletion in $K\downarrow$ would be one of the most obvious examples of the effect of cities on radiation. This reduction is expected because the polluted urban atmosphere will reduce the atmospheric transmissivity for shortwave radiation. Unfortunately observational studies find this hard to prove (OKE, 1979a) because nearby rural measurements may well be contaminated by downwind transport of these urban pollutants. Table II (from OKE, 1988) shows typical reductions in incoming shortwave radiation. These values tend to fall into two groups: (a) large reductions for those cities with a sizeable industrial base where SPM (suspended particulate matter) is common; here reductions of up to 30% in $K\downarrow$ on individual days have been observed with typical annual reductions of around 10%; (b) other cities where SPM levels are low, but photochemical smog exists, reveal much smaller reductions in incoming shortwave radiation, typically less than 5%. Observations show that the reduction in $K\downarrow$ is variable both spatially (i.e. from city to city) and temporally (i.e. diurnally and seasonally). Causes of these variations include (OKE, 1979a):

TABLE II

ATTENUATION OF INCOMING SHORTWAVE RADIATION (FROM OKE, 1988)

Author (Year)	City	Data Period	Attenuation (%)
PASZYNSKI (1972)	Upper Silesian industrial district		12
PROBALD (1974)	Budapest, Hungary	1955–1968	9
HESS et al. (1978)	Cracow, Poland	1976–1978	12
HAY (1984)	Vancouver, Canada	1979–1981	1–2
SANDERSON et al. (1973)	Detroit-Windsor, USA	1970–1971	9
ROUSE et al. (1973)	Hamilton, Canada	1970–1972	12
YAMASHITA (1979)	Tokyo, Japan	1967–1968	12–14
SEKIHARA (1973)	Tokyo, Japan	1961–1971	10–15
PETERSON and STOFFEL (1980)	St. Louis, USA	1975–1977	3
METHOD and CARLSON (1982)	St. Louis, USA	1975–1976	7
CHOW and CHANG (1981)	Shanghai, China		15
WANG and LIU (1982)	Hangchow, China		9
ESTOURNEL et al. (1983)	Toulouse, France	1979	3
PETERSON et al. (1978)	Los Angeles, USA	1973	6–8

Note: based on studies since 1970 only; includes data for studies with very different periods of observation (3 months–13 years) and cloud conditions (some only use clear skies).

- geographical location and time of day/year (determine the optical pathlength);
- averaging period;
- industrial and manufacturing activities and mode of transport;
- wind direction in relation to the location of industries, major highways, etc.;
- sky cover (viz. cloudy or clear).

Examples of such variability are found in the studies presented in Table II. A recent study by STANHILL and KALMA (1993) reported an annual decline in $K{\downarrow}$ of 1.08% over the 35 years of measurements at Kowloon (Hong Kong). They attributed most of this to reduced direct beam radiation resulting from increased emissions of aerosols from fossil fuel combustion. This follows the results of STANHILL and MORESHET (1992) who showed the influence of vehicle pollution on depletion of $K{\downarrow}$ in Israel. Cities with a large, old and inefficient vehicle fleet are likely to find similar reductions in $K{\downarrow}$. Such variability means that it is impossible to generalise about the size of the reduction in shortwave radiation in cities. In terms of future climates, cities in rapidly industrialising nations will likely have the greatest reductions in $K{\downarrow}$ if fossil fuels remain the major primary energy source. Warmer UBL temperatures, resulting from greenhouse warming, have the potential to increase photochemical smog formation which may further reduce $K{\downarrow}$.

These studies refer to the receipt of shortwave radiation at UCL level. Atmospheric constituents (including water vapour, aerosols and SPM) will selectively absorb radiation which further alters the radiation budget both of the UCL and the UBL, as described for the globe in Chapter 10 by ANDREAE. Some studies suggest that radiative heating from absorption within the UBL is important. Others agree with LYONS and FORGAN (1975) who noted that atmospheric pollutants tend to scatter, rather than absorb, the solar beam. Diffuse radiation, rather than radiative heating of the UBL, is then increased and accounts for 60–80% of the attenuated direct beam radiation (OKE, 1988). METHOD and CARLSON (1982) found that maximum heating rates due to absorption by aerosols were more than 3°C/day near the surface, but averaged over the day were only 1°C/day near the ground and closer to zero at the top of the urban convective boundary layer. They concluded that heating effects due to aerosols were not large and any heating was slightly less than the cooling that resulted from increased longwave emissions (see below). Numerical simulations (YOSHIDA, 1991) showed that the UBL temperature structure was influenced by aerosols during the daytime only. Under calm conditions, where radiation scattering is dominant, urban temperatures were lowered and the urban-induced thermal circulation (see below) weakened. This had the effect of enhancing the aerosol concentrations even further. Urban-induced circulations were not affected when radiation absorption, rather than scattering, dominates. In these conditions near surface temperatures decreased and upper temperatures either increased or were unchanged.

Ultraviolet (UV) radiation is readily absorbed by ozone which is created in the UBL as a part of the photochemical smog cycle. Thus we would expect to find reduced levels of UV radiation beneath urban atmospheres, but estimates are highly variable in space and time; 10–15% reductions in Tokyo (Japan) and up to 50% on individual days in Los Angeles (USA) in the autumn (OKE, 1976, 1979b).

While aerosols may reduce $K\downarrow$, they can increase the emission of longwave radiation because of their effect on UBL temperature and emissivity; an urban-induced "greenhouse effect" is created. As with the depletion of shortwave radiation, this effect varies from city to city, e.g. WHITE et al. (1978) found no difference in $L\downarrow$ between urban St. Louis (USA) and the surrounding rural area whereas ROUSE et al. (1973) observed large daytime differences in the industrial city of Hamilton (Canada). There is evidence from many of these studies that a reduction in $K\downarrow$ due to aerosol absorption yields an increase in $L\downarrow$, hence there is little change to the total incoming radiation (OKE, 1988). Increases in $L\downarrow$ are not always due to aerosol effects on atmospheric emissivity, however, but result from a warmer urban atmosphere arising from the heat island effect (e.g. ESTOURNEL et al., 1983; OKE and FUGGLE, 1972; SUCKLING, 1981). The general consensus is that urban values of $L\downarrow$ are larger, primarily because of the warmer atmosphere (OKE, 1988) but atmospheric aerosols may also play a role. Large downwelling longwave fluxes are often offset by increases in emitted longwave radiation from the UCL (i.e. $L\uparrow$) hence differences between urban and rural net longwave radiation will be small.

Cities therefore influence the receipt of radiation because of the composition and temperature of the urban atmosphere. The factors discussed, plus the variability of pollution emissions (amount, type and timing) both between cities and within individual cities, means that it is difficult to generalise an urban effect. In general, the shortwave radiation received at the UCL may be reduced in intensity, be more diffuse and be depleted in some wavelengths. Reductions in $K\downarrow$ will be offset by increases in $L\downarrow$.

The altered nature of the urban surface also has a strong influence on the receipt and absorption of radiation because its convoluted, three-dimensional nature can result in shading and multiple reflection of the solar beam once it enters the urban canyon. Multiple reflections lead to greater shortwave trapping and absorption and so the albedo of the UCL is reduced. Urban albedos are also lower because of the dark nature of many materials used in urban construction (e.g. asphalt), however radiative trapping means that any three-dimensional surface will absorb more radiation than a horizontal one constructed of the same material. AIDA (1982) found this difference to be of the order 20% based on an ex-

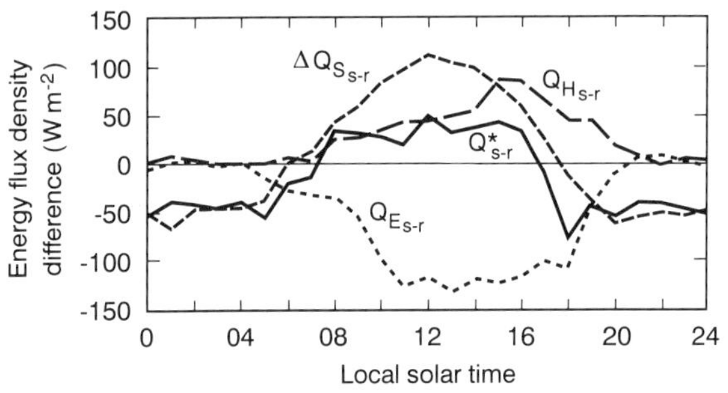

Fig. 7. Measured differences between suburban and rural energy balance components (from CLEUGH and OKE, 1986). s, suburban; r, rural.

building density and high occurrence of concrete rooftops which intercept large amounts of solar radiation at this tropical latitude (ca. 19°N).

In general urban latent heat fluxes are smaller than their rural counterparts, in proportion to the reduction in vegetation. However, even in drought conditions Q_E can still act as an important energy sink, by day, in suburban areas. These non-trivial latent heat fluxes are maintained by external water use (irrigation of gardens, lawns and parks). Considerable day-to-day variability in β is also a common feature in suburban land use (e.g. KALANDA et al., 1980; OKE and MCCAUGHEY, 1983; CLEUGH and OKE, 1986; GRIMMOND et al., 1993). Some of this arises from variations in rainfall and irrigation and hence surface moisture availability (GRIMMOND and OKE, 1986). OKE and MCCAUGHEY (1983) suggested that variability in β could be related to differences in incoming radiation, however neither of these factors provides a complete account. CLEUGH and OKE (1986) noted that such temporal variability was absent at their "control" site and argued that the turbulent flux partitioning is also linked to changes in the saturation deficit (i.e. dryness) of the overlying UBL. This is enabled by the roughness of the urban surface which enhances turbulent mixing and thus strong "coupling" between the UBL and surface fluxes. Aerodynamically rough surfaces, such as forest canopies and cities, are better coupled to the UBL because turbulence is enhanced. CLEUGH and OKE (1986) and GRIMMOND et al. (1993) suggest that this coupling enables synoptic control on turbulent flux partitioning which could explain their observed day-to-day variability of β. Smoother surfaces, such as pasture in flat terrain, are not so well coupled to the PBL so surface turbulent fluxes are primarily determined by moisture supply and available energy (i.e. the net radiation minus the heat storage).

Airflow

Modified surface heating and increased friction lead to perturbations in the mean wind velocity vectors and the turbulence within the UCL and UBL. This section describes the perturbations to mean flow in the UCL and UBL first, followed by a brief overview of changes in turbulence. The mean wind and its turbulent structure influence the sizes of the turbulent fluxes and the diffusion and dispersion of atmospheric pollutants released into the UBL.

Within the UCL, the presence of buildings dominates the pattern of flow. The nature of airflow around individual buildings has been well documented in the literature (e.g. MERONEY, 1982). The introduction of sharp-edged and inflexible buildings into a moving airstream decreases the velocity of the approach flow. The streamlines are displaced and converge above

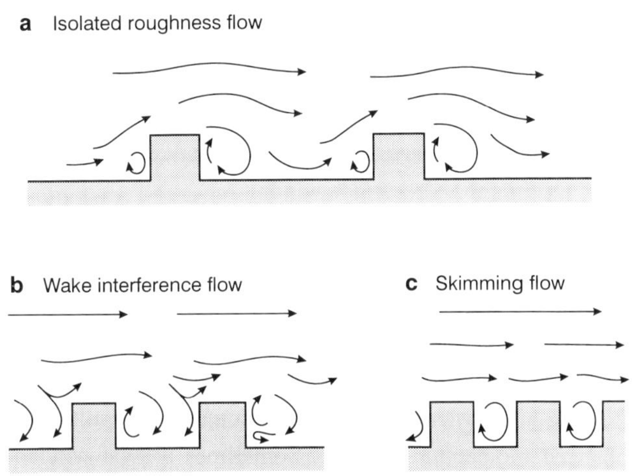

Fig. 8. Hypothesised effects of building spacing on airflow. The airflow is from left to right across the page and the urban canyons are assumed to be infinitely long in the cross wind direction (after OKE, 1987).

the building, yielding a jet of high velocity air above and a recirculating eddy located to the lee of the building. Airflow approaching from other wind directions will complicate this basic flow regime.

The effect of an agglomeration of buildings will depend on their orientation and spacing. In the UCL, the important scaling variables for flow are thus the aspect ratio, the ratio of canyon length to height and building orientation to the airflow. The overall morphology of the city, in terms of the building height, length and spacing, determines the airflow patterns in the UCL plus its effectiveness as a momentum sink. These also affect the interaction between the UCL and the UBL and the microclimate within the UCL which is where people live and breathe. Fig. 8 depicts the types of flow and, by extrapolation, the microclimate which could be expected for a variety of building densities where the approach flow is oriented normal to the long axis of the urban canyon. The circulating vortex within the urban canyon exchanges scalars from within the canyon to the atmosphere above. Its strength depends on the speed and direction of the approach flow and the aspect ratio of the urban canyon. Note that airflow oriented along canyons is usually accelerated, with frictional slowing and uplift at the canyon walls.

There are three mechanisms which alter mesoscale airflow in the UBL (AUER, 1981). Firstly, the changed density and characteristics of the roughness elements alter the friction of the urban surface and its effectiveness as a momentum sink. These surface modifications lead to, secondly, variations in the vertical transport of momentum via thermal and mechanical mixing. Lastly, the altered thermal state of the UBL leads to pressure and stability changes. Which of these effects dominates depends on the regional flow regime, atmospheric stability and the scale of interest. Calm conditions and/or increased distance from the aerodynamically-rough urban surface favour the influence of buoyancy, free convection and thus thermal effects. Closer proximity to the rough urban surface and/or stronger airflow allow frictional influences to dominate.

When winds are light, the urban heat island produces a low pressure area at ground level and hence convergent flow (SHREFFLER, 1979). This is accompanied by uplift in the UBL ban and desert sites were significant; the authors suggest that this could be the result of a growing urban heat island. In comparison to other arid cities, Kuwait City's heat island is very small. The authors argue that this is a result of the lack of trees; the similarity of thermal properties of construction materials between the city buildings and the desert; lower building heights; and the moderating influence of the Arabian Gulf.

These studies show that urban heat islands exist in tropical cities, but they are less pronounced and the timing of the peak heat island strength is altered compared to mid-latitude cities. Heat island strengths and occurrence are greater in the dry season.

Changes to surface energy partitioning in urban land use should be reflected in surface temperatures. For this reason, plus the wealth of spatial information provided, several studies (e.g. PRICE, 1979; VUKOVICH, 1983; ROTH et al., 1989b) have investigated the spatial expression of the urban heat island using surface temperatures measured from satellites. GALLO et al. (1993), for example, found a linear relationship between the normalised difference vegetation index (NDVI) and observed urban-rural minimum air temperature differences for 37 cities in the USA.

Satellite-derived temperatures, however, do not always reveal the same spatial and temporal patterns as air temperature surveys. Several common features found in satellite studies were summarised, and confirmed, by ROTH et al. (1989b). They noted, firstly, that heat island intensities were largest by day and smallest by night, almost the reverse of the diurnal cycle of the UCL air temperature heat island. Secondly, the spatial temperature structure was closely linked to the land use patterns (i.e. features such as parks, shopping centres) during the day, but not at night. ROTH et al. found that the warmest urban surfaces (by day) were not in the city core but in areas with large flat-topped buildings or extensive pavement. Some of these discrepancies may simply be linked to the difference between a radiative surface temperature and near-surface air temperatures where the latter depend on the surface energy balance plus heating and cooling in an air volume. Satellites may also preferentially view horizontal surfaces such as pavements, rooftops and treetops. The amount of the three dimensional active urban surface "viewed" or sampled by a satellite sensor then depends on its view angle and the ratio of active to plan area of the urban landscape being sampled. GOWARD (1981) showed that rooftops have much lower thermal admittance values and thus will show a greater surface temperature amplitude which may explain the satellite thermal patterns. The discrepancy between the UCL air temperature heat island and that viewed from a satellite simply illustrates our limited knowledge of urban processes. But it also means that satellite-based observations of urban climate processes must be interpreted with care.

Humidity

Much less is known about urban effects on humidity. Humidity is expected to differ between urban and rural locations because of the following factors:

- smaller input of vapour as a result of a reduction in vegetation in the UCL and hence smaller transpiration rates (but note that the reduced atmospheric stability at night may allow transpiration and evaporation to continue into the night);
- emissions of water vapour by industry and transport;
- intercepted water on impervious surfaces may lead to enhanced evaporation following rainfall;

- anthropogenic control of water application;
- juxtaposition of wet and dry surfaces means microscale advection can enhance evaporation.

Like the thermal anomaly, urban humidity effects need to be separated into the UCL and UBL effects. Cities are often asserted to have lower relative humidities because of the misconception that urban areas are devoid of water. Most measurements of relative humidity in the UCL support these assertions but only because the relative humidity index depends both on atmospheric water vapour content and temperature. In fact, relative humidity is more sensitive to temperature changes than to humidity changes, so lower relative humidities often simply reflect the extra warmth of cities.

Studies comparing actual humidity show that cities are often drier than their surroundings but there is a diurnal and seasonal dependence. HAGE's (1975) study found, in summer, that urban areas were drier than rural by day and more moist by night. Seasonally, HAGE found that the vapour density was always greater in winter. Similar results (see OKE, 1979a) have been reported from other cities, e.g. in Chicago, ACKERMAN (1987) found lower vapour pressures and dew point temperatures in the forenoon and on spring afternoons. Urban-rural humidity differences, like temperatures, were found to be very sensitive to wind speed and cloud cover. For tropical Ibadan, ADEBAYO (1991) observed slightly reduced vapour pressures, especially in the afternoon, based on 2 years of data.

The UBL climate

These thermal and moisture anomalies are not restricted to the UCL. The UCL warm, dry "island" extends upwards into the UBL and can be advected downwind. The temperature and humidity in the UBL also result from entrainment, advective influences and the effects of radiative heating and cooling. Measurements conducted in St. Louis in summer daytime conditions also showed a thermal excess and a specific humidity deficit in the UBL temperature profiles (ACKERMAN and MANSELL, 1978; TAPPER, 1990) which were displaced downwind. These thermodynamic anomalies varied inversely with wind speed (SHEA and AUER, 1978) and extended to heights of 500–1000 m (AUER, 1981); they also resulted in higher convective condensation levels. The deeper UBL slightly dilutes and reduces the heating impact. These observations were in summertime, anticyclonic conditions. In wintertime, or nocturnally, the UBL can be more moist.

The drier UBL is due to both a decreased surface evaporation flux and enhanced entrainment of dry air from the stable inversion layer above the UBL, as demonstrated by HILDEBRAND and ACKERMAN (1984). They found that Q_H decreased with height in both urban and rural boundary-layers, but urban Q_H fluxes were always larger. Note that these large fluxes, accompanied by large vertical velocities in the UBL, enhance the entrainment process. Their urban/rural comparison found large rural Q_E fluxes at the surface which decreased with height in contrast to the large Q_E flux at the top of the UBL. Evaporation in the UBL thus increases with height and leads to a drying of the UBL.

Precipitation

Two of the basic ingredients for raindrop formation; condensation nuclei and strong up-whose timing and magnitude are unpredictable. A study by BROECKER and DENTON (1990) indicates that massive reorganizations of the ocean-atmosphere system are key events that link cyclic changes in the Earth's orbit to the advance and retreat of ice sheets. What are the relationships between climate change and the mode of operation for ocean circulation systems? What sorts of surprises might be in store for us if the greenhouse warming is effective in changing the mode of operation of the ocean-atmosphere system?

In this chapter, we review lessons learned from the study of climatic changes in the past, the corresponding large-scale reorganization of the mode of operation for the ocean-atmosphere system, and the possible linkage between climate change and ocean circulation. Understanding the mechanism of global ocean circulation is crucial in seeking such linkages. The great ocean conveyor circulation of BROECKER (1991) is reviewed for this purpose. Based on analysis of air bubbles in polar ice cores, the CO_2 content in the atmosphere varied significantly with climate changes over the last 100,000 years. We summarize in this chapter the history of atmospheric CO_2 variations and review possible causes of atmospheric CO_2 variations in relation to the last glacial-to-interglacial climate change.

Changes in climate and ocean circulation in the past

The cyclic climatic changes between warm and cold periods are recorded by oxygen isotope ($\delta^{18}O$) in foraminifera shells deposited in the deep sea sediments. Fig. 1 shows the marine $\delta^{18}O$ record obtained from the benthic foraminifera from a deep sea core raised from the east equatorial Pacific (SHACKLETON et al., 1983). Sharp changes in $\delta^{18}O$ 128,000 and 12,000 years ago denote the end of the glacial period (Terminations II and I). Slower

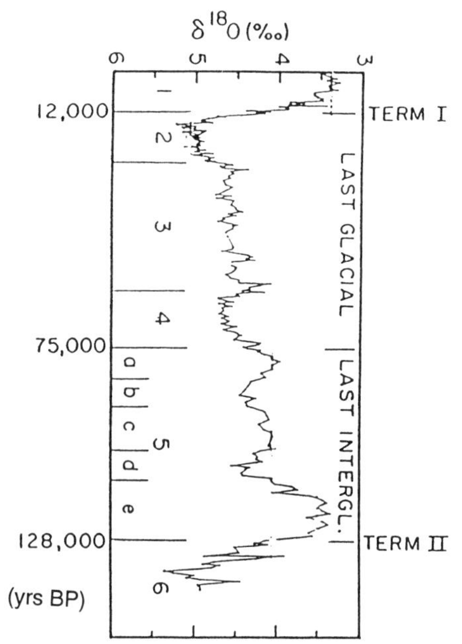

Fig. 1. Oxygen isotope record of deep sea core V19–30 from equatorial East Pacific (SHACKLETON et al., 1983). The marine isotope stages of Emiliani and the terminations are shown. The generally accepted ages for the major climate changes are also given.

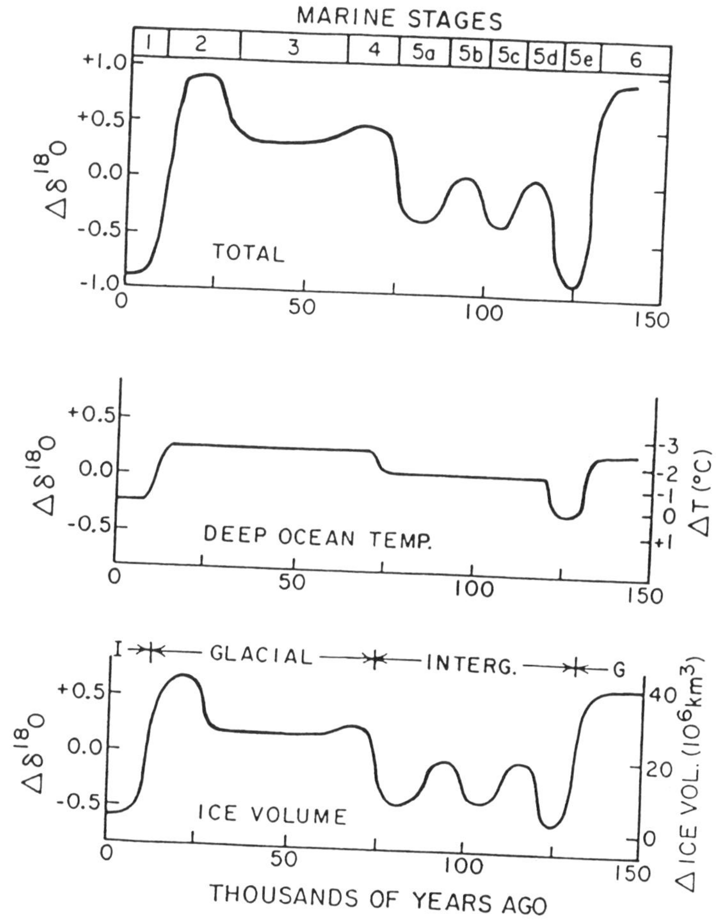

Fig. 2. Generalized total change in oxygen isotope ($\delta^{18}O$) for benthic foraminifera in the deep Pacific Ocean is given in the upper panel. The portion of $\delta^{18}O$ signal corresponding to temperature change of deep sea water is shown in the middle panel. The lower panel shows that portion of $\delta^{18}O$ change equivalent to ice volume (from BROECKER, 1993).

changes in between these two terminations represent the cooling of the Earth system until the peak of the last glacial time. The oxygen isotope ratio changes with both ice volume and deep water temperature. The increase in ice volume enriches the heavier oxygen isotope in sea water because the isotope fractionation during the evaporation of sea water preferentially sends the lighter oxygen isotope into the gas phase. This water vapour eventually precipitates from the air as snow and then becomes ice in the high latitude regions. The deep water temperature where benthic foraminifera lived also affects the oxygen isotope composition of carbonate shell, with a general lowering of $\delta^{18}O$ corresponding to warmer temperature. As shown in Fig. 2, the record given in Fig. 1 can be separated into two records related to these two factors (BROECKER, 1993). The total change in oxygen isotope ($\Delta\delta^{18}O$) is shown as a generalized smooth curve in the upper panel. The stratigraphic nomenclature used by quaternary geologists is also shown as marine stages on the top of the figure. The present interglacial (i.e. Holocene) is designated as marine stage 1. The last interval of glaciation is subdivided into three stages (2, 3 and 4). The entire last interglacial is designated as stage 5, which has subsequently been divided into five parts. The warm maxima are stages 5e, 5c and 5a, and the intervening intervals of cooler climate are 5d and 5b. The pe-

period. Although we do not have hard evidence to substantiate the fact that lower atmospheric CO_2 leads to colder climate and higher CO_2 leads to warmer climate, the strong correlation in ice core records between low atmospheric CO_2 concentration and cold periods and between high atmospheric CO_2 and warm periods suggest that the atmospheric CO_2 content must have played an important role in the positive feedback of climate changes.

As mentioned in the last section, the increase in atmospheric CO_2 content from glacial to interglacial time has been determined to be about 80 ppm (BARNOLA et al., 1987; NEFTEL et al., 1988). The hard question is what caused the glacial to interglacial CO_2 change? There are a number of possible explanations, but no one of them by itself can fully explain both CO_2 record in ice cores and carbon isotope record preserved in deep sea sediment. The answer must be "We do not yet know". In chapter 12 of this volume, the land-use change has been described as a possible factor affecting climate change. In fact, the amount of carbon residing in living wood and in soil humus at present totalled about four times that in the atmosphere as CO_2 (BROECKER and PENG, 1993). Hence, glacial to interglacial changes in the size of the organic carbon reservoir in land biosphere are potential causes for CO_2 changes. However, according to the analysis of BROECKER and PENG (1993) based on ^{13}C measurements of benthic and planktonic foraminifera from deep sea sediment, the ocean-atmosphere system had during glacial time 1.4% higher carbon content than interglacial. This is consistent with the fact that most current boreal forests were under the ice sheet during the glacial time. Extra amounts of carbon were transferred from land biosphere to the ocean-atmosphere system resulting from massive forest removal. This excess carbon would create a 25 ppm higher glacial atmospheric CO_2 content when it is in equilibrium with carbon and carbonate in the ocean. This result is opposite to observations made in ice cores.

The ocean contains about 60 times more carbon than the atmosphere. After taking into account the buffer capacity of sea water, the effective amount of carbon in the ocean, which would respond to changes in atmospheric CO_2 concentration, is still about six times the amount in the atmosphere (assuming an average buffer factor of 10). Hence, the atmospheric CO_2 is controlled to a large extent by the carbon cycle in the ocean. The three most obvious factors that affected the partial pressure of CO_2 in ocean waters as global climate changed from glacial to Holocene were temperature, salinity and biological pumping.

For each 1°C that the surface ocean is warmed, the partial pressure of CO_2 is increased by about 4.3%. An increase in the glacial CO_2 content of 200 ppm to the interglacial value of 280 ppm would require a 10°C warming. On the basis of the faunal record in deep sea sediments (CLIMAP, 1981) and oxygen isotope ratios from foraminifera (SHACKLETON, 1987; LABEYRIE et al., 1987), the average glacial to interglacial sea surface temperature warming was estimated to be about 2°C. In the absence of any other change, a 2°C warming would have introduced a 17 ppm increase in the atmospheric CO_2 content (BROECKER and PENG, 1986).

Because more fresh water was stored in continental ice caps during glacial time, the sea must have been saltier than it is now. If one assumes a 150-m lowering of sea level, one can estimate the volume of excess continental ice to have been 55×10^6 km^3. Hence, for an ocean volume of 1375×10^6 km^3, the salinity of sea water was about 4% higher during glacial time than it is now. This salinity effect applies to oceanic alkalinity and total CO_2 concentration, as well as to concentrations of oceanic major ions (such as Na^+ and Cl^-). The salinity decrease at the close of glacial time would have lowered the atmospheric CO_2 level by about 11 ppm (BROECKER and PENG, 1986). Thus, any salinity-induced CO_2 decrease would cancel out most of the temperature-induced CO_2 increase.

The biological pump removes surface water CO_2 in the form of organic matter (produced by photosynthesis) to the interior, where the organic matter is oxidized and returned to inorganic form. In so doing, the pCO_2 in surface waters is less than that in the interior. The efficiency of carbon removal is limited by the nutrient content of surface waters. If all the nutrients were completely extracted from surface waters by photosynthesis, the atmospheric CO_2 content would be about 150 ppm. On the other hand, if the ocean were without biology, the atmospheric CO_2 would be about 470 ppm. The pre-industrial CO_2 content of 280 ppm implies that the nutrients NO_3 and PO_4 were not efficiently utilized in surface waters. The lower CO_2 content of 200 ppm in the glacial period could be caused by a more efficient biological pump. In fact, a polar nutrient hypothesis was proposed simultaneously by three research groups (KNOX and MCELROY, 1984; SARMIENTO and TOGGWEILER, 1984; SIEGENTHALER and WENK, 1984), who showed that the polar seas have an influence on the atmospheric CO_2 content far beyond that expected based on their small area. Because warm surface ocean waters are currently free of the nutrients NO_3 and PO_4, the increase in the biological pumping capacity needed to draw down the atmospheric CO_2 content could not be attributed to enhanced plant productivity in warm waters. By contrast, because polar surface waters are currently rich in these nutrient constituents, the potential exists for stripping them through enhanced polar biological activity and thereby drawing down the CO_2 content of the atmosphere and also the CO_2 concentration of the warm surface ocean. Unfortunately, specific predictions made by this hypothesis were not supported by the marine sediment record. For example, the change in PO_4 and NO_3 in polar surface oceans should be recorded in the shells of planktonic foraminifera grown in polar surface waters. BOYLE (1988a,b) found no significant difference between the glacial and the interglacial Cd/Ca ratio in deep sea cores (CO_2 and PO_4 are assumed linearly related in the ocean at all times (BOYLE, 1986)) either from the northern Atlantic or from the Antarctic. The polar nutrient hypothesis also predicted an increase in the $^{13}C/^{12}C$ ratios for polar surface waters during glacial time. Instead, the foraminifera record shows a decrease (LABEYRIE and DUPLESSY, 1985; CHARLES and FAIRBANKS, 1990). Finally, any lowering of atmospheric CO_2 content by the biological pumping effect must be accompanied by a significant reduction in O_2 in the deep ocean. The absence of anaerobic sediments in the glacial record suggests that the rather large O_2 reductions called for by the polar nutrient hypothesis did not occur.

On the basis of Cd measurements on benthic foraminifera, BOYLE (1988b) discovered that the maximum nutrient level in the water column shifted from the current shallow depth to a greater depth in the glacial Atlantic. He showed that this downshift, when coupled with a process of returning to a balanced oceanic $CaCO_3$ budget (i.e. $CaCO_3$ compensation), would lead to an increase in the alkalinity of the entire ocean. This alkalinity increase would in turn decrease the atmospheric CO_2. Under the assumption that the downward shift in the Atlantic nutrients was an ocean wide phenomenon, Boyle's calculation could account for a drop of ~40 ppm in atmospheric CO_2 content during glacial time. Besides being too small to explain the entire atmospheric CO_2 change, the $CaCO_3$ compensation required by this mechanism has a time constant of several thousand years, making this mechanism too slow to account for the rapidity of the CO_2 changes reflected in the Greenland ice core record.

Following Boyle's lead in suggesting that an alkalinity change may have driven the glacial to interglacial change in atmospheric CO_2, BROECKER and PENG (1989) proposed a polar alkalinity hypothesis. This hypothesis builds on the basic concept of the polar nutrient hypothesis, namely that the partial pressure of CO_2 in polar surface waters, especially in the Antarctic ocean, controls the CO_2 partial pressure of both the warm surface ocean and atmosphere. It also requires an increase in the alkalinity of polar surface waters, such as that produced by Boyle's nutrient-deepening hypothesis. Four constraints were obtained for reconstructing the composition of waters in the glacial Antarctic ocean. First, the pCO_2 of Antarctic surface waters was 200 μatm instead of 280 μatm, according to the ice core data and in line with the fact that Antarctic surface waters dictate the atmospheric CO_2 concentration. Second, the NO_3 and PO_4 contents of the Antarctic surface waters remained the same as they are today, according to Boyle's Cd results. Third, in accord with the near constancy of the oceanic lysocline (defined as the depth of ocean at which the rate of $CaCO_3$ dissolution increases significantly), deep waters in the Antarctic, Pacific and Indian oceans had nearly the same CO_3^{2-} ion content as they have today. Fourth, in accord with $^{13}C/^{12}C$ data on benthic forams, the salinity-normalized ΣCO_2 of glacial deep Antarctic water was nearly the same as that in glacial deep Pacific water. The reconstructed composition of glacial Antarctic waters had the salinity-normalized alkalinity of ~75 μeq kg^{-1} higher and the salinity-normalized ΣCO_2 in deep water of ~100 μmol kg^{-1} higher than the interglacial time. The alkalinity change is responsible for an 80 μatm change in atmospheric CO_2 partial pressure between the glacial time and the Holocene. Unfortunately, this hypothesis also cannot explain the large drop in the $\delta^{13}C$ value for the planktonic foraminifera from Antarctic sediments of glacial age.

Part of the alkalinity increases that the polar alkalinity hypothesis requires for glacial surface water in the Antarctic could be a natural consequence of the demise of the NADW during glacial time. The cessation of NADW stopped the flow of low-alkalinity deep water into the circumpolar region. Because little $CaCO_3$ is produced by organisms living in Antarctic surface waters, the alkalinity rise experienced by circumpolar deep water must also have occurred in Antarctic surface water. Coupled with an increase in the degree of utilization of the nutrients brought to the surface by the upwelling in the Antarctic, this alkalinity rise brought about a drop in the pCO_2 of surface waters in the Antarctic and hence in the warm surface ocean and atmosphere. What are mechanisms in the ocean-atmosphere system that can cause the turning on and shutting off of the NADW? BROECKER'S (1991) ocean conveyor circulation theory offers some possible answers to this question.

Role of ocean conveyor circulation

The most important feature of the ocean conveyor circulation is the production of deep water in the northern Atlantic. Waters in the vicinity of Iceland are cooled and their density is increased to the point that they sink into the abyss. As shown in Fig. 8, the newly formed deep water, known as NADW, flows southward from Iceland to the tip of Africa with its characteristic property of high salinity, low nutrient content and high $^{14}C/^{12}C$ ratio. This flow forms the lower limb of the conveyor. In addition, there is the northward flowing wedge of Antarctic Bottom Water (AABW) from southern Atlantic that underrides the

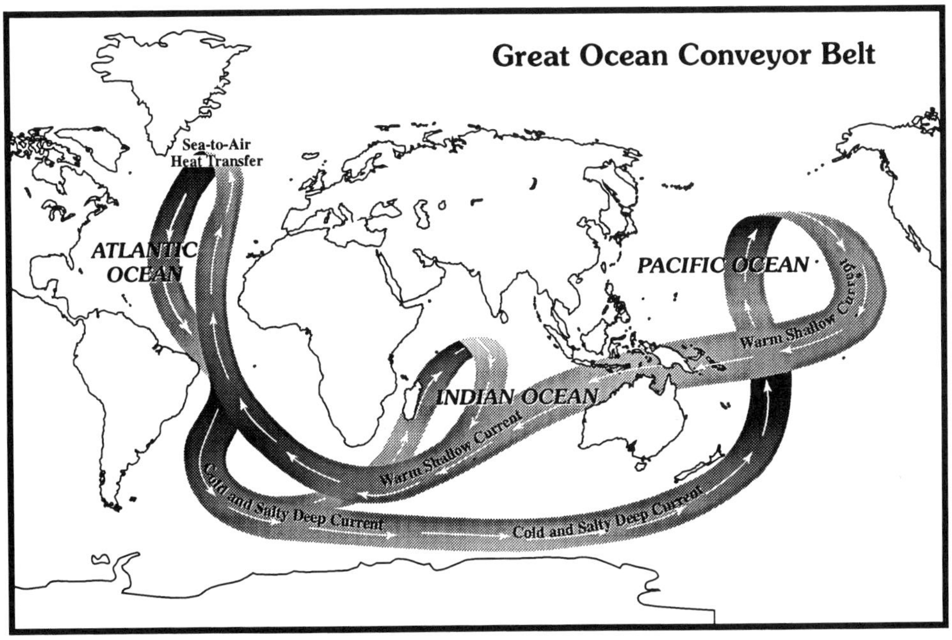

Fig. 8. The diagram of the great ocean conveyor (BROECKER, 1991).

NADW mass and is mixed upward into NADW, increasing the flow of the lower limb. Southward of 30°S, this lower limb joins the rapidly flowing deep Antarctic circumpolar current, which mixes the NADW exiting the Atlantic with new deep water produced around the Antarctic continent and also with old deep waters recirculated back into the Antarctic from the deep Pacific and Indian oceans. This current is the great mix-master of the world ocean. Some of this lower limb water branches northward into deep Indian and Pacific oceans. The upwelling of these lower branches in the northern Indian and Pacific oceans forms the upper limbs of the conveyor, which returns along the upper ocean through the Indonesian archipelago and around the tip of Africa to the Atlantic Ocean. However, in reality, the upwelling is widely spread, with a large amount taking place in the Antarctic. Hence, there is a significant return flow through the Antarctic via the Drake Passage into the South Atlantic, which is not portrayed in the conveyor diagram shown in Fig. 8.

The flux of NADW into the deep Atlantic can be estimated by measuring the radiocarbon in samples of deep water from the Atlantic Ocean. The radiocarbon measurements give the mean residence time of the deep water. The flux is then estimated by dividing the volume of water contained in the deep Atlantic with its mean residence time. Based on radiocarbon measurements made during the GEOSECS programme, corrected for the contribution of radiocarbon made by AABW to the conveyor's lower limb and adjusted for the impact of temporal changes in the $^{14}C/^{12}C$ ratio for atmospheric CO_2, BROECKER (1991) estimated that the current flux of NADW is close to 20 Sv (1 Sv = 1 $\times$ 10^6 m^3 of water per second). This immense magnitude of flux is 20 times the combined flows of all the world's rivers and is even larger than the combined rainfall for the entire globe.

BROECKER (1991) suggested that the conveyor is propelled by the excess salt left behind in the Atlantic as the result of vapour export (BROECKER et al., 1985). Because of the difference in the circulation patterns, surface waters in the North Atlantic are on average warmer than those in the North Pacific. As a result, more water is evaporated from the North Atlantic than from the North Pacific and in turn there is a net transport of water vapour through the atmosphere from the Atlantic to the Pacific. Therefore, the North Atlantic water becomes saltier than the North Pacific water. The salinity difference between surface waters in the northern Atlantic and those at comparable latitudes in the Pacific ranges from 2 to 3 g/l. The enrichment of salt in the North Atlantic must somehow be compensated by a flow of salt through the sea from the Atlantic to the Pacific. Formation of NADW by the sinking of dense surface waters as a result of winter cooling of saltier water in the North Atlantic initiates this compensation process. Broecker estimated that the rate of vapour loss from the Atlantic basin is 0.35 ± 0.12 Sv. Adopting a 20 Sv NADW outflowing flux and taking the salinity of this lower limb water mass to be ~34.9‰, the combined return flow must have a salinity of ~34.3‰ (20 × 34.9/20.35). As can be seen, the salinity contrast between seawater (34.9‰) and water vapour (0.0‰) is ~60 times that of the waters being traded between the Atlantic and the remainder of the ocean (i.e. 34.9–34.3 or 0.6‰). It is for this reason that only a 0.35 Sv vapour loss can force a mighty 20 Sv deep ocean flow.

The important role of conveyor circulation in climate change is the transport of heat from lower latitudes to higher latitudes in the northern Atlantic basin. The warmer surface return flow from lower latitudes has to be cooled and its density has to increase to a point at which it can sink and become the conveyor's lower limb. The amount of heat released into the atmosphere is equal to the product of the conveyor's flux and temperature reductions required for making NADW. The water temperature of the conveyor's upper limb averages ~10°C, and that of NADW averages ~3°C. Thus, each cubic centimetre of upper-limb water releases 29.3 Joules to the atmosphere when it becomes the lower-limb deep water. With a 20 Sv deep flow, the formation of NADW gives off 16.7×10^{21} J each year, which is equal to 35% of the heat received from the sun by the Atlantic north of 40°N latitude. This extra heat is responsible for Europe's exceptionally mild winters. The immediate question is: what would happen if the conveyor ceased to operate? Model simulations (Rind et al., 1986; Manabe and Stauffer, 1988) indicate that a very different temperature distribution pattern would be created and thus a different climate system would result. The temperature of surface waters in the northern Atlantic would become ~5°C colder if the conveyor were inoperative.

One potential threat to the operation of conveyor circulation is the addition of fresh water to the northern Atlantic. Dilution of seawater may potentially weaken the conveyor because of its negative effect on the production of NADW. Continuous dilution of seawater would eventually halt the formation of NADW and, thus, would shut down the conveyor. If this happened, fresh water would pool at the surface of the northern Atlantic and create a severe barrier to deep-water formation. This phenomenon is currently seen in the northern Pacific. Using an oceanic GCM, Maier-Reimer and Mikolajewicz (1989) demonstrate that addition of enough excess fresh water to the source region of NADW can kill the model's thermohaline circulation abruptly, on a time scale of about a few decades.

Broecker (1991) argued that changes in thermohaline circulation in the Atlantic ocean were the main causes for the abrupt and large climatic changes experienced by the northern Atlantic basin during the last glacial period. His conclusion is based on the observation of the oxygen isotope record from Greenland ice cores (Dansgaard et al., 1971; Hammer et al., 1985), which reflects changes in air temperature. The turning on and off of the conveyor could cause 5–8°C changes in air temperature over Greenland and Europe. As shown in Fig. 6, Greenland air temperature during glacial time varied with magnitude and abruptness expected if the conveyor were to turn on and off on a millennia time scale. BROECKER et al. (1990) show that when the northern end of the Atlantic basin was surrounded by ice sheets during glacial time, stable operation of the Atlantic circulation system was not possible because the ice sheets constitute a tremendous source of fresh water. The heat released during the operation of the conveyor tends to melt the ice, adding a large amount of fresh water to the ocean. In addition, salt is efficiently transported out of this region by the conveyor. Hence, fresh water dilution and salt export reduce the density of waters in the Atlantic. This process continues until the density is too low to produce NADW. Thus, the conveyor stops. Without the conveyor, the salt export and meltwater dilution are reduced to the point at which salt enrichment is once again possible by water vapour export. The increase in salt content leads to a rise in water density, which eventually creates an environment for the generation of NADW and turns on the conveyor. Apparently, this cycle repeats over and over again in glacial time. In contrast, during the nine or ten thousand years after the Younger Dryas cold period, the ever changing glacial conveyor became firmly locked in the 'on' position as demonstrated by nearly constant Greenland air temperature, as shown in the isotope record (Fig. 5).

To explain the temperature cycles characterized by abrupt warmings followed by gradual coolings as recorded in the Greenland ice core, BROECKER (1991) offers a salt oscillator hypothesis. The abrupt warmings result from initiation of the conveyor, which runs vigorously at the beginning. The reason for this vigour is that to overcome the freshwater pool present in the northern Atlantic when the conveyor is inoperative, the salinity of the Atlantic water would have to rise above the level required for steady-state operation of the conveyor. Thus, when the conveyor turns on, the buoyancy contrast between deep water produced in the northern Atlantic and that present in the remainder of the world ocean will be extraordinarily large. This excess density will drive the conveyor at an unusually high rate. As a consequence, a greater amount of heat will be released to the atmosphere over the North Atlantic. Hence, the atmosphere warms up abruptly in the beginning. However, once operative, the strength of conveyor will steadily weaken because the combination of dilution with meltwater and export of excess salt by the conveyor will reduce the buoyancy contrast between deep waters inside and outside of the Atlantic. As the strength of the conveyor gradually wanes, the amount of heat given off to the atmosphere over the northern Atlantic also will decrease, causing air temperature to drop accordingly. Eventually the conveyor will shut down and cut off the supply of heat from the ocean. The atmospheric temperature cycle generated in this way resembles that observed in the Greenland ice core record.

Future surprises

We have learned from the preceding discussion that the operation of conveyor circulation has played an important role in determining air temperature changes in the northern Atlantic basin. A shutdown of the Atlantic conveyor belt cooled the North Atlantic and its adjacent revealed by the Camp Century Ice Core. In: K. K. TUREKIAN (Editor), *The Late Cenozoic Glacial Ages*. Yale University Press, New Haven, CT, pp. 37–56.

DANSGAARD, W., WHITEAND, J. W. C. and JOHNSON, S. J., 1989. The abrupt termination of the Younger Dryas climate event. *Nature,* 339: 532–533.

DANSGAARD, W., JOHNSEN, S. J., CLAUSEN, H. B., DAHL-JENSEN, D., GUNDESTRUP, N. S., HAMMER, C. U., HVIDGERG, C. S., STEFFENSEN, J. P., SVEINBJORNSDOTTIR, A. E., JOUSEL, J. and BOND, G., 1993. Evidence for general instability of past climate from a 250-kyr ice-core record. *Nature,* 364: 218–220.

DELMAS, R. J., ASCENCIO, J.-M. and LEGRAND, M., 1980. Polar ice evidence that atmospheric CO_2 20,000 yr BP was 50% of present. *Nature*, 284: 155–157.

GRIP (GREENLAND ICE-CORE PROJECT) MEMBERS, 1993. Climate instability during the last interglacial period recorded in the GRIP ice core. *Nature*, 364: 203–207.

HAMMER, C. U., CLAUSEN, H. B., DANSGAARD, W., NEFTEL, A., KRISTINSDOTTIR, P. and JOHNSON, E., 1985. Continuous impurity analysis along the Dye 3 deep core. In. C. C. LANGWAY, H. OESCHGER and W. DANSGAARD (Editors), *Greenland Ice Core. Geophysics, Geochemistry and the Environment.* Am. Geophys. Union, Washington, DC, pp. 90–94.

JOHNSEN, S. J., CLAUSEN, H. B., DANSGAARD, W., FUHRER, K., GUNDESTRUP, N., HAMMER, C. U., IVERSEN, P., JOUSEL, J., STAUFFER, B. and STEFFENSEN, J. P., 1992. Irregular glacial interstadials recorded in a new Greenland ice core. *Nature*, 359: 311–313.

JOUZEL, J., LORIUS, C., PETIT, J. R., GENTHON, C., BARKOV, N. I., KOTLYAKOV, V. M. and PETROV, V. M., 1987. Vostok ice core: a continuous isotope temperature record over the last climatic cycle (160,000 years). *Nature,* 329: 403–408.

JOUZEL, J., BARKOV, N. I., BARNOLA, J. M., BENDER, M., CHAPPELLAZ, J., GENTHON, C., KOTLYAKOV, V. M., LIPENKOV, V., LORIUS, C., PETIT, J. R., RAYNAUD, D., RAISBECK, G., RITZ, C., SOWERS, T., STIEVENARD, M., YIOU, F. and YIOU, P., 1993. Extending the Vostok ice-core record of palaeoclimate to the penultimate glacial period. *Nature,* 364: 407–412.

KEELING, C. D., BACASTOW, R. B., CARTER, A. F., PIPER, S. C., WHORF, T. P., HEIMANN, M., MOOK, W. G. and ROELOFFZEN, H., 1989. A three-dimensional model of atmospheric CO_2 transport based on observed winds. 1. Analysis of observational data. *Geophys. Mono.* Am. Geophys. Union, Washington, DC, pp. 165–231.

KNOX, F. and MCELROY, M., 1984. Changes in atmospheric CO_2. influence of marine biota at high latitudes. *J. Geophys. Res.*, 89: 4629–4637.

LABEYRIE, L. D. and DUPLESSY, J. C., 1985. Changes in oceanic $^{13}C/^{12}C$ ratio during the last 140,000 years. high latitude surface water records. *Paleogeogr., Paleoclimatol., Paleoecol.*, 50: 217–240.

LABEYRIE, L. D., DUPLESSY, J. C. and BLANC, P. L., 1987. Variations in mode of formation and temperature of oceanic deep waters over the past 125,000 years. *Nature*, 327: 477–482.

LEHMAN, S. J. and KEIGWIN, L. D., 1992. Sudden changes in North Atlantic circulation during the last deglaciation. *Nature*, 356: 757–762.

MAIER-REIMER, E. and MIKOLAJEWICZ, U., 1989. Experiments with an OGCM on the cause of the Younger Dryas. In: A. AYALA-CASTANARES, W. WOOSTER and A. YANEZ-ARANCIBIA (Editors), *Oceanography.* UNAM Press, Mexico, pp. 87–100.

MANABE, S. and STAUFFER, R. J., 1988. Two stable equilibria of a coupled ocean-atmosphere model, *J. Climate*, 1: 841–866.

MIX, A. C. and FAIRBANKS, R. G., 1985. North Atlantic surface-ocean control of Pleistocene deep-ocean circulation. *Earth Planet. Sci. Lett.,* 73: 231–243.

NEFTEL, A., OESCHGER, H., SCHWANDER, J., STAUFFER, B. and ZUMBRUNN, R., 1982. Ice core sample measurements give atmospheric CO_2 content during the past 40,000 yr. *Nature*, 295: 220–223.

NEFTEL, A., MOOR, E., OESCHGER, H. and STAUFFER, B., 1985. Evidence from polar ice cores for the increase in atmospheric CO_2 in the past two centuries. *Nature*, 315: 45–47.

NEFTEL, A., OESCHGER, H., STAFFELBACH, T. and STAUFFER, B., 1988. CO_2 record in the Byrd ice core 50,000–5,000 years BP. *Nature*, 331: 609–611.

PEARMAN, G. I., ETHERIDGE, D., DE SILVA, F. and FRASER, P. J., 1986. Evidence of changing concentrations of atmospheric CO_2, N_2O and CH_4 from air bubbles in Antarctic ice. *Nature*, 320: 248–250.

RAYNAUD, D. and BARNOLA, J. M., 1985. An Antarctic ice core reveals atmospheric CO_2 variations over the past few centuries. *Nature*, 315: 309–311.

RIND, D., PETEET, D., BROECKER, W. S., MCINTYRE, A. and RUDDIMAN, W., 1986. The impact of cold North Atlantic sea surface temperatures on climate. Implications for the Younger Dryas cooling (11–10k). *Climatol. Dynam.*, 1: 3–33.

SARMIENTO, J. L. and TOGGWEILER, R., 1984. A new model for the role of the oceans in determining atmospheric pCO_2 . *Nature*, 308: 621–624.

SCHLOSSER, P., BONISCH, G., RHEIN, M. and BAYER, R., 1991. Reduction of deep water formation in the Greenland Sea during the 1980s. Evidence from tracer data. *Science*, 251: 1054–1056.

SHACKLETON, N. J., 1987. Oxygen isotopes, ice volume and sea level. *Quaternary Sci. Rev.*, 6: 183–190.

SHACKLETON, N. J., IMBRIE, J. and HALL, M. A., 1983. Oxygen and carbon isotope record of East Pacific core V19–30: implications for the formation of deep water in the late Pleistocene North Atlantic. *Earth Planet. Sci. Lett.*, 65: 233–244.

SIEGENTHALER, U. and WENK, T., 1984. Rapid atmospheric CO_2 variations and ocean circulation. *Nature*, 308: 624–626.

STAUFFER, B., HOFER, H., OESCHGER, H., SCHWANDER, J. and SIEGENTHALER, U., 1984. Atmospheric CO_2 concentrations during the last glaciation. *Ann. Glaciol.*, 5: 160–164.

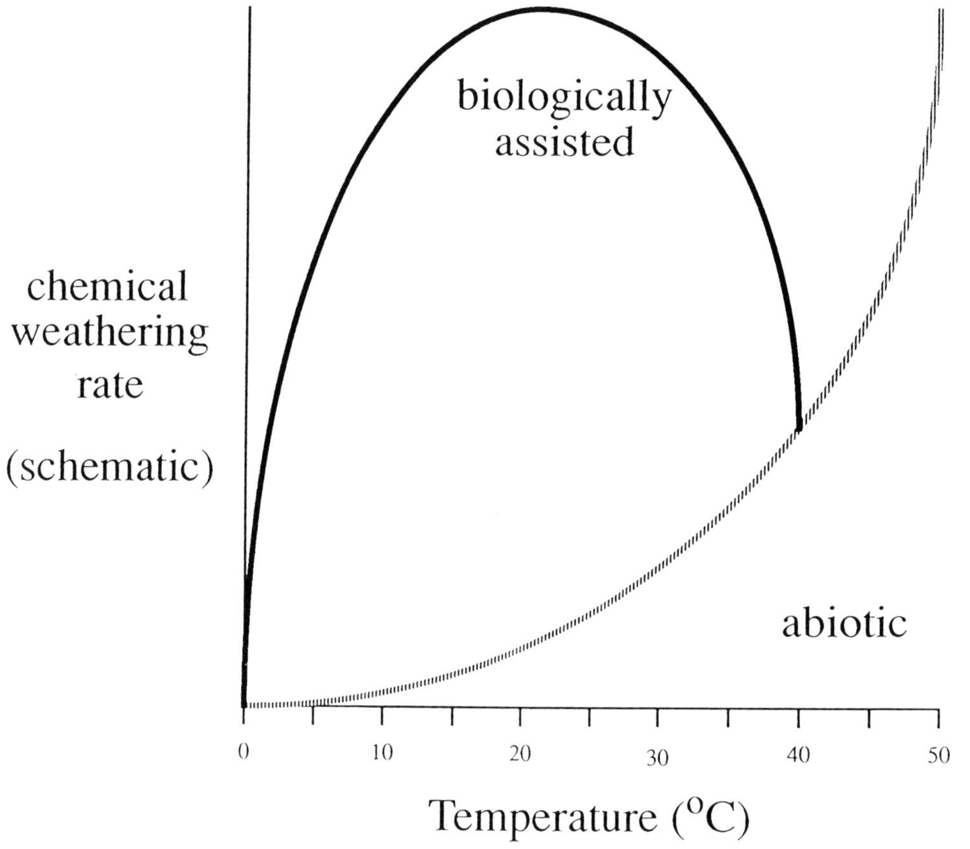

Fig. 4. Idealization of the effect of temperature changes on rates of chemical weathering. Solid line: Biologically assisted chemical weathering rate, displaying the characteristic parabolic response of living systems to temperature change. The biota generate high soil pCO_2 values and release organic acids to the soil, both of which presumably increase rates of chemical weathering. The biota also play an integral role in the development of soil structure, including its ability to retain water. Dashed line: Weathering rates on an abiotic Earth. The diagram is meant to include both the direct effect of temperature on mineral dissolution rates and the indirect effect of temperature on the intensity of the hydrologic cycle. The magnitude of the biological effect is shown only schematically, but likely ranges from a factor of 10 (BERNER, 1993) to 1,000 (SCHWARTZMANN and VOLK, 1989).

metastable; the regional climate changes associated with deforestation (temperature increase, soil moisture decrease) may prevent the re-establishment of the rain-forest ecosystem.

Thus the temperature dependency of silicate weathering rates might look something like Fig. 4. Within the range of organism tolerance, biotically enhanced weathering rates are greater than abiotic rates, and the planet is cooled. However, when the system enters positive feedback (above 25°C) the regulatory mechanism switches to the abiotic one, which demands substantially warmer temperatures to sustain a given weathering rate. In the next section we discuss the implications of this hypothesis for considerations of future climates.

Geophysiology of future climates

As in medicine, predictions about the future from geophysiology must be couched in terms of substantial uncertainty. Nevertheless, the physician who has studied the life history of the patient, looking for similar illnesses and reactions in the past, has a basis for making a prognosis. So too does the geophysiologist who has studied the operation of the climate system through Earth history.

Perhaps the most important question concerns the consequences of continued and accelerated consumption of fossil fuels and release of CO_2 to the atmosphere. HOFFERT and COVEY (1992) have studied the correlation through Earth history between proxy indicators of paleo-pCO_2 and climate, and have concluded that a good correlation exists. Warm climates typically coincide with high pCO_2 values. Current estimates are that the amount of carbon dioxide stored in fossil fuels equals about 4200 Gtonnes (IPCC, 1991), or about seven times the pre-industrial atmospheric CO_2 content.

WALKER and KASTING (1992), using a simple box model of the global carbon cycle, predicted that even with very conservative fossil fuel burning scenarios, the result in all cases will be the same: eventual, complete combustion of the recoverable fossil fuels will lead to an atmospheric CO_2 level some six to seven times higher than the pre-industrial value when combustion ceases. The recovery then will take thousands of years. Even after 10,000 years atmospheric pCO_2 might be as high as 600 ppm, more than double the pre-industrial level. The onset of the next glacial period may well be delayed, if not altogether prevented.

From proxy indicators of ancient CO_2 levels (e.g. FREEMAN and HAYES, 1992) it appears that the Earth has not experienced such high levels of atmospheric carbon dioxide (7× the present level) for 100 million years, a time when the ice-caps were small, if present, globally averaged temperatures were some 10°C greater and sea level was considerably higher (BARRON and WASHINGTON, 1985; Chapter 3 by BARRON). By analogy, the persistence of the fossil fuel CO_2 pulse, for time-scales that approach the response time of continental ice sheets, should cause considerable warming and glacial melting. In the transition to this warmer state, sea-ice melting in the North Atlantic might cause temporary shutdowns of North Atlantic Deep Water formation, and thus brief returns to cooler conditions (see Chapter 14 by PENG).

Global warming is likely to reduce, at least temporarily, marine algal productivity and thus CCN production, by expanding the warm-water sphere at the expense of the cold-water sphere where vertical mixing is vigorous. This stage will be temporary. Because reduced productivity will lead to reduced sedimentation of the nutrient phosphate with organic matter, this will allow the phosphate content of the oceans to increase. Within a few hundred thousand years, the phosphate content will have risen sufficiently high so that the upwelling flux of phosphate is sufficient to sustain productivities comparable to today, despite lower rates of vertical transport (cf. BROECKER and PENG, 1982).

On very long time-scales (tens to hundreds of millions of years) the ineluctable increase in solar luminosity presents the most significant of climate forcings, threatening the existence of life on the planet. For most of Earth history, the increase in solar luminosity has been offset by a regulated decrease in atmospheric CO_2 levels, from perhaps 10% 4 billion years ago to the pre-industrial value of 0.028%. During the last glacial period, CO_2 levels fell to 180 ppm, approaching the region where net photosynthesis by C3 plants is severely limited by CO_2. LOVELOCK and WHITFIELD (1982) calculated that within the next 100 million years, CO_2 levels would have to fall below the compensation point for photosynthesis to maintain an equable climate. CALDEIRA and KASTING (1992) redressed this issue, using more recent CO_2–greenhouse effect relationships, and concluded that the biosphere could survive for another billion years before plant growth failed.

However, the present temperature of the Earth, 15°C, is well below the optimum for plant growth (around 20–25°C). It might be argued that there is ample room for warming before the temperatures rise enough to threaten even the contemporary mainstream biota. The capacity of organisms to adapt would suggest that temperatures of even 60°C might not be inconsistent with life.

Consider first adaptation. It is true that organisms can adapt to harsh environments. Bacteria live in hot springs at temperatures close to boiling, in saturated salt solutions, and in extremely acidic solutions. Closer examination shows that the ecosystems of these extreme environments are sparse and the price of adaptation has been high in terms of the ability of vigorous growth possessed by the mainstream biota. It could be said that organisms living in harsh environments are rather like the eccentrics within a wealthy society, living a subsidised existence on the benefit of global welfare. A significant evolutionary change would be needed before they could be expected to run the system.

The major molecules of organisms, proteins, lipo-protein and lipo-polysaccharide complexes, and nucleic acids, are stable only at temperatures below 50°C. The upper temperature limit for even a warm-blooded mammal is below 50°C; at 50°C a second degree burn takes only 60°seconds contact time. Organisms require internal temperatures low enough to permit the survival of significant molecules for days at least. In practice a sustained ambient temperature of 40°C is the upper limit for mainstream life.

More important than the temperature and environmental limits of the organisms themselves is the sensitivity of the whole system. We have seen earlier in this chapter that ecological optima are typically different from physiological optima. From this generality it is reasonable to conclude that the climate system operating today, which depends to a large part on the acceleration of chemical weathering rates provided by plants, will fail long before the upper temperature limit for mainstream life is reached. Failure of this system could spell the end for mainstream life if, as we have suggested here, the planet is operating at a cooler global temperature because of the presence of the biota. However, through natural selection, a new mechanism of planetary cooling may evolve to fill the role currently played by C3 plants.

Actually, one such mechanism has already evolved, perhaps in response to the decline of atmospheric CO_2 levels in the last few tens of millions of years: C4 photosynthesis. C4 plants maintain elevated internal CO_2 concentrations and are thus able to continue photosynthesising at very low external CO_2 concentrations. The C4 pathway is primarily restricted to grasses today; no algae, bryophytes, ferns, gymnosperms, and few angiosperms utilise it (KREBS, 1978). However, this innovation may become more widely utilised by photosynthesisers in the future (LOVELOCK, 1989).

In conclusion, the Earth is approaching a critical point in its climatic and evolutionary history when the mainstream mechanisms for climate regulation will be overcome by external stresses. The climatic oscillations of the Quaternary are viewed from a geophysiological perspective as an indication of this impending failure of the climate system, with positive feedback in the CO_2 and albedo subsystems dominating. The extreme sensitivity of the climate system to weak astronomical forcing should be taken as a warning for the future. Our destruction of the tropical rain-forests, so important to the climate stability of the region and to the diversity of life on the planet (LOVELOCK, 1992), may generate as many surprises for the future as will the injection of mass quantities of carbon dioxide into the atmosphere. Earth history tells us that the planet will survive our tamperings, but it may indeed be a very different world in the future.

Acknowledgements

This work was supported by Gaia Charity using funds generously donated by Knut Kloster of the World City Corporation, and to LRK by Shell Research Ltd., UK and the Geologic Record of Global Change Program at the National Science Foundation.

References

ADAMS, J. M., FAURE, H., FAURE-DENARD, L., MCGLADE, J. M. and WOODWARD, F. I., 1990. Increases in terrestrial carbon storage from the Last Glacial Maximum to the present. *Nature*, 348: 711–714.

ANDERSON, T. L. and CHARLSON, R. J., 1991. Geophysiology of natural marine sulfate aerosols. *Atmos. Environ.*, 11: 2445–2447.

BANNISTER, P., 1976. *Introduction to Physiological Plant Ecology.* Wiley, New York, 273 pp.

BARNOLA, J. M., RAYNAUD, D., KOROTKEVICH, Y. S. and LORIUS, C., 1987. Vostok ice core provides 160,000-year record of atmospheric CO_2. *Nature*, 329: 408–414.

BARRON, E. J. and WASHINGTON, W. M., 1985. Warm Cretaceous climates: high atmospheric CO_2 as a plausible mechanism. In: E. T. SUNDQUIST and W. S. BROECKER (Editors), *The Carbon Cycle and Atmospheric CO_2; Natural Variations Archean to Present.* American Geophysical Union, Washington, DC, pp. 546–553.

BERGER, W. H., FISCHER, K., LAI, C. and WU, G., 1987. *Ocean Productivity and Organic Carbon Flux. I. Overview and Maps of Primary Production and Export Production.* University of California, San Diego, CA, SIO Reference 87-30, 1-67.

BERNER, R. A., 1993. Paleozoic atmospheric CO_2: importance of solar radiation and plant evolution. *Science*, 261: 68–70.

BERNER, R. A., LASAGA, A. C. and GARRELS, R. M., 1983. The carbonate-silicate geochemical cycle and its effect on atmospheric carbon dioxide over the past 100 million years. *Am. J. Sci.*, 283: 641–683.

BONAN, G. B., POLLARD, D. and THOMPSON, S. L., 1992. Effects of boreal forest vegetation on global climate. *Nature*, 359: 716–718.

BOYLE, E. A., 1988. Cadmium chemical tracer of deep water paleoceanography. *Paleoceanography*, 3: 471–489.

BRICAUD, A. and STRAMSKI, D., 1990. Spectral absorption coefficients of living phytoplankton and nonalgal biogenous matter: a comparison between the Peru upwelling area and the Sargasso Sea. *Limnol. Oceanogr.*, 35: 562–582.

BROECKER, W. S. and PENG, T.-H., 1982. *Tracers in the Sea.* Eldigio Press, Palisades, NY, 690 pp.

BROECKER, W. S. and PENG, T.-H., 1989. The cause of the glacial to interglacial atmospheric CO_2 changes: a polar alkalinity hypothesis. *Global Biogeoch. Cycles*, 3: 215–239.

CALDEIRA, K. and KASTING, J. F., 1992. The life span of the biosphere revisited. *Nature*, 360: 721–723.

CHARLES, C. D. and FAIRBANKS, R. G., 1990. Glacial to interglacial changes in the isotopic gradients of southern ocean surface waters. In: U. BLEIL and J. THIEDE (Editors), *Geological History of the Polar Oceans, Arctic versus Antarctic.* Kluwer, Boston, MA, 823 pp.

BENJAMIN, S. G., *see* CARLSON, T. N. and BENJAMIN, S. G.
BENJAMIN, S. G., *see* CARLSON, T. N. et al.
BENTON, M. J., 119, 137
BERBERA, J. J., *see* SELF, S. et al.
BERGER, A, GALLÉE H., FICHEFET, TH., MARSIAT, I. and TRICOT, C., 44, 45, 47, 60
BERGER, A., 5, 9, 17, 24, 25, 26, 28, 29, 31, 39, 44, 46, 60, 231, 233
BERGER, A. and LOUTRE, M. F., 26, 27, 29, 60
BERGER, A., FICHEFET, TH., GALLÉE H., TRICOT, C. and VAN YPERSELE, J. P., 50, 60
BERGER, A., FICHEFET, TH., GALLÉE, H., MARSIAT, I., TRICOT, C. and VAN YPERSELE, J. P., 44, 45, 60
BERGER, A., GALLÉE, H. and MÉLICE, J. L., 55, 56, 59, 60
BERGER, A., GALLÉE, H. and TRICOT, C., 60
BERGER, A., GUIOT, J., KUKLA, G. and PESTIAUX, P., 42, 49, 60
BERGER, A., LOUTRE, M. F. and TRICOT, C., 29, 50, 60
BERGER, A., *see* ADEM, J. et al.
BERGER, A., *see* GALLÉE, H. et al.
BERGER, A., *see* IMBRIE, J. et al.
BERGER, A., *see* KUKLA, G. et al.
BERGER, A., *see* LOUTRE, M. F. et al.
BERGER, A., *see* MARSIAT, I. and BERGER, A.
BERGER, A., *see* PESTIAUX, P. et al.
BERGER, A., *see* SANTER, B. et al.
BERGER, A., *see* SHACKLETON, N. J. et al.
BERGER, A., *see* TRICOT, C. and BERGER, A.
BERGER, A., TRICOT, C., GALLÉE, H. and LOUTRE, M. F., 52, 53, 60
BERGER, D., *see* SCOTTO, J. et al.
BERGER, D., *see* URBACH, F. et al.
BERGER, W. H., 553
BERGER, W. H., FISCHER, K., LAI, C. and WU, G., 545, 551
BERGGREN, W. A., *see* CORLISS, B. H. et al.
BERGSTRÖM, H., *see* HÖGSTRÖM, U. et al.
BERGTHORSSON, P., 193, 233
BERNER, R. A., 75, 80, 91, 126, 137, 539, 547, 548, 551
BERNER, R. A., LASAGA, A. C. and GARRELS, R. M., 75, 84, 91, 103, 137, 538, 547, 551
BERNER, W., OESCHGER, H. and STAUFFER, B., 524, 533
BERNSTEIN, R. E., *see* BETZER, P. R. et al.
BERNSTEIN, R. L., *see* BETZER, P. R. et al.
BERNSTEIN, R. L., *see* COAKLEY JR., J. A. et al.
BERRESHEIM, H., ANDREAE, M. O., AYERS, G. P. and GILLETT, R. W., 381, 390
BERRESHEIM, H., ANDREAE, M. O., AYERS, G. P., GILLETT, R. W., MERRILL, J. T., DAVIS, V. J. and CHAMEIDES, W. L., 381, 390
BERRESHEIM, H., ANDREAE, M. O., IVERSON, R. L. and LI, S.-M., 381, 390
BERRESHEIM, H., *see* ANDREAE, M. O. et al.
BERRESHEIM, H., *see* TALBOT, R. W. et al.
BERRIOR, A., *see* HARTMANN, D. L. et al.
BERRY, B. J. L., 437, 469
BERRY, J. C., *see* TRENBERTH, K. E. et al.
BERRY, W. B. N., *see* WILDE, P. et al.
BERTRAND, H., *see* SEBAI, A. et al.
BESSE, J., *see* COURTILLOT, V. et al.
BESSE, J., *see* VANDAMME, D. et al.
BETTS, A. K., BALL, J. H., BELJAARS, A. C. M., MILLER, M. J. and VITERBO, P., 269, 275
BETZER, P. R., CARDER, K. L., DUCE, R. A., MERRILL, J. T., TINDALE, N. W., UEMATSU, M., COSTELLO, D. K., YOUNG, R. W., FEELY, R. A., BRELAND, J. A., BERNSTEIN, R. E. and GRECO, A. M., 350, 390
BEVAN, A., *see* PARKER, D. E. et al.
BHALME, H. N. and MOOLEY, D. A., 199, 234
BHANDARI, N., SHUKLA, P. N. and AZMI, R. J., 120, 137
BHANDARI, N., SHUKLA, P. N. and CINI CASAGNOLI, G., 134, 137
BHANDARI, N., SHUKLA, P. N. and PANDEY, J., 120, 137
BHRUMRALKAR, C. M., *see* SEAMAN, N. L. et al.
BICE, D. M., NEWTON, C. R., MCCAULEY, S., REINERS, P. W. and MCROBERTS, C. A., 115, 119, 137
BICKLE, M. J., *see* MCKENZIE, D. P. and BICKLE, M. J.
BIDIGARE, R. R., *see* SMITH, R. C. et al.
BIGG, E. K., 372, 390
BIGG, E. K., *see* AYERS, G. P. et al.
BILSBORROW, R. E.and OKOTH–OGENDO, H. W. O., 435, 469
BINGEMER, H. G., ANDREAE, M. O., ANDREAE, T. W., ARTAXO, P., HELAS, G., JACOB, D. J., MIHALOPOULOS, N. and NGUYEN, B. C., 381, 390
BINGEMER, H. G., *see* ANDREAE, M. O. et al.
BIRCHFIELD, G. E., 37, 60
BIRCHFIELD, G. E. and WEERTMAN, J., 31, 60
BIRCHFIELD, G. E., WEERTMANN, J. and LUNDE, A. T., 40, 60
BIRD, T., *see* COOK, E. R. et al.
BIRKS, J. W., *see* CRUTZEN, P. J. and BIRKS, J. W.
BISCHOF, W., *see* BOLIN, B. and BISCHOF, W.
BISHOP, L., *see* BOJKOV, R. D. et al.
BISHOP, L., *see* STOLARSKI, R. S. et al.
BJERKNES, J., 214, 225, 234
BJÖRN, L. O., *see* MADRONICH, S. et al.
BLACKBURN, T., 154, 182
BLACKMON, M. L., *see* DESER, C. and BLACKMON, M. L.
BLACKMON, M. L., *see* WALLACE, J. M. et al.
BLAD, B. L., *see* STARKS, P. J. et al.
BLANC, P. L., *see* DUPLESSY, J. C. et al.

BLANC, P. L., , *see* LABEYRIE, L. D. et al.
BLANC, P. L., *see* LOUTRE, M. F. et al.
BLANCHARD, D. C., 351, 390
BLANCHET, J. P., 451, 469
BLANCHET, J. P., *see* CESS, R. D. et al.
BLIFFORD, I. H., *see* GILLETTE, D. A. and BLIFFORD, I. H.
BLOOMFIELD, P., 231, 234
BLOOMFIELD, P., *see* STOLARSKI, R. S. et al.
BLUMTHALER, M. and AMBACH, W., 425, 430
BLUTH, G. J. S., SCHNETZLER, C. C., KRUEGER, A. J. and WALTER, L. S., 352, 357, 391
BOATMAN, J. F., *see* GALLOWAY, J. N. et al.
BOATMAN, J. F., *see* WHELPDALE, D. M. et al.
BOATMAN, J. F., WELLMAN, D. L., VAN VALIN, C. C., GUNTER, R. L., RAY, J. D., SIEVERING, H., KIM, Y., WILKINSON, S. K. and LURIA, M., 381, 391
BOCLET, D., *see* ROCCHIA, R. et al.
BOECKELMANN, K., *see* HOLSER, W. T. et al.
BOER, G. J., 283, 284, 285, 286, 292, 293, 296, 310
BOER, G. J. and SARGENT, N. E., 286, 288, 293, 310
BOER, G. J., *see* CESS, R. D. et al.
BOER, G. J., *see* GATES, W. L. et al.
BOERSMA, A., 112
BOERSMA, A. and SHACKLETON, N. J., 109, 137
BOERSMA, A., *see* SHACKLETON, N. J. and BOERSMA, A.
BOERSMA, A., SHACKLETON, N. J., HALL, M. and GIVEN, Q., 109, 112, 137
BOJKOV, R. D., BISHOP, L., HILL, W. J., REINSEL, G. C. and TIAO, G. C., 339, 344
BOJKOV, R. D., *see* WANG, W.-C. et al.
BOJKOV, R., *see* STOLARSKI, R. S. et al.
BOLIN, B. and BISCHOF, W., 230, 234
BOLIN, B. and CHARLSON, R. J., 347, 391
BONAN, G. B., POLLARD, D. and THOMPSON, S. L., 438, 452, 469, 541, 551
BONANI, G., *see* BOND, G. et al.
BONANI, G., *see* BROECKER, W. S. et al.
BOND, G., HEINRICH, H., BROECKER, W. S., LABEYRIE, L. D., MCMANUS, J., ANDREWS, J., HUON, S., JANTSCHIK, R., CLASEN, S., SIMET, C., TEDESCO, K., KLAS, M.BONANI, G. and IVY, S., 25, 60
BOND, G., *see* BROECKER, W. S. et al.
BOND, G., *see* DANSGAARD, W. et al.
BONISCH, G., *see* SCHLOSSER, P. et al.
BONNEL, B., *see* FOUQUART, Y. et al.
BONSANG, B., *see* NGUYEN, B. C. et al.
BONTE, P., *see* ROCCHIA, R. et al.
BONY, S. and LE TREUT, H., 254, 275
BOOTH, C. R., *see* LUBIN, D. et al.
BOOTH, C. R., *see* STAMNES, K. et al.
BORNSTEIN, R. D., 479, 510
BORNSTEIN, R. D. and JOHNSTON, D. S., 496, 510
BORNSTEIN, R. D., *see* LOOSE, T. and BORNSTEIN, R. D.
BOSART, L. F., *see* KESSLER, R. W. et al.
BOSTON, P. J., *see* SCHNEIDER, S. H. and BOSTON, P. J.
BOTTJER, D. J., *see* SCHUBERT, J. K. and BOTTJER, D. J.
BOTTOMLEY, M., FOLLAND, C. K., HSIUNG, J., NEWELL, R. E. and PARKER, D. E., 152, 153, 158, 182
BOTZET, M., *see* BENGTSSON, L. et al.
BOUCHER, N. P., *see* SMITH, R. C. et al.
BOULTON, G. S., 25, 60
BOULTON, G. S., SMITH, G. D., JONES, A. S. and NEWSOME, J., 47, 60
BOVILL, E., 441, 469
BOWEN, M., *see* STAMNES, K. et al.
BOWNE, N. E. and BALL, J. T., 496, 510
BOWYER, P. A., *see* CIPRIANO, R. J. et al.
BOYLE, E. A., 527, 528, 533, 547, 551
BOYLE, E. A. and KEIGWIN, L., 88, 91
BOYLE, E. A., *see* IMBRIE, J. et al.
BRADLEY, R. S., 17, 22, 33, 60, 168, 176, 182, 195, 205, 220, 234
BRADLEY, R. S. and JONES, P. D., 168, 169, 176, 177, 182, 183, 195, 196, 197, 234
BRADLEY, R. S., DIAZ, H. F., EISCHEID, J. K., JONES, P. D., KELLY, P. M. and GOODESS, C. M., 155, 166, 183
BRADLEY, R. S., DIAZ, H. F., KILADIS, G. N. and EISCHEID, J. K., 174, 183, 234
BRADLEY, R. S., KEIMIG, F. T. and DIAZ, H. F., 228, 234
BRADLEY, R. S., KELLY, P. M., JONES, P. D., GOODESS, C. M. and DIAZ, H. F., 154, 183
BRADLEY, R. S., *see* DIAZ, H. F. and BRADLEY, R. S.
BRADLEY, R. S., *see* DIAZ, H. F. et al.
BRADLEY, R. S., *see* JONES, P. D. and BRADLEY, R. S.
BRADLEY, R. S., *see* JONES, P. D. et al.
BRAHAM JR., R. R., 372, 373, 380, 388, 391, 504, 510
BRAND, U., 125, 137
BRANSTATOR, G. W., *see* MADDEN, R. A. et al.
BRANSTATOR, G. W., *see* MEEHL, G. A. and BRANSTATOR, G. W.
BRANSTATOR, G. W., *see* TRENBERTH, K. E. et al.
BRASS, G., SALTZMAN, E., SLOAN, J., SOUTHAM, J., HAY, W., HOLZER, W. and PETERSON, W., 73, 91
BRASSEUR, G. P. and GRANIER, C., 413, 430
BRASSEUR, G. P. and HITCHMAN, M. H., 413, 430
BRASSEUR, G. P. and SOLOMON, S., 405, 430
BRASSEUR, G. P., HITCHMAN, M. H., WALTERS, S., DYMEK, M., FALISE, E. and PIRRE, M., 413, 430

BRASSEUR, G. P., *see* GRANIER, C. and BRASSEUR, G. P.
BRASSEUR, G. P., *see* HAUGLUSTAINE, D. A. et al.
BRASSEUR, G. P., *see* HITCHMAN, M. H. and BRASSEUR, G. P.
BRAZEL, A. J., *see* NASRALLAH, H. A. et al.
BRAZEL, S. W., *see* BALLING JR., R. C. and BRAZEL, S. W.
BRAZIUNAS, T. F., *see* STUIVER, M. and BRAZIUNAS, T. F.
BRELAND, J. A., *see* BETZER, P. R. et al.
BRÉMOND, M.-P., *see* CACHIER, H. et al.
BRENCHLEY, P. J., 126, 137
BRESCHER, R., *see* OSTLUND, H. G. et al.
BRETAGNON, P., *see* LOUTRE, M. F. et al.
BRETHERTON, F. P., *see* SODEN, B. J. and BRETHERTON, F. P.
BREWER, P. G., BROECKER, W. S., JENKINS, W. J., RHINES, P. B., ROOTH, C. G., SWIFT, J. H., TAKAHASHI, T. and WILLIAMS R. T., 532, 533
BREZA, J. R., *see* ZACHOS, J. C. et al.
BRIAT, M., *see* PETIT, J.-R. et al.
BRICAUD, A. and STRAMSKI, D., 541, 551
BRIDGE, D., *see* SIMKIN, T. et al.
BRIEGEL, L. M., 305, 310
BRIEGLIEB, B., 413, 430
BRIEGLEB, B. P., *see* KIEHL, J. T. and BRIEGLEB, B. P.
BRIEGLEB, B. P., *see* SHINE, K. P. et al.
BRIFFA, K. R. and JONES, P. D., 183
BRIFFA, K. R., JONES, P. D., BARTHOLIN, T. S., ECKSTEIN, D., SCHWEINGRUBER, F. H., KARLÉN, W., ZETTERBERG, P. and ERONEN, M., 194, 234
BRIFFA, K. R., *see* JONES, P. D. and BRIFFA, K. R.
BRISKIN, M. and HARRELL, J., 31, 61
BROCCOLI, A. J. and MANABE, S., 35, 36, 61
BROCCOLI, A. J., *see* MANABE, S. and BROCCOLI, A. J.
BROCK, C. A., *see* RADKE, L. F. et al.
BROECKER, W. S., 226, 234, 389, 391, 517, 518, 519, 520, 522, 526, 527, 528, 529, 530, 531, 532, 533
BROECKER, W. S. and DENTON, G. H., 31, 41, 61, 531, 533
BROECKER, W. S. and PENG, T.-H., 524, 529, 533, 547, 549, 551
BROECKER, W. S., ANDREE, M., WOLFI, W., OESCHGERI, H., BONANII, C., KENNETTI, J. and PETEET, D., 37, 43, 61
BROECKER, W. S., BOND, G., KLAS, M., BONANI, G. and WOLFI, W., 533
BROECKER, W. S., FLOWER, B. P., TRUMBORE, S., WOLFI, W., KENNETT, J. P., TELLER, J. T. and BONANI, G., 24, 43, 61
BROECKER, W. S., PETEET, D. M. and RIND, D., 43, 61, 88, 91, 195, 234, 389, 391, 533
BROECKER, W. S., *see* BOND, G. et al.
BROECKER, W. S., *see* BREWER, P. G. et al.
BROECKER, W. S., *see* RIND, D. et al.
BROGNIEZ, G., *see* FOUQUART, Y. et al.
BROOKS, R. R., *see* WOLBACH, W. S. et al.
BROWELL, E. V., *see* ANDREAE, M. O. et al.
BROWN, J. F., *see* GALLO, K. P. et al.
BROWN, J., *see* KUKLA, G. et al.
BROWN, L. R., 439, 469
BROWN, L. R., *see* ECKHOLM, E. and BROWN, L. R.
BROWN, S., *see* HOUGHTON, R. A. et al.
BROWNING, S. R., *see* HERMAN, B. M. et al.
BRUGMAN, W. A., *see* VISSCHER, H. and BRUGMAN, W. A.
BRUNDTLAND, G. H., 439, 440, 441, 469
BRUNE, W. H., *see* ANDERSON, J. G. et al.
BRUNELL, R., *see* GUNST, R. F. et al.
BRUNNER, J. S., *see* STREET-PERROTT, F. A. et al.
BRYAN, F., 88, 91, 226, 234
BRYAN, F., *see* MANABE, S. et al.
BRYAN, K., *see* GORDON, A. L. et al.
BRYAN, K., *see* MANABE, S. et al.
BRYAN, K., *see* STOUFFER, R. J. et al.
BRYANT, H. K., GLANCY, R. T. and AUER, A. H., 496, 510
BRYSON, R. A., 231, 234, 244, 347, 391
BRYSON, R. A. and SWAIN, A. M., 194, 234
BRYSON, R. A., IRVING, W. M. and LARSON, J. A., 194, 234
BRYSON, R. A., *see* BAERREIS, D. A. and BRYSON, R. A.
BRYSON, R. A., *see* KUTZBACH, J. E. and BRYSON, R. A.
BUAT-MÉNARD, P., *see* CACHIER, H. et al.
BUAT-MENARD, P., *see* DUCE, R. A. et al.
BUCK, C. F., *see* MAYEWSKI, P. A. et al.
BUCKLEY, B., *see* COOK, E. R. et al.
BUDYKO, M., 230, 234
BUDYKO, M. and RONOV, A., 75, 91
BUFFETAUT, E., *see* ROCCHIA, R. et al.
BUGGISCH, W., 114, 124, 125, 138
BUISSON, M., *see* FABRY, C. and BUISSON, M.
BUJA, L. E., *see* TRENBERTH, K. E. et al.
BÜRGERMEISTER, S. and GEORGII, H.-W., 381, 391
BURIEZ, J. C., *see* FOUQUART, Y. et al.
BUSALACCHI, A. J., *see* LAU, K.-M. and BUSALACCHI, A. J.
BUSCHBACHER, R., *see* UHL, C. and BUSCHBACHER, R.
BUTCHART, N., *see* AUSTIN , J. et al.
BUTLER, J. L., *see* SANDBERG, C. A. et al.
BUTTERWORTH, I. J., *see* ALLAN, R. J. et al.

CABRAL, G. M. R., *see* SHUTTLEWORTH, W. J. et al.
CACHIER, H. and DUCRET, J., 451, 469
CACHIER, H., BRÉMOND, M.-P. and BUAT-MÉNARD, P., 353, 391
CALDEIRA, K. and KASTING, J. F., 550, 551

CALDEIRA, K. and RAMPINO, M. R., 104, 105, 132, 133, 138
CALDEIRA, K., RAMPINO, M. R., VOLK, T. and ZACHOS, J. C., 104, 138
CALDEIRA, K., *see* RAMPINO, M. R. and CALDEIRA, K.
CALDWELL, M. M., CAMP, L. B., WARNER, C. W. and FLINT, S. D., 420, 430
CALDWELL, M. M., *see* MADRONICH, S. et al.
CALHOUN, J. A., *see* BATES, T. S. et al.
CALLANDER, B. A., *see* HOUGHTON, J. T. et al.
CALVERT, J., *see* FEHSENFELD, F. et al.
CAMP, L. B., *see* CALDWELL, M. M. et al.
CAMP, W. V., *see* POLLACK, J. B. et al., 241
CAMPBELL, I. H., CZAMANSKE, G. K., FEDORENKO, V. A., HILL, R. I. and STEPANOV, V., 127, 138
CAMPBELL, L. D., *see* STANLEY, S. M. and CAMPBELL, L. D.
CANE, M. A., 214, 234
CANE, M. A. and ZEBIAK, S. E., 211, 214, 234
CAO, H. X., 183
CAO, H. X., MITCHELL, J. F. B. and LAVERY, J. R., 164
CAPPETTA, H., *see* COURTILLOT, V. et al.
CAPUTO, M., 125, 138
CARDER, K. L., *see* BETZER, P. R. et al.
CAREY, S., *see* SIGURDSSON, H. et al.
CARISSIMO, B. C., OORT, A. H. and VONDER HAAR, T. H., 85, 91
CARLSON, T. N. and BENJAMIN, S. G., 367, 388, 391
CARLSON, T. N. and CAVERLY, T. S., 367, 388, 391
CARLSON, T. N., DODD, J. K., BENJAMIN, S. G. and COOPER, J. N., 492, 510
CARLSON, T. N., *see* METHOD, T. and CARLSON, T. N.
CARMACK, E. C., *see* AAGAARD, K. and CARMACK, E. C.
CARNEGGIE, D., *see* RUNNING, S. W. et al.
CARPENTER, T. H., *see* RASMUSSEN, E. M. and CARPENTER, T. H.
CARTER, A. F., *see* KEELING, C. D.
CARTLIDGE, J. E., *see* CORFIELD, R. M. et al.
CASIER, J-G., *see* CLAEYS, P. et al.
CASSINIS, G., 120, 138
CASTILLO, R. C., *see* PUESCHEL, R. F. et al.
CAVERLY, T. S., *see* CARLSON, T. N. and CAVERLY, T. S.
CAYAN, D. R. and DOUGLAS, A. V., 497, 510
CAYAN, D. R., *see* DOUGLAS, A. V. et al.
CAYAN, D. R., *see* EBBESMEYER, C. C. et al.
CAYAN, D. R., *see* NAMIAS, J. and CAYAN, D. R.
CAYAN, D. R., *see* VENRICK, E. L. et al.
CERF, A., *see* FOUQUART, Y. et al.
CERLING, T. E., 75, 80, 91
CESS, R. D., 275
CESS, R. D. and VULIS, I. L., 265, 275
CESS, R. D., HARRISON, E. F., MINNIS, P., BARKSTROM, B. R., RAMANATHAN, V. and KWON, T. Y., 319, 344
CESS, R. D., POTTER, G. L., BLANCHET, J. P., BOER, G. J., DEL GENIO, A. D., DÉQUE, M., DYMNIKOV, V., GALIN, V., GATES, W. L., GHAN, S. J., KIEHL, J. T., LACIS, A. A., LE TREUT, H., LI, Z.-X., LIANG, X.-Z., MCAVANEY, B. J., MELESHKO, V. P., MITCHELL, J. F. B., MORCRETTE, J.-J., RANDALL, D. A., RIKUS, L., ROECKNER, E., ROYER, J. F., SCHLESE, U., SHEININ, D. A., SLINGO, A., SOKOLOV, A. P., TAYLOR, K. E., WASHINGTON, W. M., WETHERALD, R. T., YAGAI, I. and ZHANG, M. H., 260, 275
CESS, R. D., POTTER, G. L., BLANCHET, J. P., BOER, G. J., GHAN, S. J., KIEHL, J. T., LE TREUT, H., LI, Z. X., LIANG, X. Z., MITCHELL, J. F. B., MORCRETTE, J. J., RANDALL, D. A., RICHES, M. R., ROECKNER, E., SCHLESE, U., SLINGO, A., TAYLOR K. E., WASHINGTON, W. M., WETHERALD, R. T. and YAGAI, I., 39, 61
CESS, R. D., POTTER, G. L., GATES, W. L., MORCRETTE, J.-J. and CORSETTI, L., 254
CESS, R. D., *see* CHARLSON, R. J. et al.
CESS, R. D., *see* COAKLEY JR., J. A. et al.
CESS, R. D., *see* FALKOWSKI, P. G. et al.
CESS, R. D., *see* HARRISON, E. F. et al.
CESS, R. D., *see* OWEN, T. et al.
CESS, R. D., *see* RAMANATHAN, V. et al.
CESS, R., *see* FALKOWSKI, P. G. et al.
CESS, R., *see* OWEN, T. et al.
CHAHINE, M., 453, 469
CHAHINE, M. T. and SUSSKIND, J., 263, 275
CHAI, Z., *see* XU, D. et al.
CHAI, Z-F., *see* WANG, J-X. et al.
CHAMEIDES, W. L., *see* BERRESHEIM, H. et al.
CHAMLEY, H., *see* ROBERT, C. and CHAMLEY, H.
CHANDLER, M., RIND, D. and TUEDY, R., 84, 91
CHANDLER, M., *see* RIND, D. and CHANDLER, M.
CHANG, A. T. C. and WILHEIT, T. T., 263, 275
CHANG, A. T. C., FOSTER, J. L. and HALL, D. K., 272, 275
CHANG, A. T. C., *see* FOSTER, J. L. and CHANG, A. T. C.
CHANG, A. T. C., *see* WILHEIT, T. J. et al.
CHANG, C. C., *see* CHOW, S. D. and CHANG, C. C.
CHANG, H.-R. and WEBSTER, P. J., 305, 310
CHANG, H.-R., *see* WEBSTER, P. J. and CHANG, H.-R.
CHANG, L. A., *see* WU, M. L. C. and CHANG, L. A.
CHANGERY, M. G., *see* KARL, T. R. et al.
CHANGNON JR., S. A, 504
CHANGNON JR., S. A., 477, 504, 506, 510
CHANGNON JR., S. A. and HUFF, F. A., 504, 510
CHANGNON JR., S. A., HUFF, F. A., SHICKEDANZ, P. T. and VOGEL, J. L., 504, 510

CROWLEY, T., 80
CROWLEY, T. J., 34, 61, 92, 228, 235
CROWLEY, T. J. and MITCHELL, J. F. B., 33
CROWLEY, T. J. and NORTH, G. R., 90, 92, 99, 138
CROWLEY, T. J., CROWLEY, J. and NORTH, G. R., 61
CROWLEY, T. J., KIM, K. Y., MENGEL, J. G. and SHORT, D. A., 34, 61
CROWLEY, T. J., *see* NORTH, G. R. and CROWLEY, T. J.
CROWLEY, T. J., *see* SHORT, D. A. et al.
CROWLEY, T., *see* MAIER-REIMER, E. et al.
CROZAT, G., 356, 392
CRUTZEN, P. J., 101, 102, 138, 401, 413, 430
CRUTZEN, P. J. and ANDREAE, M. O., 376, 392
CRUTZEN, P. J. and BIRKS, J. W., 347, 392
CRUTZEN, P. J., *see* GRAEDEL, T. E. and CRUTZEN, P. J.
CRUTZEN, P. J., *see* HAO, W. M. et al.
CRUTZEN, P. J., *see* LANGNER, J. et al.
CRUTZEN, P. J., *see* SANHUEZA, E. et al.
CRUTZEN, P. J., *see* SCHARFFE, D. et al.
CRUTZEN, P. J., *see* SEILER, W. and CRUTZEN, P. J.
CRUZ, F., *see* JAUREGUI, E. et al.
CRUTZEN, P. J., *see* ZIMMERMAN, P. R. et al.
CUBASCH, U., *see* GATES, W. L. et al.
CULLIS, C. F. and HIRSCHLER, M. M., 357, 392
CUNNINGHAM, G. L., *see* SCHLESINGER, W. H. et al.
CURRIE, R. G., 199, 201, 202, 235
CURRIE, R. G. and FAIRBRIDGE, R. W., 199, 235
CURRIE, R. G. and O'BRIEN, D. P., 199, 235
CURRY, J. A. and EBERT, E. E., 257, 276
CZAMANSKE, G. K., *see* CAMPBELL, I. H. et al.

D'HONDT, S., KELLER, G. and STALLARD, R. F., 118, 139
D'HONDT, S., *see* SIGURDSSON, H. et al.
DABBERDT, W. F. and DAVIS, P. A., 437, 469, 492, 510
DAHL, J., *see* VENKATESAN, M. I. and DAHL, J.
DAHL-JENSEN, D., *see* DANSGAARD, W. et al.
DAHL-JENSEN, D., *see* TAYLOR, K. C. et al.
DAHNI, R. R., *see* MCAVANEY, B. J. et al.
DAHNI, R. R., *see* POWER, S. B. et al.
DALRYMPLE, G. B., *see* SHARPTON, V. L. et al.
DALU, G., 277
DAMON, P. E., 200
DAMON, P. E., LERMAN, J. C. and LONG, A., 235
DAMON, P. E., *see* JIRIKOVIC, J. and DAMON, P. E.
DANARD, M., GRAY, M. and LYV, G., 46, 61
DANSGAARD, W. et al., 226, 235
DANSGAARD, W., JOHNSEN, S. J., CLAUSEN, H. B., DAHL-JENSEN, D., GUNDESTRUP, N. S., HAMMER, C. U., HVLDBORG, C. S., STEFFENSEN, J. P., SVELNBJÖRNSDOTTIR, A. E., JOUZEL, J. and BEND, G., 25, 58, 61, 517, 534
DANSGAARD, W., JOHNSON, S. J., CLAUSEN, H. B. and LANGWAY JR., C. C., 522, 523, 530, 533
DANSGAARD, W., *see* HAMMER, C. U. et al.
DANSGAARD, W., *see* JOHNSEN, S. J. et al.
DANSGAARD, W., *see* MAYEWSKI, P. A. et al.
DANSGAARD, W., WHITEAND, J. W. C. and JOHNSON, S. J., 517, 534
DAOHAN, C., LINZHONG, L. and JIAQING, Z., 102, 138
DAO-YI, X., QIN-WEN, Z., YI-YIN, S., YAN, Z., ZHI-FANG, C. and JIN-WEN, H., 120, 138
DARNELL, W. L., *see* GUPTA, S. K. et al.
DARNELL, W. L., STAYLOR, W. F., GUPTA, S. K., RITCHEY, N. A. and WILBER, A. C., 265, 276
DAUM, P. H., *see* TEN BRINK, H. M. et al.
DAVE, J. V., *see* SEKERA, Z. and DAVE, J. V.
DAVENPORT, S. A. et al., 102, 139
DAVIDSON, C. I., *see* MAYEWSKI, P. A. et al.
DAVIDSON, K. A., *see* TWOMEY, S. et al.
DAVIES, R. E., *see* DE GRUIJL, F. R. et al.
DAVIES, R. E., *see* URBACH, F. et al.
DAVIES, R., *see* COAKLEY JR., J. A. and DAVIES, R.
DAVIES, T. D., *see* GOODESS, C. M. et al.
DAVIS, P. A., *see* DABBERDT, W. F. and DAVIS, P. A.
DAVIS, V. J., *see* BERRESHEIM, H. et al.
DE ABREU SÁ, L. D., *see* SHUTTLEWORTH, W. J. et al.
DE ANGELIS, M., BARKOV, N. I. and PETROV, V. N., 24, 61, 351, 392
DE BEAULIEU, J. L., *see* GUIOT, J. et al.
DE DILVA, F., *see* ETHERIDGE, D. M. et al.
DE GRUIJL, F. R., *see* MADRONICH, S. and DE GRUIJL, F. R.
DE GRUIJL, F. R., STERENBORG, H. J. C. M., FORBES, P. D., DAVIES, R. E., COLE, C., KELFKENS, H., VAN WEELDEN, H., SLAPER, H. and VAN DER LEUN, J. C, 420, 430
DE MORA, S. J., *see* ANDREAE, M. O. et al.
DE MORAES, J. C., *see* SHUTTLEWORTH, W. J. et al.
DE OLIVEIRA, A. E., *see* SALATI, E. et al.
DE PAULA SILVA FILHO, V., *see* SHUTTLEWORTH, W. J. et al.
DE SILVA, F., *see* PEARMAN, G. I. et al.
DE SILVA, S. L., WOLFF, J. A. and SHARPTON, V. L., 134, 139
DE VRIES, T. J., 223, 235
DEAN, J. S., 194, 235
DEAN, W. E., *see* ZACHOS, J. C. et al.
DEBLONDE, G. and PELTIER, W. R., 44, 50, 61
DEERING, D. W., *see* AHMAD, S. P. and DEERING, D. W.
DEL GENIO, A. D., *see* CESS, R. D. et al.
DEL GENIO, A. D., *see* FU, R. et al.
DELANY, A. C., *see* PROSPERO, J. M. et al.
DE LAUBENFELS, M. W., 95, 139
DELGENIO, A., *see* HANSEN, J. et al.
DELMAS, R. J., ASCENCIO, J.-M. and LEGRAND, M., 524, 534

DELMAS, R. J., *see* LEGRAND, M. R. et al.
DELMAS, R. J., *see* RAYNAUD, D. et al.
DE LUISI, J., 425, 430
DELWORTH, T., MANABE, S. and STOUFFER, R. J., 226, 235
DEMARIA, M. and KAPLAN, J., 310, 311
DEMARIA, M., *see* SILVA DIAS, P. L. et al.
DEMORE, W. B., GOLDEN, D. M., MOLINA, M. J., HAMPSON, R. F., KURYLO, M. J., HOWARD, C. J. and RAVISHANKARA, A. R., 430
DENN, F. M., *see* MINNIS, P. et al.
DENTON, G. H. and ARMSTRONG, R. L., 82, 92
DENTON, G. H. and HUGHES, T. J., 61
DENTON, G. H., *see* BROECKER, W. S. and DENTON, G. H.
DENTON, G. H., *see* HUGHES, T. J. et al.
DENTON, G. H., *see* PORTER, S. C. and DENTON, G. H.
DÉQUE, M., *see* CESS, R. D. et al.
DEQUE, M., *see* ROYER, J. F. et al.
DEROME, J., *see* MICHAUD, R. and DEROME, J.
DÉSALMAND, F., 376, 392
DESER, C. and BLACKMON, M. L., 227, 235
DESER, C. and WALLACE, J. M., 211, 222, 235
DESER, C., *see* WRIGHT, P. B. et al.
DESHLER, T., HOFMANN, D. J., HEREFORD, V. and SUTTER, C. B., 407, 416, 430
DETTWILLER, J., 151, 183
DEWEY, K. F., 278
DIAZ, H. F., 199, 200, 235
DIAZ, H. F. and BRADLEY, R. S., 198, 231, 235
DIAZ, H. F. and KILADIS, G. N., 220, 221, 235
DIAZ, H. F. and MARKGRAF, V., 207, 222, 227, 235
DIAZ, H. F. and PULWARTY, R. S., 218, 222, 223, 235
DIAZ, H. F., BRADLEY, R. S. and EISCHEID, J. K., 155, 166, 183
DIAZ, H. F., *see* BRADLEY, R. S. et al.
DIAZ, H. F., *see* FU, C. et al.
DIAZ, H. F., *see* HUGHES, M. K. and DIAZ, H. F.
DIAZ, H. F., *see* JONES, P. D. et al.
DIAZ, H. F., *see* KARL, T. R. et al.
DIAZ, H. F., *see* KILADIS, G. N. and DIAZ, H. F.
DIAZ, H. F., *see* PARTHASARATHY, B. et al.
DICKEY, J. O., *see* HIDE, R. et al.
DICKINS, J. M., 119, 139
DICKINSON, R. E., 270, 276, 435, 464, 469
DICKINSON, R. E. and HENDERSON-SELLERS, A., 437, 461, 463, 465, 469
DICKINSON, R. E. and KENNEDY, P. J., 461, 469, 541, 552
DICKINSON, R. E., HENDERSON-SELLERS, A. and KENNEDY, P. J., 467, 469
DICKINSON, R. E., HENDERSON-SELLERS, A., KENNEDY, P. J. and WILSON, M. F., 467, 469
DICKINSON, R. E., PINTY, B. and VERSTRAETE, M. M., 265, 276
DICKINSON, R. E., *see* HENDERSON-SELLERS, A. et al.
DICKINSON, R. E., *see* PENNER, J. E. et al.
DICKSON, C. R., *see* ANGELL, J. K. et al.
DICKSON, R. R., 227
DICKSON, R. R., MEINCKE, J., MALMBERG, S.-A. and LEE, A. J., 88, 92, 235
DICKSON, S. M., *see* ERICKSON, D. J. and DICKSON, S. M.
DIETACHMAYER, G. S., *see* HOLLAND, G. J. and DIETACHMAYER, G. S.
DIFFEY, B. L., *see* MCKINLAY, A. F. and DIFFEY, B. L.
DIPASQUALE, R. C., *see* WHITLOCK, C. H. et al.279
DIRMEYER, P. A., 470
DIRMHIRN, I. and EATON, F. D., 46, 62
DLUGOKENCKY, E. J., MASAIRE, K. A., LANG, P. M., TANS, P. P., STEELE, L. P. and NISBET, E. G., 322, 344
DOBSON, G. M. B., 408, 430
DODD, J. K., *see* CARLSON, T. N. et al.
DOEHNE, E., *see* MARGOLIS, S. V. et al.
DOELLING, D. R., *see* MINNIS, P. et al.
DONALL, E. G., *see* SEAMAN, N. L. et al.
DONESO, L., *see* SANHUEZA, E. et al.
DONN, W. L. and SHAW, D. M., 32, 62, 75, 92
DONNELL, C. A., 154, 183
DONOSO, L., *see* SCHARFFE, D. et al.
DONOVAN, S. K., 114, 139
DORSEY, H. G., *see* OSTLUND, H. G. et al.
DOUGLAS, A. V., CAYAN, D. R. and NAMIAS, J., 168, 183, 226, 235
DOUGLAS, A. V., *see* CAYAN, D. R. and DOUGLAS, A. V.
DOUGLAS, I., 440, 470
DOUGLAS, R. G., *see* SAVIN, S. et al.
DOUROJEANNI, M. J., *see* SALATI, E. et al.
DOUTHITT, C. B., *see* RICH, T. H. et al.
DOWNEY, W. K., *see* JOHNSON, D. R. and DOWNEY, W. K.
DOWSETT, H. J. and LOUBERE, P., 117, 139
DREESEN, R., *see* SANDBERG, C. A. et al.
DREGNE, H. E., 441, 465, 470
DREGNE, H. E., *see* TUCKER, C. J. et al.
DREVER, J. I. and ZOBRIST, J., 547, 552
DRUILHET, A., *see* ESTOURNEL, C. et al.
DUCE, R. A., LISS, P. S., MERRILL, J. T., ATLAS, A. L., BUAT-MENARD, P., HICKS, B. B., MILLER, J. M., PROSPERO, J. M., ARIMOTO, R., CHURCH, T. M., ELLIS, W., GALLOWAY, J. N., HANSEN, L., JICKELLS, T. D., KNAP, A. H. and REINHARDT, K. H., 350, 364, 380, 392
DUCE, R. A., *see* BETZER, P. R. et al.
DUCE, R. A., *see* PROSPERO, J. M. et al.
DUCE, R. A., *see* UEMATSU, M. et al.
DUCRET, J., *see* CACHIER, H. and DUCRET, J.
DUDEK, M. P., *see* WANG, W.-C. et al.

DUDEK, M. P., WANG, W.-C. and LIANG, X.-Z., 336, 337, 345
DUDHA, A., *see* TAYLOR, F. W. et al.
DUFFY, A., *see* IMBRIE, J. et al.
DULANEY, W., *see* GOWARD, S. N. et al.
DUMEIL, L., *see* BARNETT, T. P. et al.
DUNN, D.A., *see* MOORE, T. C. et al.
DUNNING, G. R. and HODYCH, J. P., 127, 139
DUPLESSY, J. C., LABEYRIE, L. and BLANC, P. L., 43, 62
DUPLESSY, J. C., LABEYRIE, L., ARNOLD, M., PATERNE, M., DUPRAT, J. and VAN WEERING, T. C. E., 47
DUPLESSY, J. C., *see* LABEYRIE, L. D. and DUPLESSY, J. C.
DUPLESSY, J. C., *see* LABEYRIE, L. D. et al.
DUPLESSY, J. C., *see* PESTIAUX, P. et al.
DUPLESSY, J. C., SHACKLETON, N. J., FAIRBANKS, R. G., LABEYRIE, L., OPPO, D. and KALLEL, N., 88, 92
DUPLESSY, J. CL., LABEYRIE, L. D., ARNOLD, M., PATERNE, M., DUPRAT, J. and VAN WEERING, T. C. E., 24, 62
DUPRAT, J., *see* DUPLESSY, J. CL. et al.
DURBIDGE, T. B., *see* HENDERSON-SELLERS, A. et al.
DURBIDGE, T. B., *see* MCGUFFIE, K. et al.
DURKEE, P. A., 377, 392
DURKEE, P. A., PFEIL, F., FROST, E. and SHEMA, R., 364, 392
DURKEE, P. A., *see* COAKLEY JR., J. A. et al.
DUTTON, E. G. and CHRISTY, J. R., 205, 236
DYE, D. G., *see* GOWARD, S. N. et al.
DYMEK, M., *see* BRASSEUR, G. P. et al.
DYMNIKOV, V., *see* CESS, R. D. et al.
DZIETARA, S. and JANICOT, S., 457, 470

EAST, C. *see* OKE, T. R. and EAST, C.
EASTERLING, D. R., *see* KARL, T. R. et al.
EATON, F. D., *see* DIRMHIRN, I. and EATON, F. D.
EATON, F. D., *see* WHITE, J. M. et al.
EBBESMEYER, C. C., CAYAN, D. R., MCLAIN, D. R., NICHOLS, F. H., PETERSON, D. H. and REDMOND, K. T., 227, 236
EBERT, E. E., *see* CURRY, J. A. and EBERT, E. E.
EBINGER, C. J., 127
EBINGER, C. J., YEMANE, T., WOLDEGABRIEL, G., ARONSON, J. L. and WALTER, R. C., 139
ECKHOLM, E. and BROWN, L. R., 435, 470
ECKSTEIN, D., *see* BRIFFA, K. R. et al.
EDDY, J. A., 175, 183, 200, 236
EDDY, J. A., GILLILAND, R. L. and HOYT, D. V., 199, 236
EDDY, J. A., GILMAN, D. L., and TROTTER, D. E., 236
EDDY, J. A., GILMAN, P. A. and TROTTER, D. E., 200
EDDY, J. A., *see* SANTER, B. et al.
EHHALT, D. H., RUDOLPH, J. and SCHMIDT, U., 356, 392
EHRMANN, W. U., 119
EHRMANN, W. U. and MACKENSEN, A., 139
EHRMANN, W. U., *see* MACKENSEN, A. and EHRMANN, W. U.
EINFELD, W., WARD, D. E. and HARDY, C. C., 353, 392
EISCHEID, J. K., *see* BRADLEY, R. S. et al.
EISCHEID, J. K., *see* DIAZ, H. F. et al.
EISCHEID, J. K., *see* PARTHASARATHY, B. et al.
EKDAHL, C. A., *see* KEELING, C. D. et al.
EL-BADRY, M. A., 435, 470
ELBERT, W., *see* ANDREAE, M. O. et al.
ELLIOT, D. H., *see* HEIMAN, A. et al.
ELLIOTT, W. P., *see* GAFFEN, D. J. et al.
ELLIS, H. T. and PUESCHEL, R. F., 379, 392
ELLIS, W., 154, 183
ELLIS, W., *see* DUCE, R. A. et al.
ELSAESSER, H. W., MACCRACKEN, M. C., WALTON, J. J. and GROTCH, S. L., 158, 183
ELSBERRY, R. L., *see* FIORINO, M. J. and ELSBERRY, R. L.
ELSNER, J. B. and TSONIS, A. A., 158, 179, 183
ELSOM, D. M. and MEADEN, G. T., 504, 510
ELSON, L. S., *see* WATERS, J. W. et al.
ELSTON, W. E. and TWIST, D., 128, 139
ELY, G. A., *see* ROCHE, A. E. et al.
EMANUEL, K. A., 310, 311
EMANUEL, W. R., SHUGART, H. H. and STEVENSON, M. P., 466, 470
EMERY, W. J., *see* PICKARD, G. L. and EMERY, W. J.
EMERY, W. J., *see* ROTH, M. et al.
EMILIANI, C., 22, 62, 100, 139
EMILIANI, C., KRAUS, E. B. and SHOEMAKER, E. M., 100, 139
ENFIELD, D. B., 207, 222, 223, 236
ENFIELD, D. B. and CID, S. L., 199, 222, 227, 236
ENGARDT, M. and RODHE, H., 384, 392
ENGELHARDT, W. V., MATTHAI, S. K. and WALZEBUCK, J., 122, 139
ENGELKEMEIR, A., *see* CRAIG, H. et al.
ENSOR, D. S., PORCH, W. M., PILAT, M. J. and CHARLSON, R. J., 347, 392
ENTEKHABI, D., *see* NICHOLSON, S. E. and ENTEKHABI, D.
EPHRAUMS, J. J., 2, 6, 15
EPHRAUMS, J. J., *see* HOUGHTON, J. J. et al.
EPHRAUMS, J. J., *see* HOUGHTON, J. T. et al.
EPSHTEYN, O. G., 73, 92
ERICKSON III, D. J., *see* GHAN, S. J. et al.
ERICKSON, D. J. and DICKSON, S. M., 102, 139
ERONEN, M., *see* BRIFFA, K. R. et al.
ESBENSEN, S. K. and KUSHNIR, V., 210, 236
ESCH, M., *see* BENGTSSON, L. et al.

ESCRITOR, F., *see* HUNTLEY, M. E. et al.
ESHET, Y., 117, 120, 121, 139
ESTES, R. and HUTCHISON, J. H., 82, 92
ESTOURNEL, C., VEHIL, R., GUEDALIA, D., FONTAN, J. and DRUILHET, A., 486, 487, 510
ETHERIDGE, D. M., PEARMAN, G. I. and DE SILVA, F., 447, 470
ETHERIDGE, D., PATTERSON, G. R. and PLATT, C. M. R., 364, 392
ETHERIDGE, D., *see* PEARMAN, G. I. et al.
EUROPEAN TIMBER TRENDS and PROSPECTS, 444, 470
EVANS, J. L., 310, 311
EVANS, J. L., *see* LIGHTHILL, J. et al.
EVANS, J. L., *see* RYAN, B. F. et al.
EYRE, J. R., 263, 276
EYRE, J. R., KELLY, G. A., MCNALLY, A. P., ANDERSSON, E. and PERSSON, A., 263, 276

FABRE, J., *see* PETIT-MAIRE, J. R. et al.
FABRY, C. and BUISSON, M., 408, 431
FAIRALL, C. W., KEPERT, J. D. and HOLLAND, G. J., 310, 311
FAIRBANKS, R. G., 24, 62
FAIRBANKS, R. G., *see* BARD, E. et al.
FAIRBANKS, R. G., *see* CHARLES, C. D. and FAIRBANKS, R. G.
FAIRBANKS, R. G., *see* COLE, J. E. et al.
FAIRBANKS, R. G., *see* DUPLESSY, J. C. et al.
FAIRBANKS, R. G., *see* GUILDERSON, TH. P. et al.
FAIRBANKS, R. G., *see* MILLER, K. G. et al.
FAIRBANKS, R. G., *see* MIX, A. C. and FAIRBANKS, R. G.
FAIRBRIDGE, R. W., *see* CURRIE, R. G. and FAIRBRIDGE, R. W.
FALISE, E., *see* BRASSEUR, G. P. et al.
FALKOWSKI, P. G., KIM, Y., KOLBER, Z., WILSON, C., WIRICK, C. and CESS, R., 377, 392, 545, 552
FALL, R., *see* FEHSENFELD, F. et al.
FAN, D., YANG, R. and HUANG, Z., 126, 139
FAO, 434, 440, 445, 449, 470, 472
FAO/UNEP, 464, 470
FARMAN, J. C., GARDINER, B. G. and SHANKLIN, J. D., 399, 406, 428, 431
FARMER, G., *see* JONES, P. D. et al.
FARRAR, E., *see* BAKSI, A. K. and FARRAR, E.
FARRELL, B. and WATTERSON, I. G., 305, 311
FASTHOOK, J. L., *see* HUGHES, T. J. et al.
FAURE, H., *see* ADAMS, J. M. et al.
FAWCETT, P. J., *see* BARRON, E. J. et al.
FEARS, T., *see* SCOTTO, J. et al.
FEDORENKO, V. A., *see* CAMPBELL, I. H. et al.
FEELY, R. A., *see* BETZER, P. R. et al.
FEGLEY, B., *see* PRINN, R. G. and FEGLEY, B. et al.
FEHSENFELD, F. and LIU, S. C., 400, 431
FEHSENFELD, F., CALVERT, J., FALL, R., GOLDAN, P., GUENTHER, A. B., HEWITT, C. N., LAMB, B., LIU, S., TRAINER, M., WESTBERG, H. and ZIMMERMAN, P., 356, 392
FELDSTEIN, S. B., *see* KILADIS, G. N. and FELDSTEIN, S. B.
FELLOWS, L., *see* MALINGREAU, J. P. et al.
FELS, E. and KELLER, R., 440, 470
FELTON, E. A., *see* RICH, T. H. et al.
FENG, J. Z. and PETZOLD, D. E., 497, 510
FENIET-SAIGNE, C., *see* LEGRAND, M. R. et al.
FENNER, J. M., *see* CORLISS, B. H. et al.
FENNESSY, M., *see* SUD, Y. C. and FENNESSY, M.
FERAUD, G., 144
FEREK, R. J., *see* ANDREAE, M. O. et al.
FERGUSON, W. S., GRIFFIN, J. J. and GOLDBERG, E. D., 350, 392
FERMAN, M. A., *see* WOLFF, G. R. et al.
FERRARE, R. A., FRASER, R. S. and KAUFMAN, Y. J., 363, 392
FERRARE, R. A., *see* KAUFMAN, Y. J. et al.
FERRIER, B., *see* SIMPSON, J. S. et al.
FICHEFET, TH., *see* BERGER, A. et al.
FICHEFET, TH., *see* GALLÉE, H. et al.
FIORINO, M. J. and ELSBERRY, R. L., 305, 311
FISCH, G., *see* SHUTTLEWORTH, W. J. et al.
FISCHER, A. G., 75, 92
FISCHER, K., *see* BERGER, W. H. et al.
FISHER, R., *see* GRAETZ, D. et al.
FISHMAN, J., 319, 339, 345
FITZGERALD, J. W. and SPYERS-DURAN, P. A., 373, 392
FITZHARRIS, B. B., *see* SALINGER, M. J. et al.
FLAGAN, R. C., *see* PANDIS, S. N. et al.
FLAGAN, R. C., *see* ZHANG, S.-H. et al.
FLEMING, T. H., *see* HEIMAN, A. et al.
FLETCHER, J. O., *see* FU, C. et al.
FLINT, S. D., *see* CALDWELL, M. M. et al.
FLOHN, H. and KAPALA, A., 228, 236
FLOHN, H., KAPALA, A., KNOCHE, H. R. and MÄCHEL, H., 200, 236
FLOHN, H., *see* HENSE, A. et al.
FLOHN, H., *see* SANTER, B. et al.
FLOHN, H.FLOHN, H., KAPALA, A., KNOCHE, H. R. and MÄCHEL, H., 228
FLOWER, B. P., *see* BROECKER, W. S. et al.
FLOWER, D. A., *see* BARATH, F. T. et al.
FLOWER, D. A., *see* WATERS, J. W. et al.
FLOWERS, E. C., *see* PETERSON, J. T. et al.
FLYGER, H., HEIDAM, N. Z., HANSEN, K., MEGAW, W. J., WALTHER, E. G. and HOGAN, A. W., 381, 392
FOLAND, K. A., *see* HEIMAN, A. et al.
FOLLAND, C. K. and PARKER, D. E., 153, 183
FOLLAND, C. K., KARL, T. R. and VINNIKOV, K. YA., 155, 157, 158, 160, 164, 170, 171, 183, 437, 470

FOLLAND, C. K., KARL, T. R., NICHOLLS, N., NYENZI, B. S., PARKER, D. E. and VINNIKOV, K. YA., 155, 157, 158, 160, 164, 170, 171, 184
FOLLAND, C. K., PALMER, T. N. and PARKER, D. E., 38, 62, 166, 183, 229, 236
FOLLAND, C. K., *see* BOTTOMLEY, M. et al.
FOLLAND, C. K., *see* PARKER, D. E. and FOLLAND, C. K.
FOLLAND, C. K., *see* PARKER, D. E. et al.
FOLLAND, C. K., *see* ROWELL, D. P. et al.
FONTAN, J., *see* ESTOURNEL, C. et al.
FORBES, P. D., *see* DE GRUIJL, F. R. et al.
FORGAN, B. W., 364, 393
FORGAN, B. W., *see* LYONS, T. J. and FORGAN, B. W.
FORSYTH D., *see* LUYENDYK, B. et al.
FORTELIUS, C. and HOLOPAINEN, E., 268, 276, 286, 311
FOSTER, J. L. and CHANG, A. T. C., 272, 276
FOSTER, J. L., *see* CHANG, A. T. C. et al.
FOUKAL, P. and LEAN, J., 199, 236
FOUQUART, Y. and ISAKA, H., 375, 393
FOUQUART, Y., BONNEL, B., BROGNIEZ, G., BURIEZ, J. C., SMITH, L., MORCRETTE, J. J. and CERF, A., 367, 393
FOUQUART, Y., BONNEL, B., CHAOUI ROQUAI, M., SANTER, R. and CERF, A., 367, 393
FOWELL, S. J., *see* OLSEN, P. E. et al.
FOX, P. T., *see* JAMES, R. W. and FOX, P. T.
FRAEDRICH, K. and BANTZER, C., 194, 236
FRAKES, L. A. and FRANCES, J. E., 73, 92
FRAKES, L. A., FRANCIS, J. E. and SYTKUS, J. I., 71, 73, 92
FRAKES, L. A., *see* HAYES, D. E. and FRAKES, L. A.
FRAMJI, K. K. and MAHAJAN, I. K., 439, 470
FRANCES, J. E., *see* FRAKES, L. A. and FRANCES, J. E.
FRANCES, J. E., *see* FRAKES, L. A. et al.
FRANCEY, R., *see* COOK, E. R. et al.
FRANKE, R. and CHASIN, B., 441, 470
FRANKIGNOUL, C., 226, 236
FRANKIGNOUL, C. and HASSELMANN, K., 43, 62
FRASER, 447
FRASER, J. R., 300, 311
FRASER, J. R., *see* MCAVANEY, B. J. et al.
FRASER, P. J., *see* PEARMAN, G. I. and FRASER, P. J.
FRASER, P. J., *see* PEARMAN, G. I. et al.
FRASER, R. S., KAUFMAN, Y. J. and MAHONEY, L., 379, 393
FRASER, R. S., *see* FERRARE, R. A. et al.
FRASER, R. S., *see* KAUFMAN, Y. J. et al.
FREDERICK, J. E., *see* LUBIN, D. et al.
FREDERIKSEN, J. S., 309, 311
FREEMAN, K. H. and HAYES, J. M., 75, 92, 549, 552
FRERKING, M. A., *see* BARATH, F. T. et al.
FRIIS-CHRISTENSEN, E. and LAASEN, K., 175, 179, 184, 200, 236
FRISBIE, P. R., *see* HUDSON, J. G. and FRISBIE, P. R.
FROIDEVAUX, L., *see* WATERS, J. W. et al.
FROST, E., *see* DURKEE, P. A. et al.
FU, C., DIAZ, H. F. and FLETCHER, J. O., 211, 236
FU, C., *see* KARL, T. R. et al.
FU, R., DEL GENIO, A. D. and ROSSOW, W. B., 269, 276
FUGGLE, R. F., 487
FUGGLE, R. F., *see* OKE, T. R. and FUGGLE, R. F.
FUGLISTER, F. J., *see* GILMAN, D. L. et al.
FUHRER, K., *see* JOHNSEN, S. J. et al.
FUJIWHARA, S., 305, 311
FULTON, S. R., *see* SCHUBERT, W. H. et al.
FUNG, I. Y., *see* PRENTICE, K. C. and FUNG, I. Y.
FUNG, I., *see* HANSEN, J. E. et al.
FUNG, I., *see* KAUFMAN, Y. J. et al.
FUNG, I., *see* MATTHEWS, E. et al.
FYFE, W., *see* APSIMON, H. et al.

GAFFEN, D. J., 169, 184
GAFFEN, D. J., BARNETT, T. P. and ELLIOTT, W. P., 228, 236
GAGE, K. S. and REID, G. C., 199, 236
GAL-CHEN, T., *see* TAYLOR, B. L. et al.
GALIN, V., *see* CESS, R. D. et al.
GALL, R., YOUNG, K., SCHOTLAND, R. and SHMITZ, J., 153, 184
GALLÉE, H., 56, 58, 62
GALLÉE, H., BERGER, A. and SHACKLETON, N. J., 47, 49, 62
GALLÉE, H., *see* BERGER, A. et al.
GALLÉE, H., VAN YPERSELE, J. P., FICHEFET, TH., MARSIAT, I., TRICOT, C. and BERGER, A., 45, 47, 48, 52, 58, 59, 62
GALLÉE, H., VAN YPERSELE, J. P., FICHEFET, TH., TRICOT, C. and BERGER, A., 45, 58, 62
GALLIMORE, R. G., *see* KUTZBACH, J. E. and GALLIMORE, R. G.
GALLO, K. P., MCNAB, A. L., KARL, T. R., BROWN, J. F., HOOD, J. J. and TARPLEY, J. D., 502, 510
GALLO, K. P. *see* KARL, T. R. et al.
GALLO, K., *see* OHRING, G. et al.
GALLOWAY, J. N., 358, 393
GALLOWAY, J. N., KEENE, W. C., PSZENNY, A. A. P., WHELPDALE, D. M., SIEVERING, H., MERRILL, J. T. and BOATMAN, J. F., 381, 393
GALLOWAY, J. N., *see* CHURCH, T. M. et al.
GALLOWAY, J. N., *see* DUCE, R. A. et al.
GAMMON, R. H., *see* BATES, T. S. et al.
GANAPATHY, R., 95, 115, 118, 139
GANOR, E. J., *see* PUESCHEL, R. F. et al.
GARADZHA, M. P. and NEZVAL, YE. I., 425, 431

GARCIA, R. R., *see* HAMILTON, K. and GARCIA, R. R.
GARDINER, B. G., *see* FARMAN, J. C. et al.
GARP, 8, 9, 17
GARRELS, R. M. *see* BERNER, R. A. et al.
GARSTANG, M., *see* ANDREAE, M. O. et al.
GARSTANG, M., *see* SWAP, R. et al.
GARSTANG, M., *see* TALBOT, R. W. et al.
GASH, J. H. C., *see* SHUTTLEWORTH, W. J. et al.
GASPAR, PH., *see* ADEM, J. et al.
GATES, W. L., 35, 62, 321, 345
GATES, W. L., CUBASCH, U., MEEHL, G. A., MITCHELL, J. F. B. and STOUFFER, R. J., 321, 327, 331, 345
GATES, W. L., MITCHELL, J. F. B., BOER, G. J., CUBASCH, U. and MELESHKO, V. P., 33, 62
GATES, W. L., *see* CESS, R. D. et al.
GAUDRY, A., *see* NGUYEN, B. C. et al.
GAULT, D. E., *see* SCHULTZ, P. H. and GAULT, D. E.
GAVIN, J., KUKLA, G. and KARL, T., 545, 552
GAVIN, J., *see* KUKLA, G. et al.
GAVIN, J., *see* PLANTICO, M. S. et al.
GAZA, R. S., *see* KESSLER, R. W. et al.
GELDSETZER, H. H. J., GOODFELLOW, W. D., MCLAREN, D. J. and ORCHARD, M. J., 125, 139
GELDSETZER, H. H. J., *see* GOODFELLOW, W. D. et al.
GELDSETZER, H. H. J., *see* WANG, K. et al.
GELEYN, J. F., *see* TANRE, D. et al.
GELLER, M. A. and ALPERT, J. C., 200, 236
GENTHON, C., BARNOLA, J. M., RAYNAUD, D., LORIUS, CL., JOUZEL, J., BARKOV, N. I., KOROTKEVITCH, Y. S. and KOTLYAKOV, V. M., 52, 62
GENTHON, C., *see* JOUZEL, J. et al.
GEORGII, H.-W., *see* BÜRGERMEISTER, S. and GEORGII, H.-W.
GERLACH, T. M. and GRAEBER, E. J., 133, 139
GERMAIN, C., *see* LEGRAND, M. R. et al.
GERSTEL, J., THUNELL, R. C., ZACHOS, J. C. and ARTHUR, M. A., 110, 139
GERSTL, S. A. and ZARDECKI, A., 98, 139
GEYH, M. A., *see* RÖTHLISBERGER, F. and GEYH, M. A.
GHAN, S. J., PENNER, J. E. and TAYLOR, K. E., 545, 552
GHAN, S. J., *see* CESS, R. D. et al.
GHAN, S. J., *see* COVEY, C. et al.
GHAN, S. J., TAYLOR, K. E., PENNER, J. E. and ERICKSON III, D. J., 378, 393
GHIL, M. and LE TREUT, H., 41, 62
GHIL, M. and VAUTARD, R., 179, 184, 201, 236
GHIL, M., *see* LE TREUT, H. et al.
GHIL, M., *see* VAUTARD, R. and GHIL, M.
GIBSON, G. G., *see* HARRISON, E. F. et al.
GIBSON, G. G., *see* MINNIS, P. et al.
GIES, H. P., *see* ROY, C. T. et al.
GILL, A. E., 282, 288, 305, 311
GILLE, J. C., 431
GILLE, J. C. and LYJAK, L. V., 411, 431
GILLE, J. C. and RUSSELL III, J. M., 411, 431
GILLETT, R. W., *see* AYERS, G. P. et al.
GILLETT, R. W., *see* BERRESHEIM, H. et al.
GILLETTE, D. A. and BLIFFORD, I. H., 381, 382, 393
GILLILAND, R. L., 175, 184
GILLILAND, R. L., *see* EDDY, J. A. et al.
GILMAN P. A., *see* EDDY, J. A. et al.
GILMAN, D. L., FUGLISTER, F. J. and MITCHELL JR., J. M., 191, 192, 236
GILMAN, D. L., *see* EDDY, J. A. et al.
GILMORE, J. S., *see* ORTH, C. J. et al.
GILMOUR, I., *see* WOLBACH, W. S. et al.
GILMOUR, I., WOLBACH, W. S. and ANDERS, E., 98, 101, 139, 140
GIORGI, F., 321
GIORGI, F. and MEARNS, L. O., 345
GIVEN, Q., *see* BOERSMA, A. et al.
GLANCY, R. T., *see* BRYANT, H. K. et al.
GLANTZ, M. H., 229, 236, 438, 470
GLANTZ, M. H., KATZ, R. W. and NICHOLLS, N., 207, 236
GLASS, B. P., 118, 140
GLEN, W., 96, 129, 134, 140
GLOERSEN, P., *see* PARKINSON, C. L. and GLOERSEN, P.
GOBLE, R., *see* YERSEL, M. and GOBLE, R.
GODINEZ, L., *see* JAUREGUI, E. et al.
GODOWITCH, J. M., 496, 510
GODOWITCH, J. M., *see* CHING, J. K. S. et al.
GODOWITCH, J. M., *see* CLARKE, J. F. et al.
GOETZ, S. J., *see* HALL, F. G. et al.
GOLDAN, P., *see* FEHSENFELD, F. et al.
GOLDBERG, B., *see* CORRELL, D. L. et al.
GOLDBERG, E. D., *see* FERGUSON, W. S. et al.
GOLDBERG, R., *see* RIND, D. et al.
GOLDEN, D. M., *see* DEMORE, W. B. et al.
GOLDMAN, A., *see* MANKIN, W. G. et al.
GOLDSTEIN, W., *see* MOHNEN, V. A. et al.
GOODESS, C. M., PALUTIKOF, J. P. and DAVIES, T. D., 56, 59, 62
GOODESS, C. M., *see* BRADLEY, R. S. et al.
GOODESS, C. M., *see* JONES, P. D. et al.
GOODESS, C. M., *see* KELLY, P. M. et al.
GOODESS, C. M., *see* SEAR, C. B. et al.
GOODFELLOW, W. D., 114
GOODFELLOW, W. D. and MCLAREN, D. J., 125
GOODFELLOW, W. D., GELDSETZER, H. H. J., MCLAREN, D. J., ORCHARD, M. J. and KLAPPER, G., 124, 140
GOODFELLOW, W. D., NOWLAN, G. S., MC-CRACKEN, A. D., LENZ, A. C. and GREGOIRE, D. C., 114, 116, 125, 126, 140

GOODFELLOW, W. D., *see* GELDSETZER, H. H. J. et al.
GOODFELLOW, W. D., *see* MCLAREN, D. J. and GOODFELLOW, W. D.
GOODRICH, V. R., *see* CORRELL, D. L. et al.
GORDON, A. L., ZEBIAK, S. E. and BRYAN, K., 226, 236
GORE, A., 4, 17
GORNITZ, V., 459, 470
GORNITZ, V. and SOLOW, A., 231, 236
GORNITZ, V., *see* HENDERSON-SELLERS, A. and GORNITZ, V.
GORSE, J., 465, 470
GOUGH, D., 81, 92
GOUVEIA, A. D., *see* SATHYENDRANATH, S. et al.
GOW, A. J., *see* ALLEY, R. B. et al.
GOW, A. J., *see* TAYLOR, K. C. et al.
GOWARD, S. N., 492, 502, 510
GOWARD, S. N., MARKHAM, B., DYE, D. G., DULANEY, W. and YANG, J., 269, 276
GRAEBER, E. J., *see* GERLACH, T. M. and GRAEBER, E. J.
GRAEDEL, T. E. and CRUTZEN, P. J., 447, 470
GRAEME, E., *see* ROY, C. T. et al.
GRAETZ, D., FISHER, R. and WILSON, M. F., 440, 470
GRAHAM, N., 168, 184
GRAHAM, N. E. and BARNETT, T. P., 209, 237
GRAHAM, N. E., MICHAELSEN, J. and BARNETT, T. P., 214, 237
GRAHAME, N. S., *see* MITCHELL, J. F. B. et al.
GRAM, M. B., *see* BARATH, F. T. et al.
GRANIER, C. and BRASSEUR, G. P., 415, 431
GRANIER, C., *see* BRASSEUR, G. P. and GRANIER, C.
GRANIER, C., *see* HAUGLUSTAINE, D. A. et al.
GRANIER, C., *see* MADRONICH, S. and GRANIER, C.
GRANT, K., *see* MILLER, A. J. et al.,
GRANT, W. B., 425, 431
GRANT, W. B. et al., 413, 431
GRAS, J. L., 372, 393
GRASSL, H., 360, 367, 368, 393
GRASSL, H. and LEVKOV, L., 368, 393
GRATZ, A. J., NELLIS, W. J. and HINSEY, N. A., 134, 140
GRAUSTEIN, W. C., *see* SAVOIE, D. L. et al.
GRAY, M., *see* DANARD, M. et al.
GRAY, W. M., 310, 311
GRAY, W. M., *see* LANDSEA, C. W. and GRAY, W. M.
GRAY, W. M., *see* LIGHTHILL, J. et al.
GRAY, W. M., SHEAFFER, J. D. and KNAFF, J. A., 213, 237
GRAYBILL, D. A. and SHIYATOV, S. G., 194, 237
GRECO, A. M., *see* BETZER, P. R. et al.
GRECO, A. M., *see* SWAP, R. et al.
GRECO, S., *see* SWAP, R. et al.
GREENBERG, J. P., *see* ZIMMERMAN, P. R. et al.
GREENLAND ICE-CORE PROJECT MEMBERS (GRIP), 25, 62, 226, 237, 389, 393, 534
GREENWALD, T. J., *see* STEPHENS, G. L. and GREENWALD, T. J.
GREENWOOD, D. R., *see* WING, S. L. and GREENWOOD, D. R.
GREGOIRE, D. C., *see* GOODFELLOW, W. D. et al.
GREGORY R. T., *see* RICH, T. H. et al.
GREGORY, G. L., *see* ANDREAE, M. O. et al.
GREGORY, J. M., 327, 333, 334, 345
GREGORY, J. M. and MITCHELL, J. F. B., 179, 180, 184
GRIEVE, R. A. F., 118, 140
GRIFFIN, J. J., *see* FERGUSON, W. S. et al.
GRIFFIS, K. and CHAPMAN, D. J., 98, 140
GRIFFITHS, D. C., *see* TEGART, W. J. MCG. et al.
GRIMM, E. C., JACOBSON, G. L., WATTS, W. A., HANSEN, B. C. and MAASCH, K. A., 25, 62
GRIMMOND, C. S. B., 491, 493, 511
GRIMMOND, C. S. B. and OKE, T. R., 493, 494, 498, 511
GRIMMOND, C. S. B., OKE, T. R. and CLEUGH, H. A., 481, 493, 494, 500, 511
GRIMMOND, C. S. B., *see* SCHMID, H. P. et al.
GRINSPOON, D., *see* ZAHNLE, K. and GRINSPOON, D.
GROISMAN, P. Y., KARL, T. R. and KNIGHT, R. W., 58, 62
GROISMAN, P. YA., KOKNAEVA, V. V., BELOKRYLOVA, T. A. and KARL, T. R., 164, 184
GROISMAN, P. YA., *see* JONES, P. D. et al.
GROISMAN, P. YA., *see* VINNIKOV, K. YA. et al.
GROOTES, P. M., *see* ALLEY, R. B. et al.
GROOTES, P. M., STUIVER, M., WHITE, J. W. C., JOHNSEN, S. and JOUZEL, J., 25, 62
GROSSMAN, A., *see* SHINE, K. P. et al.
GROTCH, S. L., *see* ELSAESSER, H. W. et al.
GROVE, A. T., *see* STREET, F. A. and GROVE, A. T.
GROVE, J. M., 169, 184, 196, 237
GRUBER, A., *see* OHRING, G. et al.
GRUSZCZYNSKI, M., HALAS, S., HOFFMAN, A. and MALKOWSKI, K., 120, 140
GRUSZCZYNSKI, M., HALAS, S., HOFFMAN, A., MALKOWSKI, K. and VEIZER, J., 120, 140
GUARD, C. P., *see* LIGHTHILL, J. et al.
GUEDALIA, D., *see* ESTOURNEL, C. et al.
GUENTHER, A. B., *see* FEHSENFELD, F. et al.
GUENTHER, P. R., *see* KEELING, C. D. et al.
GUETTER, P. J., *see* KUTZBACH, J. E. and GUETTER, P. J.
GUILDERSON, TH. P., FAIRBANKS, R. G. and RUBENSTONE, J. L., 24, 36, 62
GUIOT, J., PONS, A., DE BEAULIEU, J. L. and REILLE, M., 22, 62
GUIOT, J., *see* BERGER, A. et al.
GUMERMAN, G. J., 194, 237

GUNDESTRUP, N. S., *see* DANSGAARD, W. et al.
GUNDESTRUP, N. S., *see* JOHNSEN, S. J. et al.
GUNDESTRUP, N. S., *see* TAYLOR, K. C. et al.
GUNST, R. F., BASU, S. and BRUNELL, R., 159, 184
GUNTER, R. L., *see* BOATMAN, J. F. et al.
GUPTA, S. K., DARNELL, W. L. and WILBER, A. C., 265, 276
GUPTA, S. K., *see* DARNELL, W. L. et al.
GUTZLER, D. F., 228, 237
GUTZLER, D. F. and HARRISON, D. E., 218, 237
GUTZLER, D. S., *see* WALLACE, J. M. and GUTZLER, D. S.

HACK, J. J., *see* HURRELL, J. et al.
HACK, J. J., *see* SCHUBERT, W. H. et al.
HACKMAN, C. H., *see* KAYE, J. A. and HACKMAN, C. H.
HADLEY CENTER, 33, 63
HAGE, K. D., 503, 511
HAGGERTY, B. M., *see* RAMPINO, M. R. and HAGGERTY, B. M.
HAHN, C. J., *see* WARREN, S. G. et al.
HAHN, D. G., *see* MANABE, S. and HAHN, D. G.
HAKKARINEN, I. M., *see* ADLER, R. F. et al.
HALAS, S., *see* GRUSZCZYNSKI, M. et al.
HALES, J. M., *see* CHARLSON, R, J. et al.
HALL, D. K., *see* CHANG, A. T. C. et al.
HALL, D., *see* RUNNING, S. W. et al.
HALL, F. G., HUEMMRICH, K. F., GOETZ, S. J., SELLERS, P. J. and NICKESON, J. E., 270, 276
HALL, F. G., *see* SELLERS, P. J. et al.
HALL, M. A., 553
HALL, M. A., *see* SHACKLETON, N. J. et al.
HALL, M., *see* BOERSMA, A. et al.
HALL, M., *see* SANCETTA, C. et al.
HALL, N. M. J., HOSKINS, B. J., VALDES, P. J. and SENIOR, C. A., 304, 311
HALL, W. D., *see* WISCOMBE, W. J. et al.
HALLAM, A., 114, 119, 140
HALLETT, J., HUDSON, J. G. and ROGERS, C. F., 376, 393
HALLETT, J., *see* HUDSON, J. G. et al.
HALLETT, J., *see* ROGERS, C. F. et al.
HALPERT, M. S., *see* ROPELEWSKI, C. F. and HALPERT, M. S.
HALTINER, G. J. and WILLIAMS, R. T., 282, 311
HAMELIN, B., *see* BARD, E. et al.
HAMILL, P., *see* MCCORMICK, M. P. et al.
HAMILTON, K., 205, 237
HAMILTON, K. and GARCIA, R. R., 222, 237
HAMMER, C. U., CLAUSEN, H. B. and DANSGAARD, W., 177, 184
HAMMER, C. U., CLAUSEN, H. B., DANSGAARD, W., NEFTEL, A., KRISTINSDOTTIR, P. and JOHNSON, E., 523, 530, 534
HAMMER, C. U., *see* DANSGAARD, W. et al.
HAMMER, C. U., *see* JOHNSEN, S. J. et al.
HAMMER, C. U., *see* TAYLOR, K. C. et al.
HAMMOND, A. L., 477, 511
HAMPSON, R. F., *see* DEMORE, W. B. et al.
HANES, J., *see* SEBAI, A. et al.
HANSEN, A. D. A., ARTZ, R. S., PSZENNY, A. A. P. and LARSON, R. E., 393
HANSEN, A. D. A., *see* WHELPDALE, D. M. et al.
HANSEN, A. R., *see* SALTZMAN, B. et al.
HANSEN, B. C., *see* GRIMM, E. C. et al.
HANSEN, J., 35, 318
HANSEN, J. and LACIS, A., 80, 92, 326, 345, 352, 359, 362, 382, 393
HANSEN, J. and LEBEDEFF, S., 158
HANSEN, J. E., JOHNSON, D., LACIS, A., LEBEDEFF, S., LEE, P., RIND, D. and RUSSELL, G., 437, 470
HANSEN, J. E., LACIS, A., RIND, D., RUSSELL, G., STONE, P., FUNG, I., RUEDY, R. and LERNER, J., 35, 63, 466, 470
HANSEN, J. E., *see* WANG, W.-C. et al.
HANSEN, J., FUNG, I., LACIS, A., RIND, D., LEBEDEFF, S., RUEDY, R., RUSSELL, G. and STONE, P., 175, 184
HANSEN, J., LACIS, A., RUEDY, R. and SATO, M., 176, 184, 205, 207, 237, 352, 393
HANSEN, J., LACIS, A., RUEDY, R., SATO, M. and WILSON, H., 17, 80, 92, 336, 345, 352, 384, 393
HANSEN, J., RIND, D., DELGENIO, A., LACIS, A., LEBEDEFF, S., PRATHER, M., RUEDY, R. and KARL, T., 334, 345
HANSEN, J., ROSSOW, W. and FUNG, I., 336, 345
HANSEN, J., SATO, M., LACIS, A. and RUEDY, R., 336, 345
HANSEN, J., *see* CHARLSON, R, J. et al.
HANSEN, J., *see* LACIS, A. et al.
HANSEN, J., *see* LORIUS, C. et al.
HANSEN, J.E and LEBEDEFF, S., 184
HANSEN, K., *see* FLYGER, H. et al.
HANSEN, L., *see* DUCE, R. A. et al.
HAO, W. M., LIU, M.-H. and CRUTZEN, P. J., 451, 471
HAO, W. M., *see* SANHUEZA, E. et al.
HAO, W. M., *see* SCHARFFE, D. et al.
HARDY, C. C., *see* EINFELD, W. et al.
HARDY, C. C., *see* WARD, D. E. and HARDY, C. C.
HARMS, D. E., RAMAN, S. and MADALA, R. V., 263, 276
HARNACK, R. P. and LANDSBERG, H. E., 504, 511
HARRELL, J., *see* BRISKIN, M. and HARRELL, J.
HARRINGTON, C. R., 176, 184
HARRIS, W. M., *see* BARATH, F. T. et al.
HARRISON, D. E. and SCHOPF, P. S., 211, 237
HARRISON, D. E., *see* GUTZLER, D. F. and HARRISON, D. E.
HARRISON, E. F., 211
HARRISON, E. F., MINNIS, P., BARKSTROM, B. R. and GIBSON, G. G., 254, 276

HARRISON, E. F., MINNIS, P., BARKSTROM, B. R., RAMANATHAN V., CESS, R. D. and GIBSON, G. G., 256, 276
HARRISON, E. F., *see* CESS, R. D. et al.
HARRISON, E. F., *see* MINNIS, P. et al.
HARRISON, E. F., *see* RAMANATHAN, V. et al.
HARRISON, S. P., *see* STREET-PERROTT, F. A. and HARRISON, S. P.
HARRISS, R. C., *see* ANDREAE, M. O. et al.
HARRISS, R. C., *see* TALBOT, R. W. et al.
HART, T. L., 456, 471
HARTLEY, W. N., 408, 431
HARTMANN, D. L., RAMANATHAN, V., BERRIOR, A. and HUNT, G. E., 254, 276
HARTMANN, D. L., *see* KLEIN, S. A. and HARTMANN, D. L.
HARTMANN, D., *see* RAMANATHAN, V. et al.
HARTZELL, F. Z., 154, 184
HARVEY, L. D. D., 52, 54, 63, 448, 471
HARVEY, L. D. D. and SCHNEIDER, S. H., 32, 63
HARWOOD, D. M., 82, 92
HARWOOD, R. S., *see* WATERS, J. W. et al.
HASSAN, F. A., 194, 237
HASSELMANN, K., 43, 63, 175, 184
HASSELMANN, K., *see* FRANKIGNOUL, C. and HASSELMANN, K.
HASSELMANN, K., *see* HERTERICH, K. and HASSELMANN, K.
HASSELMANN, K., *see* SAUSEN, R. et al.
HASTENRATH, S., 229, 237
HASTENRATH, S. and HELLER, L., 218, 237
HAUGLUSTAINE, D. A., GRANIER, C., BRASSEUR, G. P. and MAGIE, G., 319, 336, 339, 345, 418, 419, 431
HAUGLUSTAINE, D. A., *see* SHINE, K. P. et al.
HAUSCHILD, H., *see* RUDOLF, B. et al.
HAY, J. E., 486, 511
HAY, J. E., *see* SALINGER, M. J. et al.
HAY, W., *see* BRASS, G. et al.
HAYDER, M. E., *see* WATTS, R. G. and HAYDER, M. E.
HAYES JR., D. R., *see* CORRELL, D. L. et al.
HAYES, D. E. and FRAKES, L. A., 82, 92
HAYES, J. M., *see* FREEMAN, K. H. and HAYES, J. M.
HAYS, J. D., IMBRIE, J. and SHACKLETON, N. J., 31, 40, 63
HAYS, J. D., *see* IMBRIE, J. et al.
HAYS, J. D., *see* MARTINSON, D. G. et al.
HAYS, P. B., *see* WALKER, J. C. G. et al.
HAYWARD, T. L., *see* VENRICK, E. L. et al.
HE, Q. X., *see* HSU, K. J. et al.
HEAMAN, L. M., *see* LECHEMINANT, A. N. and HEAMAN, L. M.
HEATH, D. F., KRUEGER, A. J. and PARK, H., 409, 431
HEATHCOTE, R. L., 468, 471
HECHT, A., 21, 63
HECKLEY, W. A., KELLY, G. and TIEDTKE, M., 264, 276
HEGG, D. A., RADKE, L. F. and HOBBS, P. V., 372, 393
HEGG, D. A., *see* RADKE, L. F. et al.
HEIDAM, N. Z., *see* FLYGER, H. et al.
HEIM JR., R. R., 278
HEIM JR., R. R., *see* KARL, T. R. et al.
HEIMAN, A., 127
HEIMAN, A., FLEMING, T. H., ELLIOT, D. H. and FOLAND, K. A., 140
HEIMANN, M., 533
HEIMANN, M., *see* KEELING, C. D. et al.
HEINRICH, H., *see* BOND, G. et al.
HEINTZENBERG, J., 382, 393
HEINTZENBERG, J. and LARSSEN, S., 380, 393
HEISER, M. D., *see* SELLERS, P. J. et al.
HELAS, G., *see* BINGEMER, H. G. et al.
HELD, I. M., 192, 226, 237, 307, 308, 311
HELD, I. M., *see* STEPHENSON, D. and HELD, I. M.
HELD, I. M., *see* SUAREZ, M. J. and HELD, I. M.
HELLDEN, U., 436, 441, 452, 471
HELLER, L., *see* HASTENRATH, S. and HELLER, L.
HENDERSON-SELLERS, A., 5, 17, 18, 157, 164, 184, 231, 237, 435, 453, 459, 465, 466, 467, 471
HENDERSON-SELLERS, A. and GORNITZ, V., 438, 440, 441, 452, 455, 459, 461, 471
HENDERSON-SELLERS, A. and MCGUFFIE, K., 8, 10, 12, 17, 232, 237, 379, 393, 467, 471, 545, 552
HENDERSON-SELLERS, A. and PITMAN, A. J., 440, 471
HENDERSON-SELLERS, A. and WILSON, M. F., 452, 471
HENDERSON-SELLERS, A., DICKINSON, R. E., DURBIDGE, T. B., KENNEDY, P. J., MCGUFFIE, K. and PITMAN, A. J., 435, 457, 459, 461, 471
HENDERSON-SELLERS, A., DURBIDGE, T. B., PITMAN, A. J., DICKINSON, R. E., KENNEDY, P. J. and MCGUFFIE, K., 231, 237
HENDERSON-SELLERS, A., *see* DICKINSON, R. E. and HENDERSON-SELLERS, A.
HENDERSON-SELLERS, A., *see* DICKINSON, R. E. et al.
HENDERSON-SELLERS, A., *see* HENDERSON-SELLERS, B. et al.
HENDERSON-SELLERS, A., *see* JONES, P. A. and HENDERSON-SELLERS, A.
HENDERSON-SELLERS, A., *see* MCGUFFIE, K. et al.
HENDERSON-SELLERS, A., *see* SHINE, K. P. et al.
HENDERSON-SELLERS, B., HENDERSON-SELLERS, A., BENBOW, S. M. P. and MCGUFFIE, K., 6, 17
HENG-YI, W., *see* SMITH, E. A. et al.
HENSE, A., KRAHE, P. and FLOHN, H., 228, 237
HEREFORD, V., *see* DESHLER, T. et al.
HERMAN, B. M., BROWNING, S. R. and RABINOFF, R., 389, 393

HERMAN, J. R., *see* STOLARSKI, R. S. et al.
HERTERICH, H., *see* LAUTENSCHLAGER, M. and HERTERICH, H.
HERTERICH, K. and HASSELMANN, K., 43, 63
HERTOGEN, J., *see* SMIT, J. and HERTOGEN, J.
HESS, M. et al., 486, 511
HEUER, K., 95, 140
HEUSSER, L., *see* SANCETTA, C. et al.
HEWITT, C. N., *see* FEHSENFELD, F. et al.
HICKEY, L. J., 111, 140
HICKEY, L. J., *see* JOHNSON, K. R. and HICKEY, L. J.
HICKS, B. B., *see* DUCE, R. A. et al.
HIDE, R., DICKEY, J. O., MARCUS, S. L., ROSEN, R. D. and SALSTEIN, D. A., 292, 311
HILDEBRAND, A. R. et al., 96, 140
HILDEBRAND, P. H. and ACKERMAN, B., 496, 503, 511
HILL, G. F., *see* ANDREAE, M. O. et al.
HILL, R. I., *see* CAMPBELL, I. H. et al.
HILL, W. J., *see* BOJKOV, R. D. et al.
HINSEY, N. A., *see* GRATZ, A. J. et al.
HIPPEL, D. V., *see* SUBAK, S. et al.
HIRSCHLER, M. M., *see* CULLIS, C. F. and HIRSCHLER, M. M.
HIRST, A. C., 211, 237
HIRST, A. C., *see* BATTISTI, D. S. and HIRST, A. C.
HITCHMAN, M. H. and BRASSEUR, G. P., 413, 431
HITCHMAN, M. H., *see* BRASSEUR, G. P. and HITCHMAN, M. H.
HITCHMAN, M. H., *see* BRASSEUR, G. P. et al.
HJELMFELT, M. R., 479, 511
HOBBS, P. V., 372, 394
HOBBS, P. V. and RADKE, L. F., 376, 394
HOBBS, P. V., RADKE, L. F. and SHUMWAY, S. E., 372, 380, 394
HOBBS, P. V., *see* HEGG, D. A. et al.
HOBBS, P. V., *see* KING, M. D. et al.
HOBBS, P. V., *see* RADKE, L. F. et al.
HODYCH, J. P., *see* DUNNING, G. R. and HODYCH, J. P.
HOECKER, W. H., *see* ANGELL, J. K. et al.
HOFER, H., *see* STAUFFER, B. et al.
HOFF, R. M., *see* BARRIE, L. A. and HOFF, R. M.
HOFFERT, M. I. and COVEY, C., 52, 63, 80, 92, 549, 552
HOFFMAN, A., *see* GRUSZCZYNSKI, M. et al.
HOFFMAN, D. J., *see* CHARLSON, R, J. et al.
HOFFMAN, M. E., *see* SWAID, H. and HOFFMAN, M. E.
HOFMANN, D. J., 389, 394
HOFMANN, D. J., *see* CHARLSON, R. J. et al.
HOFMANN, D. J., *see* DESHLER, T. et al.
HOGAN, A. W., *see* FLYGER, H. et al.
HÖGSTRÖM, U., BERGSTRÖM, H. and ALEXANDERSSON, H., 496, 511
HOLBEN, B. N., *see* KAUFMAN, Y. J. et al.
HOLDEN, J. R., *see* BARATH, F. T. et al.
HOLDRIDGE, L. R., 466, 471
HOLLAND, G. J., 289, 290, 304, 305, 309, 311
HOLLAND, G. J. and DIETACHMAYER, G. S., 305, 311
HOLLAND, G. J. and LANDER, M., 311
HOLLAND, G. J., MCBRIDE, J. L. and NICHOLLS, N. N., 310, 311
HOLLAND, G. J., *see* FAIRALL, C. W. et al.
HOLLAND, G. J., *see* LANDER, M. and HOLLAND, G. J.
HOLLAND, G. J., *see* LIGHTHILL, J. et al.
HOLLAND, G. J., *see* RITCHIE, E. A. and HOLLAND, G. J.
HOLLAND, G. J., *see* RITCHIE, E. A. et al.
HOLLAND, G. J., *see* SIMPSON, J. S. et al.
HOLLAND, G. J., *see* WANG, Y. and HOLLAND, G. J.
HOLLAND, H. D., *see* KASTING, J. F. et al.
HOLLINGSWORTH, A., HORN, J. and UPPALA, S., 264, 276
HOLLOWAY JR., J. L., *see* MANABE, S. and HOLLOWAY JR., J. L.
HOLLOWAY, J. M., *see* HUSAR, R. B. et al.
HOLOPAINEN, E., *see* FORTELIUS, C. and HOLOPAINEN, E.
HOLSER, W. T. and MAGARITZ, M., 120, 140
HOLSER, W. T. and SCHÖNLAUB, H. P., 122
HOLSER, W. T., SCHÖNLAUB, H. P., BOECKELMANN, K. and MAGARITZ, M., 120, 140
HOLSER, W. T., *see* MAGARITZ, M. et al.
HOLTON, J. R., 202, 237, 282, 286, 288, 289, 290, 311
HOLTON, J. R., *see* WEBSTER, P. J. and HOLTON, J. R.
HOLZER, W., *see* BRASS, G. et al.
HONG, Y., *see* APSIMON, H. et al.
HONRATH, R. E., *see* MAYEWSKI, P. A. et al.
HOOD, J. J., *see* GALLO, K. P. et al.
HOOD, L. L., *see* MCCORMICK, M. P. and HOOD, L. L.
HOPPEL, W. A., 372, 380, 394
HOREL, J. D. and WALLACE, J. M., 204, 220, 237, 297, 304, 311
HORN, J., *see* HOLLINGSWORTH, A. et al.
HOSKINS, B. J., 289, 298, 311
HOSKINS, B. J. and KAROLY, D., 297, 305, 311
HOSKINS, B. J. and VALDES, P. J., 304, 312
HOSKINS, B. J., HSU, H. H., JAMES, I. N., MASUTANI, P. D, SARDESHMUKH, P. D. and WHITE, G. H., 283, 312
HOSKINS, B. J., MCINTYRE, M. E. and ROBERTSON, A. W., 289, 311
HOSKINS, B. J., *see* HALL, N. M. J. et al.
HOSKINS, B. J., *see* SARDESHMUKH, P. D. and HOSKINS, B. J.
HOU, H., *see* WANG, K. et al.

HOUGHTON, J. J., JENKINS, G. J. and EPHRAUMS, J. J., 317, 319, 320, 321, 323, 324, 325, 329, 330, 331, 345
HOUGHTON, J. T., 17
HOUGHTON, J. T., CALLANDER, B. A. and VARNEY, S. K., 2, 18, 33, 57, 63, 177, 180, 184, 307, 312, 317, 319, 320, 321, 323, 324, 325, 326, 327, 333, 334, 335, 344, 345, 358, 359, 362, 384, 394, 437, 444, 446, 465, 467, 471
HOUGHTON, J. T., JENKINS, G. J. and EPHRAUMS, J. J., 2, 6, 15, 18, 33, 57, 63, 177, 180, 184, 348, 359, 362, 384, 394, 437, 446, 447, 449, 450, 465, 467, 471
HOUGHTON, R. A., 447, 448, 464, 465, 471
HOUGHTON, R. A. and SKOLE, D. L., 447, 448, 471
HOUGHTON, R. A., SCHLESINGER, W. H., BROWN, S. and RICHARDS, J. F., 448, 471
HOUSE, M. R., *see* BECKER, R. T. et al.
HOUSLEY, R. A., *see* CORFIELD, R. M. et al.
HOVINE, S., 43, 63
HOWARD, C. J., *see* DEMORE, W. B. et al.
HOWARD, K. W., *see* NEGRI, A. J. et al.
HOWARD, L., 478, 511
HOWARD, W. R., *see* IMBRIE, J. et al.
HOYT, D. V., 236
HSIUNG, J. and NEWELL, R. E., 229, 237
HSIUNG, J., *see* BOTTOMLEY, M. et al.
HSU, H. H., *see* HOSKINS, B. J. et al.
HSU, K. J., 102, 140
HSU, K. J. and MCKENZIE J. A., 103, 104, 114, 107, 140
HSU, K. J., MCKENZIE J. A., WEISSERT, H. et al., 140
HSU, K. J., MCKENZIE, J. A. and HE, Q. X., 102, 103, 109, 112, 140
HSU, K. J., MCKENZIE, J. A., WEISSERT, H. et al., 102, 103, 109, 112
HSU, K. J., OBERHANSLI, H. et al., 126, 140
HUANG, Z., *see* FAN, D. et al.
HUBER, B. T., *see* BARRERA, E. et al.
HUDSON, H. S., 244
HUDSON, H. S., *see* WILLSON, R. C. and HUDSON, H. S.
HUDSON, J. G., 372, 388, 394
HUDSON, J. G. and FRISBIE, P. R., 372, 394
HUDSON, J. G., HALLETT, J. and RODGERS, C. F., 376, 394
HUDSON, J. G., *see* HALLETT, J. et al.
HUDSON, J. G., *see* LEAITCH, W. R. et al.
HUDSON, J. G., *see* ROGERS, C. F. et al.
HUDSON, J. G., *see* SAX, R. I. and HUDSON, J. G.
HUEBERT, B. J. and LAZRUS, A. L., 381, 394
HUEMMRICH, K. F., *see* HALL, F. G. et al.
HUENNEKE, L. F., *see* SCHLESINGER, W. H. et al.
HUETE, A. R., *see* RUNNING, S. W. et al.
HUFF, F. A. and CHANGNON JR., S. A, 504, 511
HUFF, F. A., *see* CHANGNON JR., S. A. and HUFF, F. A.
HUFF, F. A., *see* CHANGNON JR., S. A. et al.
HUGHES, M. K. and DIAZ, H. F., 5, 18, 194, 237
HUGHES, T. J., 40, 63
HUGHES, T. J., DENTON, G. H., ANDERSON, D. M., ANDERSON, B. G., SCHILLING, D. H., FASTHOOK, J. L. and LINGLE, C. S., 47, 63
HUGHES, T. J., *see* DENTON, G. H. and HUGHES, T. J.
HULME, M., 155, 166, 185
HUMERNIK, F. M., *see* LEZBERG, E. A. et al.
HUNT JR., E. R., *see* RUNNING, S. W. et al.
HUNT, G. E., *see* HARTMANN, D. L. et al.
HUNTLEY, M. E., MARIN, V. and ESCRITOR, F., 541, 552
HUON, S., *see* BOND, G. et al.
HURRELL, J. W. and TRENBERTH, K. E., 263, 276
HURRELL, J. W., *see* TRENBERTH, K. E. et al.
HURRELL, J., HACK, J. J. and BAUMHEFNER, D. P., 321, 345
HUSAR, R. B. and STOWE, L. L., 349, 376, 379, 394
HUSAR, R. B. and WILSON JR., W. E., 379, 384, 394
HUSAR, R. B., HOLLOWAY, J. M. and PATTERSON, D. E., 379, 394
HUT, P., ALVAREZ, W. et al., 124, 140
HUTCHISON, J. H., *see* ESTES, R. and HUTCHISON, J. H.
HUYBRECHTS, PH., 47, 63
HUYBRECHTS, PH., LETREGUILLY, A. and REEH, N., 58, 63
HVIDGERG, C. S. *see* DANSGAARD, W. et al.
HVLDBORG, C. S., *see* DANSGAARD, W. et al.
HYDE, W. T. and PELTIER, W. R., 40, 41, 63
HYDE, W. T., *see* SHORT, D. A. et al.

ILYAS, M., *see* MADRONICH, S. et al.
IMBRIE, J. and IMBRIE, J. Z., 31, 42, 49, 63, 231, 237
IMBRIE, J. and IMBRIE, K. P., 30, 63
IMBRIE, J. Z., *see* IMBRIE, J. and IMBRIE, J. Z.
IMBRIE, J., BERGER, A., BOYLE, E. A., CLEMENS, S. C., DUFFY, A., HOWARD, W. R., KUKLA, G., KUTZBACH, J., MARTINSON, D. G., MCINTYRE, A., MIX, A. C., MOLFINO, B., MORLEY, J. J., PETERSON, L. C., PISIAS, N. G., PRELL, W. L., RAYMO, M. E., SHACKLETON, N. J. and TOGGWEILER, J. R., 64
IMBRIE, J., BOYLE, E. A., CLEMENS, S. C., DUFFY, A., HOWARD, W. R., KUKLA, G., KUTZBACH, J. E., MARTINSON, D. G., MCINTYRE, A., MIX, A. C., MOLFINO, B., MORLEY, J. J., PETERSON, L. C., PISIAS, N. G., PRELL, W. L., RAYMO, M. E., SHACKLETON, N. J. and TOGGWEILER, J. R., 22, 31, 37, 63
IMBRIE, J., BOYLE, E. A., CLEMENS, S. C., DUFFY, A., HOWARD, W. R., KUKLA, G., KUTZBACH, J.,

MARTINSON, D. G., MCINTYRE, A., MIX, A. C., MOLFINO, B., MORLEY, J. J., PETERSON, L. C., PISIAS, N. G., PRELL, W. L., RAYMO, M. E., SHACKLETON, N. J., BERGER, A. and TOGGWEILER, J. R., 31
IMBRIE, J., HAYS, J. D., MARTINSON, D. G., MCINTYRE, A., MIX, A. C., MORLEY, J. J., PISIAS, N. G., PRELL, W. L. and SHACKLETON, N. J., 31, 40, 47, 49, 63
IMBRIE, J., MCINTYRE, A. and MIX, A. C., 31, 63
IMBRIE, J., *see* HAYS, J. D.
IMBRIE, J., *see* MARTINSON, D. G. et al.
IMBRIE, J., *see* SANCETTA, C. et al.
IMBRIE, J., *see* SANTER, B. et al.
IMBRIE, J., *see* SHACKLETON, N. J. and IMBRIE, J.
IMBRIE, J., *see* SHACKLETON, N. J. et al.
IMBRIE, K. P., *see* IMBRIE, J. and IMBRIE, K. P.
INGRAM, W. J., *see* MITCHELL, J. F. B. et al.
INTERGOVERNMENTAL PANEL ON CLIMATE CHANGE (IPCC), 2, 18, 72, 80, 92, 227, 230, 231, 232, 238, 549, 552
IRVING, W. M., *see* BRYSON, R. A. et al.
ISAAC, G. A., *see* LEAITCH, W. R. et al.
ISAACS, R. G., *see* WANG, W.-C. et al.
ISAKA, H., *see* FOUQUART, Y. and ISAKA, H.
ISAKSEN, I. S. A., 319, 338, 345
ISAKSEN, I. S. A., RAMASWAMY, V., RODHE, H. and WIGLEY, T. M. L., 451, 472
ISAKSEN, I. S. A., *see* WANG, W.-C. and ISAKSEN, I. S. A.
ISRIC, 440, 472
IVERSEN, P., *see* JOHNSEN, S. J. et al.
IVERSEN, T., *see* TARRASON, L. and IVERSEN, T.
IVERSON, R. L., *see* BERRESHEIM, H. et al.
IVEY, J. P., *see* AYERS, G. P. et al.
IVY, S., *see* BOND, G. et al.

JABLONSKI, D., 113, 141
JACKSON, D. L., *see* TJEMKES, S. A. et al.
JACOB, D. J., *see* ANDREAE, M. O. et al.
JACOB, D. J., *see* BINGEMER, H. G. et al.
JACOB, D. J., *see* SPIRO, P. A. et al.
JACOBSON, G. L., *see* GRIMM, E. C. et al.
JAEGER, J.-J., *see* COURTILLOT, V. et al.
JAEGER, J.-J., *see* ROCCHIA, R. et al.
JAENICKE, R., 358, 364, 394, 397
JAENICKE, R., *see* PROSPERO, J. M. et al.
JAESCHKE, W. A., *see* ANDREAE, M. O. and JAESCHKE, W. A.
JAIPRAKASH, B. C., SINGH, J. and RAJU, D. S. N., 127, 141
JAMES, I. N., *see* HOSKINS, B. J. et al.
JAMES, R. W. and FOX, P. T., 153, 185
JAMES, T. C., *see* ROCHE, A. E. et al.
JANACEK, T. R., 84, 86, 92
JANACEK, T. R. and REA, D. K., 83, 84, 86, 92
JANACEK, T. R., *see* MILLER, K. G. et al.
JANACEK, T. R., *see* REA, D. K. et al.
JANETOS, A., *see* WATSON, R. T. et al.
JANETOS, A., *see* WATSON, R. W. et al.
JANICOT, S., *see* DZIETARA, S. and JANICOT, S.
JANOWIAK, J. E., 156, 185, 218, 238
JANOWIAK, J. E., *see* ARKIN, P. A. and JANOWIAK, J. E.
JANTSCHIK, R., *see* BOND, G. et al.
JARNOT, R. F., *see* BARATH, F. T. et al.
JARNOT, R. F., *see* WATERS, J. W. et al.
JARRELL, W. M., *see* SCHLESINGER, W. H. et al.
JARZEN, D. M., *see* NICHOLS, D. J. et al.
JASTROW, R., *see* BALIUNAS, S. and JASTROW, R.
JAUREGUI, E., 477, 480, 481, 499, 501, 508, 511
JAUREGUI, E., GODINEZ, L. and CRUZ, F., 500, 511
JAUREGUI, E., *see* OKE, T. R. et al.
JEHANNO, C., *see* ROCCHIA, R. et al.
JENKINS, G. J., *see* HOUGHTON, J. J. et al.
JENKINS, W. J., *see* BREWER, P. G. et al.
JENNE, R. J., *see* WOODRUFF, S. D. et al.
JENNE, R. L. and MCKEE, T. B., 197, 238
JENNE, R. L., *see* WARREN, S. G. et al.
JENNINGS, S. G., *see* O'DOWD, C. D. et al.
JI, M., *see* LEETMAA, A. and JI, M.
JIANG, H., *see* RAYMOND, D. J. and JIANG, H.
JIANG, X., *see* BARNETT, T. P. et al.
JIAQING, Z., *see* DAOHAN, C. et al.
JICKELLS, T. D., *see* DUCE, R. A. et al.
JIN-WEN, H., *see* DAO-YI, X. et al.
JIRIKOVIC, J. and DAMON, P. E., 200, 238
JOHNSEN, S. J., CLAUSEN, H. B., DANSGAARD, W., FUHRER, K., GUNDESTRUP, N., HAMMER, C. U., IVERSEN, P., JOUSEL, J., STAUFFER, B. and STEFFENSEN, J. P., 25, 64, 517, 534
JOHNSEN, S. J., *see* DANSGAARD, W. et al.
JOHNSEN, S. J., *see* GROOTES, P. M. et al.
JOHNSON, A. I., *see* LAMB, H. H. and JOHNSON, A. I. et al.
JOHNSON, A. L. A. and SIMMS, M. J., 111, 119, 141
JOHNSON, D. R., 285, 286, 295, 296, 312
JOHNSON, D. R. and DOWNEY, W. K., 285, 312
JOHNSON, D. R., *see* TOWNSEND, R. D. and JOHNSON, D. R.
JOHNSON, D., *see* HANSEN, J. E. et al.
JOHNSON, E., *see* HAMMER, C. U. et al.
JOHNSON, G. D., *see* CROCKET, J. H. et al.
JOHNSON, G. T., OKE, T. R., LYONS, T. J., STEYN, D. G., WATSON, I. D. and VOOGT, J. A., 499, 511
JOHNSON, G. T., *see* OKE, T. R. et al.
JOHNSON, J. E., *see* QUINN, P. K. et al.
JOHNSON, K. R., 96, 112, 141
JOHNSON, K. R. and HICKEY, L. J., 111, 141
JOHNSON, K. R., NICHOLS, D. J., ATTREP JR, M. and ORTH, C. J., 141
JOHNSON, S. J., *see* DANSGAARD, W. et al.
JOHNSTON, D. S., *see* BORNSTEIN, R. D. and JOHNSTON, D. S.

JONES, A. S., *see* BOULTON, G. S. et al.
JONES, D. S., 117
JONES, D. S. et al., 141
JONES, D. S., MUELLER, P. A. et al., 108, 110, 141
JONES, P. A., 189
JONES, P. A. and HENDERSON-SELLERS, A., 164, 185
JONES, P. D., 18, 156, 157, 174, 185, 382, 394
JONES, P. D. and BRADLEY, R. S., 168, 185
JONES, P. D. and BRIFFA, K. R., 160, 185
JONES, P. D. and KELLY, P. M., 160, 185
JONES, P. D. and WIGLEY, T. M. L., 151, 158, 185
JONES, P. D., GROISMAN, P. YA., COUGHLAN, M., PLUMMER, N., WANG, W.-C. and KARL, T. R., 155, 185, 197, 198, 238
JONES, P. D., KELLY, P. M., GOODESS, C. M. and KARL, T. R., 479, 506, 511
JONES, P. D., RAPER, S. C. B. and WIGLEY, T. M. L., 158, 171, 185
JONES, P. D., RAPER, S. C. B., BRADLEY, R. S., DIAZ, H. F., KELLY, P. M. and WIGLEY, T. M. L., 156, 158, 171, 185, 207, 238
JONES, P. D., RAPER, S. C. B., CHERRY, B. S. G., GOODESS, C. M. and WIGLEY, T. M. L., 185
JONES, P. D., RAPER, S. C. B., CHERRY, B. S. G., GOODESS, C. M., WIGLEY, T. M. L., BRADLEY, R. S. and DIAZ, H. F., 158
JONES, P. D., RAPER, S. C. B., SANTER, B. D., CHERRY, B. S. G., GOODESS, C. M.,KELLY, P. M., WIGLEY, T. M. L., BRADLEY, R. S. and DIAZ, H. F., 185
JONES, P. D., *see* ALLAN, R. J. et al.
JONES, P. D., *see* BRADLE Y, R. S. and JONES, P. D.
JONES, P. D., *see* BRADLEY, R. S. et al.
JONES, P. D., *see* BRIFFA, K. R. and JONES, P. D.
JONES, P. D., *see* BRIFFA, K. R. et al.
JONES, P. D., *see* KARL, T. R. and JONES, P. D.
JONES, P. D., *see* KARL, T. R. et al.
JONES, P. D., *see* PARKER, D. E. et al.
JONES, P. D., *see* ROPELEWSKI, C. F. and JONES, P. D.
JONES, P. D., *see* SANTER, B. D. et al.
JONES, P. D., *see* SEAR, C. B. et al.
JONES, P. D., *see* WIGLEY, T. M. L. and JONES, P. D.
JONES, P. D., *see* WIGLEY, T. M. L. et al.
JONES, P. D., WIGLEY, T. M. L. and FARMER, G., 158, 171, 185
JONES, P. D., WIGLEY, T. M. W. and WRIGHT, P. B., 382, 394
JONES, R. *see* MCEVEDY, C. and JONES, R.
JORDÁN, R. S., 207, 238
JOUSEL, J., *see* DANSGAARD, W. et al.
JOUSEL, J., *see* JOHNSEN, S. J.
JOUSSAUME, S., 24, 36, 64
JOUSSAUME, S. and JOUZEL, J., 36, 64
JOUZEL, J. et al., 226, 238
JOUZEL, J., BARKOV, N. I., BARNOLA, J. M., BENDER, M., CHAPPELLAZ, J., GENTHON, C., KOTLYAKOV, V. M., LIPENKOV, V., LORIUS, C., PETIT, J. R., RAYNAUD, D., RAISBECK, G., RITZ, C., SOWERS, T., STIEVENARD, M., YIOU, F. and YIOU, P., 24, 38, 47, 50, 64, 520, 521, 534
JOUZEL, J., LORIUS, C., PETIT, J. R., GENTHON, C., BARKOV, N. I., KOTLYAKOV, V. M. and PETROV, V. M., 520, 534
JOUZEL, J., *see* DANSGAARD, W. et al.
JOUZEL, J., *see* GENTHON, C. et al.
JOUZEL, J., *see* GROOTES, P. M. et al.
JOUZEL, J., *see* JOHNSEN, S. J. et al.
JOUZEL, J., *see* JOUSSAUME, S. and JOUZEL, J.
JOUZEL, J., *see* LE TREUT, H. et al.
JOUZEL, J., *see* LORIUS, C. et al.
JOUZEL, J., *see* RAYNAUD, D. et al.
JULIAN, P. R., *see* MADDEN, R. A. and JULIAN, P. R.
JUSTICE, C. O., *see* RUNNING, S. W. et al.

KADLECEK, J. A., *see* PUESCHEL, R. F. et al.
KAHL, J. D. W., SEREEZE, M. C., STONE, R. S., SHIOTANI, S., KISLEY, M. and SCHNELL, R. C., 169, 185
KAIHO, K., 108, 141
KAIHO, K., *see* KAJIWARA, Y. and KAIHO, K.
KAIHO, K., *see* SAITO, T. et al.
KAJIWARA, Y. and KAIHO, K., 109, 141
KALANDA, B. D., 494
KALANDA, B. D., OKE, T. R. and SPITTLEHOUSE, D. L., 511
KÅLLBERG, P., *see* SWAP, R. et al.
KALLEL, N., *see* DUPLESSY, J. C. et al.
KALMA, J. D., 491, 511
KALMA, J. D., *see* STANHILL, G. and KALMA, J. D.
KANAMITSU, M. and KRISHNAMURTI, T. N., 457, 472
KAPALA, A., *see* FLOHN, H. and KAPALA, A.
KAPALA, A., *see* FLOHN, H. et al.
KAPLAN, J., *see* DEMARIA, M. and KAPLAN, J.
KARENTZ, D., *see* SMITH, R. C. et al.
KARL, T. R., 200, 231, 238
KARL, T. R. and JONES, P. D., 479, 506, 508, 511
KARL, T. R. and STEURER, P. M., 157, 185, 232, 238
KARL, T. R. and WILLIAMS JR., C. N., 185
KARL, T. R. and WILLIAMS, J., 156
KARL, T. R., DIAZ, H. F. and KUKLA, G., 155, 185, 197, 238, 478, 479, 497, 498, 508, 511
KARL, T. R., HEIM JR., R. R. and QUAYLE, R. G., 231, 238
KARL, T. R., JONES, P. D., KNIGHT, R. W., KUKLA, G., PLUMMER, N., RAZUVAYEV, V., GALLO, K. P., LINDESAY, J., CHARLSON, R. J. and PETERSON, T. C., 163, 164, 185
KARL, T. R., KUKLA, G., RAZUVAYEV, V. N., CHANGERY, M. G., QUAYLE, R. G., HEIM JR., R.

R., EASTERLING, D. R. and FU, C. B., 232, 238, 384, 394
KARL, T. R., *see* FOLLAND, C. K. et al.
KARL, T. R., *see* GALLO, K. P. et al.
KARL, T. R., *see* GROISMAN, P. Y. et al.
KARL, T. R., *see* JONES, P. D. et al.
KARL, T. R., *see* KUKLA, G. et al.
KARL, T. R., *see* PLANTICO, M. S. et al.
KARL, T. R., WILLIAMS JR., C. N. and YOUNG, P. J., 154, 185
KARL, T., *see* FOLLAND, C. K. et al.
KARL, T., *see* GAVIN, J. et al.
KARL, T., *see* HANSEN, J. et al.
KARLÉN, W., *see* BRIFFA, K. R. et al.
KAROLY, D. J., 156, 167, 186
KAROLY, D. J., COHEN, J. A., MEEHL, G. A., MEEHL, G. A., OORT, A. H., STOUFFER, R. J. and WETHERALD, R. T., 181, 186
KAROLY, D., *see* HOSKINS, B. J. and KAROLY, D.
KASAHARA, A., BALGOVIND, R. C. and KATZ, B. B., 264, 276
KASTING, J. F., 6, 18
KASTING, J. F., HOLLAND, H. D. and KUMP, L. R., 542, 552
KASTING, J. F., RICHARDSON, S. M., POLLACK, J. B. and TOON, O. B., 103, 141
KASTING, J. F., *see* CALDEIRA, K. and KASTING, J. F.
KASTING, J. F., *see* WALKER, J. C. G. and KASTING, J. F.
KASTING, J. F., *see* WALKER, J. C. G. et al.
KASTING, J. F., ZAHNLE, K. J. and WALKER, J. C. G., 543, 552
KASTING, J. K. and RICHARDSON, S. M., 84, 92
KATES, R. W., *see* TURNER II, B. L. et al.
KATZ, B. B., *see* KASAHARA, A. et al.
KATZ, M. E., *see* MILLER, K. G. et al.
KATZ, R. W., *see* GLANTZ, M. H. et al.
KATZ, R. W., *see* NICHOLLS, N. and KATZ, R. W.
KAUFFMAN, E. G., 114, 141
KAUFMAN, Y. J. and CHOU, M.-D., 379, 383, 385, 386, 394
KAUFMAN, Y. J. and NAKAJIMA, T., 377, 394
KAUFMAN, Y. J., FRASER, R. S. and FERRARE, R. A., 367, 394
KAUFMAN, Y. J., FRASER, R. S. and MAHONEY, R. L., 376, 394
KAUFMAN, Y. J., *see* FERRARE, R. A. et al.
KAUFMAN, Y. J., *see* FRASER, R. S. et al.
KAUFMAN, Y. J., *see* KING, M. D. et al.
KAUFMAN, Y. J., SETZER, A., WARD, D., TANRE, D., HOLBEN, B. N., MENZEL, P., PEREIRA, M. C. and RASMUSSEN, R., 367, 394
KAUFMAN, Y. J., TUCKER, C. J. and FUNG, I., 367, 394
KAUFMANN, Y. J., *see* RUNNING, S. W. et al.
KAYE, J. A. and HACKMAN, C. H., 418, 431
KEEHN, P. R., *see* NEGRI, A. J. et al.
KEELING, C. D., BACASTON, R. B. and WHORF, T. P., 539, 552
KEELING, C. D., BACASTOW, R. B., BAINBRIDGE, A. E., EKDAHL, C. A., GUENTHER, P. R., WATERMAN, L. S. and CHIN, J. F. S., 230, 238
KEELING, C. D., BACASTOW, R. B., CARTER, A. F., PIPER, S. C., WHORF, T. P., HEIMANN, M., MOOK, W. G. and ROELOFFZEN, H., 517, 534
KEENAN, T. D., *see* SIMPSON, J. S. et al.
KEENE, W. C., *see* CHURCH, T. M. et al.
KEENE, W. C., *see* GALLOWAY, J. N. et al.
KEENE, W. C., *see* WHELPDALE, D. M. et al.
KEIGWIN JR., L. D., *see* CORLISS, B. H. et al.
KEIGWIN, L., 82
KEIGWIN, L. and CORLISS, B. H., 81, 92
KEIGWIN, L. D., 92
KEIGWIN, L. D., *see* LEHMAN, S. J. and KEIGWIN, L. D.
KEIGWIN, L., *see* BOYLE, E. A. and KEIGWIN, L.
KEIL, D. J., *see* MILLER, K. G. et al.
KEIMIG, F. T., *see* BRADLEY, R. S. et al.
KELFKENS, H., *see* DE GRUIJL, F. R. et al.
KELLER, G., 103, 104, 109, 118, 141
KELLER, G. and BARRERA, E., 134, 141
KELLER, G., *see* BARRERA, E. and KELLER, G.
KELLER, G., *see* CORLISS, B. H. et al.
KELLER, G., *see* D'HONDT, S. et al.
KELLER, G., *see* MACLEOD, N. and KELLER, G.
KELLER, M., 448, 472
KELLER, M., MITRE, M. E. and STALLARD, R. F., 448, 449, 472
KELLER, R., *see* FELS, E. and KELLER, R.
KELLOGG, W. W., 347, 366, 394
KELLY, G. A., *see* EYRE, J. R. et al.
KELLY, G. A., *see* HECKLEY, W. A. et al.
KELLY, P. M. and SEAR, C. B., 176, 186
KELLY, P. M. and WIGLEY, T. M. L., 175, 186, 200, 238
KELLY, P. M., GOODESS, C. M. and CHERRY, B. S. G., 193, 238
KELLY, P. M., *see* BRADLEY, R. S. et al.
KELLY, P. M., *see* JONES, P. D. and KELLY, P. M.
KELLY, P. M., *see* JONES, P. D. et al.
KELLY, P. M., *see* SEAR, C. B. et al.
KELLY, P. M., *see* WIGLEY, T. M. L. et al.
KELLY-HANSEN, S.R., *see* BATES, T. S. et al.
KENNEDY, P. J., *see* DICKINSON, R. E. and KENNEDY, P. J.
KENNEDY, P. J., *see* DICKINSON, R. E. et al.
KENNEDY, P. J., *see* HENDERSON-SELLERS, A. et al.
KENNETT, J. P., 84, 93
KENNETT, J. P. J., *see* SHACKLETON, N. J. and KENNETT, J. P.
KENNETT, J. P., *see* BROECKER, W. S. et al.
KENNETT, J. P., *see* STOTT, L. D. and KENNETT, J. P.

KEPERT, J. D., *see* FAIRALL, C. W. et al.
KERR, J. B. and MCELROY, C. T., 427, 431
KERSHAW, A. P., *see* MCGLONE, M. S. et al.
KESSLER, A., 4, 18
KESSLER, R. W., BOSART, L. F. and GAZA, R. S., 153, 186
KEY, J. R., *see* SCHWEIGER, A. J. and KEY, J. R.
KIDD, R. B., *see* RUDDIMAN, W. F. et al.
KIDSON, J. W., 457, 472
KIDSON, J. W., *see* NEWELL, R. E. and KIDSON, J. W.
KIEHL, J. T., 8, 18
KIEHL, J. T. and BRIEGLEB, B. P., 231, 238, 343, 345, 362, 365, 368, 378, 395, 452, 472
KIEHL, J. T. and RAMANATHAN V., 254, 255, 276
KIEHL, J. T., *see* CESS, R. D. et al.
KIEHL, J. T., *see* WANG, W.-C. et al.
KILADIS, G. N. and DIAZ, H. F., 214, 217, 220, 221, 222, 238
KILADIS, G. N. and FELDSTEIN, S. B., 299, 300, 312
KILADIS, G. N. and SINHA, S. K., 216, 238
KILADIS, G. N. and VAN LOON, H., 214, 218, 221, 238
KILADIS, G. N. and WEICKMAN, K. M., 298, 312
KILADIS, G. N., *see* BRADLEY, R. S. et al.
KILADIS, G. N., *see* DIAZ, H. F. and KILADIS, G. N.
KIM, K. Y., *see* CROWLEY, T. J. et al.
KIM, Y., *see* BOATMAN, J. F. et al.
KIM, Y., *see* FALKOWSKI, P. G. et al.
KING, M. D., KAUFMAN, Y. J., MENZEL, W. P. and TANRE, D., 259, 276
KING, M. D., RADKE, L. F. and HOBBS, P. V., 377, 395
KING, M. D., *see* NAKAJIMA, T. and KING, M. D.
KING, M. D., *see* RADKE, L. F. et al.
KINGTON, J., 151, 152, 186
KIPP, N. G., *see* SANCETTA, C. et al.
KIPSTUHL, J., *see* TAYLOR, K. C. et al.
KIRCHGASSER, W. T., *see* BECKER, R. T. et al.
KIRK, E., *see* BARNETT, T. P. et al.
KIRSCHVINK, J. L., *see* MAGARITZ, M. et al.
KISLEY, M., *see* KAHL, J. D. W. et al.
KLAPPER, G., *see* GOODFELLOW, W. D. et al.
KLAS, M., *see* BOND, G. et al.
KLAS, M., *see* BROECKER, W. S. et al.
KLEIN, C., *see* WALKER, J. C. G. et al.
KLEIN, S. A. and HARTMANN, D. L., 255, 277
KLEIN, W. H., *see* CORRELL, D. L. et al.
KLEMCKE, C. H., *see* ROOSEN, G. R. et al.
KLETT, J. D., *see* PRUPPACHER, H. R. and KLETT, J. D.
KLOEZEMAN, W. G., *see* BARATH, F. T. et al.
KLOSE, G. J., *see* BARATH, F. T. et al.
KNAFF, J. A., *see* GRAY, W. M. et al.
KNAP, A. H., *see* CHURCH, T. M. et al.
KNAP, A. H., *see* DUCE, R. A. et al.
KNIGHT, R. J., *see* TAYLOR, F. W. et al.
KNIGHT, R. W., *see* GROISMAN, P. Y. et al.
KNIGHT, R. W., *see* KARL, T. R. et al.
KNOCHE, H. R., *see* FLOHN, H. et al.
KNOX, F., 527
KNOX, F. and MCELROY, M., 534
KOBLINSKY, C. J., 272, 277
KOCI, B., *see* MAYEWSKI, P. A. et al.
KOKNAEVA, V. V., *see* GROISMAN, P. YA. et al.
KOLBER, Z., *see* FALKOWSKI, P. G. et al.
KOMINZ, M. A. and PISIAS, N. G., 43, 64
KÖNIG, W., SAUSEN, R. and SIELMANN, F., 302, 312
KORNFIELD, J., *see* CHARNEY, J. G. et al.
KOROTKEVICH, Y. S., BARNOLA, J. M. et al., 551
KOROTKEVITCH, Y. S., *see* BARNOLA, J. M. et al.
KOROTKEVITCH, Y. S., *see* CHAPPELLAZ, J. et al.
KOROTKEVITCH, Y. S., *see* GENTHON, C. et al.
KORSHOVER, J., *see* ANGELL, J. K. and KORSHOVER, J.
KORSOG, P. E., *see* WOLFF, G. R. et al.
KOTLYAKOV, V. M., *see* GENTHON, C. et al.
KOTLYAKOV, V. M., *see* JOUZEL, J. et al.
KRAHE, P., *see* HENSE, A. et al.
KRASSILOV, V. A., 112, 141
KRAUS, E. B., *see* EMILIANI, C. et al.
KREBS, C. J., 550, 552
KRISHNAMURTI, T. N., *see* KANAMITSU, M. and KRISHNAMURTI, T. N.
KRISTINSDOTTIR, P., *see* HAMMER, C. U.
KRITZ, M. A., *see* ANDREAE, M. O. et al.
KRUEGER, A. J., *see* BLUTH, G. J. S. et al.
KRUEGER, A. J., *see* HEATH, D. F. et al.
KUKLA, G., BERGER, A., LOTTI, R. and BROWN, J., 42, 64
KUKLA, G., GAVIN, J. and KARL, T. R., 506, 511
KUKLA, G., *see* BERGER, A. et al.
KUKLA, G., *see* GAVIN, J. et al.
KUKLA, G., *see* IMBRIE, J. et al.
KUKLA, G., *see* KARL, T. R. et al.
KUKLA, G., *see* PLANTICO, M. S. et al.
KUKLA, G., *see* RIND, D. et al.
KUKLA, G., *see* ROBINSON, D. A. et al.
KUMANAN, I., *see* SANDERSON, M. et al.
KUMER, J. B., *see* ROCHE, A. E. et al.
KUMP, L., 106, 141, 543, 552
KUMP, L. R. and LOVELOCK, J. E., 18
KUMP, L. R. and VOLK, T., 547, 552
KUMP, L. R., , *see* KASTING, J. F. et al.
KUMP, L. R., *see* LOVELOCK, J. E. and KUMP, L. R.
KUO YANG, R. T. W., *see* QUINN, W. H. et al.
KURIHARA, Y., *see* LIGHTHILL, J. et al.
KURYLO, M. J., *see* DEMORE, W. B. et al.
KUSHNIR, V., *see* ESBENSEN, S. K. and KUSHNIR, V.
KUSHNIR, Y., 226, 227, 239
KUTZBACH, J. E., 32, 34, 38, 64, 90, 93

KUTZBACH, J. E. and BRYSON, R. A., 191, 239
KUTZBACH, J. E. and GALLIMORE, R. G., 40, 64, 90, 93
KUTZBACH, J. E. and GUETTER, P. J., 35, 37, 38, 64, 90, 93
KUTZBACH, J. E. and OTTO-BLIESNER, B. L., 38, 64
KUTZBACH, J. E. and STREET-PERROTT, F. A., 37, 64
KUTZBACH, J. E. and WRIGHT JR., H. E., 64
KUTZBACH, J. E. and WRIGHT, D. G., 35
KUTZBACH, J. E. and ZIEGLER, A. M., 90, 93
KUTZBACH, J. E., *see* IMBRIE, J. et al.
KUTZBACH, J. E., *see* PRELL, W. L. and KUTZBACH, J. E.
KUTZBACH, J. E., *see* RUDDIMAN, W. F. and KUTZBACH, J. E.
KUTZBACH, J. E., *see* WEBB, T. et al.
KWON, T. Y., *see* CESS, R. D. et al.
KYTE, F. T., 115, 141
KYTE, F. T. and WASSON, J. T., 115, 141
KYTE, F. T., *see* MARGOLIS, S. V. et al.
KYTE, F. T., *see* ZHOU, L. and KYTE, F. T.
KYTE, F. T., ZHOU, Z. and WASSON, J. T., 97, 100, 115, 117, 141

LAASEN, K., *see* FRIIS-CHRISTENSEN, E. and LAASEN, K.
LABEYRIE, L. D., 527
LABEYRIE, L. D. and DUPLESSY, J. C., 534, 547, 552
LABEYRIE, L. D., DUPLESSY, J. C. and BLANC, P. L., 24, 47, 48, 64, 526, 534
LABEYRIE, L. D., *see* BOND, G. et al.
LABEYRIE, L. D., *see* DUPLESSY, J. C. et al.
LABITZKE, K. and VAN LOON, H., 199, 202, 203, 239
LABITZKE, K., *see* VAN LOON, H. and LABITZKE, K.
LACAUX, J.-P., *see* PHAM-VAN-DINH et al.
LACIS, A. A., *see* CESS, R. D. et al.
LACIS, A. A., *see* HANSEN, J. and LACIS, A. A.
LACIS, A. A., *see* HANSEN, J. et al.
LACIS, A. A., *see* WANG, W.-C. et al.
LACIS, A. A., WUEBBLES, D. J. and LOGAN, J. A., 319, 336, 339, 345
LACIS, A., HANSEN, J. and SATO, M., 205, 239
LAI, C., *see* BERGER, W. H. et al.
LAMB, B., *see* FEHSENFELD, F. et al.
LAMB, H. H., 167, 177, 186, 193, 194, 195, 196, 205, 239
LAMB, H. H. and JOHNSON, A. I., 152, 186
LAMB, P. J., 166, 186
LAMB, P. J. and PEPPLER, R. A., 218, 239
LAMPREY, H. F., 441, 472
LANDER, M. and HOLLAND, G. J., 305, 312
LANDER, M., *see* HOLLAND, G. J. and LANDER, M.
LANDER, M., *see* RITCHIE, E. A. et al.
LANDSBERG, H. E., 498, 511
LANDSBERG, H. E., *see* HARNACK, R. P. and LANDSBERG, H. E.
LANDSEA, C. W. and GRAY, W. M., 310, 312
LANDSEA, C. W., *see* LIGHTHILL, J. et al.
LANG, P. M., *see* DLUGOKENCKY, E. J. et al.
LANGNER, J. and RODHE, H., 357, 363, 368, 380, 381, 395
LANGNER, J., RODHE, H., CRUTZEN, P. J. and ZIMMERMANN, P., 358, 379, 395
LANGNER, J., *see* CHARLSON, R. J. et al.
LANGWAY JR., C. C., *see* DANSGAARD, W. et al.
LANLY, J. P., 444, 472
LARSEN, R. J., *see* SAVOIE, D. L. et al.
LARSON, J. A., *see* BRYSON, R. A. et al.
LARSON, R. E., *see* HANSEN, A. D. A. et al.
LARSON, T. V., *see* CHARLSON, R. J. et al.
LARSSEN, S., *see* HEINTZENBERG, J. and LARSSEN, S.
LASAGA, A. C., 547, 552
LASAGA, A. C., *see* BERNER, R. A. et al.
LASSEN, K., 236
LASSEN, K., *see* FRIIS-CHRISTENSEN, E. and LASSEN, K.
LASZLO, I. and PINKER, R. T., 258, 268, 277
LASZLO, I., *see* PINKER, R. T. and LASZLO, I.
LASZLO, I., *see* WHITLOCK, C. H. et al.
LATHAM, D., *see* MELOSH, H. J. et al.
LATHAM, J. and SMITH, M. H., 351, 395
LATIF, M., *see* BARNETT, T. P. et al.
LATTER, J. H., *see* SIMKIN, T. et al.
LAU, G. K., *see* BARATH, F. T. et al.
LAU, K.-M. and BUSALACCHI, A. J., 262, 277
LAU, K.-M. and SHEU, P. J., 214, 239
LAURSEN, K. K., *see* RADKE, L. F. et al.
LAUTENSCHLAGER, M. and HERTERICH, H., 64
LAUTENSCHLAGER, M. and HERTERICH, K., 35
LAVAL, K., 437, 456, 457, 458, 472
LAVAL, K. and PICON, L., 455, 457, 461, 472
LAVAL, K., *see* POLCHER, J. and LAVAL, K.
LAVERY, J. R., *see* CAO, H. X. et al.
LAWSON, D. R. and WINCHESTER, J. W., 380, 395
LAZRUS, A. L., *see* HUEBERT, B. J. and LAZRUS, A. L.
LE HOUÉROU, H. N., 436, 441, 456, 472
LE TREUT, H., PORTES, J., JOUZEL, J. and GHIL, M., 41, 65
LE TREUT, H., *see* BONY, S. and LE TREUT, H.
LE TREUT, H., *see* CESS, R. D. et al.
LE TREUT, H., *see* GHIL, M. and LE TREUT, H.
LE TREUT, H., *see* LORIUS, C. et al.
LE, J., *see* SHACKLETON, N. J. et al.
LEAHY, G. D., *see* RETALLACK, G. J. et al.
LEAITCH, W. R., ISAAC, G. A., STRAPP, J. W., BANIC, C. M. and WIEBE, H. A., 372, 395
LEAITCH, W. R., STRAPP, J. W., ISAAC, G. A. and HUDSON, J. G., 372, 395
LEAN, D. R. S., *see* MAZUMBER, A. et al.

LEAN, J., 25, 64
LEAN, J. and RIND, D., 200, 239
LEAN, J. and ROWNTREE, P. R., 472
LEAN, J. and WARRILOW, D. A., 459, 461, 465, 472
LEAN, J., *see* FOUKAL, P. and LEAN, J.
LEAN, J., SKUMANICH, A. and WHITE, O., 25, 64, 200, 239, 253, 277
LEARY, P. N. and RAMPINO, M. R., 102, 141
LEBEDEFF, S., *see* HANSEN, J. et al.
LEBEDEFF, S., *see* HANSEN, J.E and LEBEDEFF, S.
LECHEMINANT, A. N. and HEAMAN, L. M., 129, 142
LEE, A. J., *see* DICKSON, R. R. et al.
LEE, P., *see* HANSEN, J. E. et al.
LEETMAA, A. and JI, M., 232, 239
LEGATES, D. R. and WILLMOTT, C. J., 155, 186, 260, 277
LEGG, E., *see* RUTLEDGE, G. et al.
LEGG, T. P., *see* PARKER, D. E. et al.
LEGGETT, J., *see* APSIMON, H. et al.
LEGRAND, M. R., DELMAS, R. J. and CHARLSON, R. J., 545, 552
LEGRAND, M. R., FENIET-SAIGNE, C., SALTZMAN, E. S., GERMAIN, C., BARKOV, N. I. and PETROV, V. N., 54, 65
LEGRAND, M. R., LORIUS, C., BARKOV, N. I. and PETROV, V. N., 54, 64
LEGRAND, M., *see* DELMAS, R. J. et al.
LEHMAN, S. J., 524
LEHMAN, S. J. and KEIGWIN, L. D., 534
LEHMAN, T. M., 100, 142
LEIGHTON, H. G., *see* LI, Z.-X. and LEIGHTON, H. G.
LEINEN, M., *see* REA, D. K. et al.
LEMKE, P., 43, 65
LENZ, A. C., *see* GOODFELLOW, W. D. et al.
LEOVY, C. B., *see* CHARLSON, R. J. et al.
LERMAN, J. C., *see* DAMON, P. E. et al.
LERNER, J., *see* HANSEN, J. et al.
LERNER, J., *see* MATTHEWS, E. et al.
LETREGUILLY, A., *see* HUYBRECHTS, PH. et al.
LEUENBERGER, M. and SIEGENTHALER, U., 52, 65
LEVKOV, L., *see* GRASSL, H. and LEVKOV, L.
LEVY III, H., *see* SAVOIE, D. L. et al.
LEWIS, B. L., *see* ANDREAE, M. O. et al.
LEWIS, J. S. et al., 100, 142
LEWIS, R. S., *see* ANDERS, E. et al.
LEWIS, R. S., *see* WOLBACH, W. S. et al.
LEZBERG, E. A., HUMENIK, F. M. and OTTERSON, D. A., 382, 395
LI, J-J., *see* WANG, K. et al.
LI, S. M., *see* BERRESHEIM, H. et al.
LI, S. M., *see* TALBOT, R. W. et al.
LI, S., *see* ANDREAE, M. O. et al.
LI, Z. X., *see* CESS, R. D. et al.
LI, Z., ZHAN, L., YAO, J. and ZHOU, Y., 120, 142
LI, Z.-X. and LEIGHTON, H. G., 265, 277
LI, Z.-X., *see* CESS, R. D. et al.
LIANG, X.-Z., *see* CESS, R. D. et al.
LIANG, X.-Z., *see* DUDEK, M. P. et al.
LIANG, X.-Z., *see* WANG, W.-C. et al.
LIBERTI, G. L., *see* PRABHAKARA, C. et al.
LIGHTHILL, J., HOLLAND, G. J., GRAY, W. M., LANDSEA, C., CRAIG, G., EVANS, J. L., KURIHARA, Y. and GUARD, C. P., 310, 312
LIM, G.-H., *see* WALLACE, J. M. et al.
LIN, C. A., *see* MYSAK, L. A. and LIN, C. A.
LINDESAY, J., *see* KARL, T. R. et al.
LINDZEN, R. S., *see* SUN, D. Z. and LINDZEN, R. S.
LINDZEN, R.S. and NIGAM, S., 209, 239
LINGLE, C. S., *see* HUGHES, T. J. et al.
LINZHONG, L., *see* DAOHAN, C. et al.
LIPENKOV, V., *see* JOUZEL, J. et al.
LIPPS, J. H., *see* SIGNOR III, P. W. and LIPPS, J. H.
LISS, P. S., *see* DUCE, R. A. et al.
LITT, T., *see* SANTER, B. et al.
LIU, H., *see* OORT, A. H. and LIU, H.
LIU, J., 486
LIU, J., *see* WANG, C. and LIU, J.
LIU, M.-H., *see* HAO, W. M. et al.
LIU, S. C., *see* FEHSENFELD, F. and LIU, S. C.
LIU, S., *see* FEHSENFELD, F. et al.
LIU, W. T., 272, 277
LIU, W. T., TANG, W. and WENTZ, F. J., 272, 277
LIVESEY, R. E., *see* BARNSTON, A. G. and LIVESEY, R. E.
LIVEZEY, R. E., *see* MO, K. C. and LIVEZEY, R. E.
LLOYD, C. R., 22, 65
LLOYD, J. C. R., *see* SHUTTLEWORTH, W. J. et al.
LOCKWOOD, G. W., *see* RADICK, R. R. et al.
LOGAN, J. A., 338, 345, 356, 395
LOGAN, J. A., *see* LACIS, A. A. et al.
LOGAN, J. A., *see* SPIRO, P. A. et al.
LOGAN, J. A., *see* SPIVAKOVSKY, C. M. et al.
LOHMANN, K. C., *see* ZACHOS, J. C. et al.
LONDER, R., *see* SCHNEIDER, S. H. and LONDER, R.
LONDON, J., *see* WARREN, S. G. et al.
LONG, A., *see* DAMON, P. E. et al.
LONG, A., *see* MICHAELSEN, J. and LONG, A.
LOO, M. S., *see* BARATH, F. T. et al.
LOOSE, T. and BORNSTEIN, R. D., 505, 511
LORENZ, E. N., 232, 239
LORIUS, C., BARNOLA, J. M. et al., 551
LORIUS, C., JOUZEL, J., RAYNAUD, D., HANSEN, J. and LE TREUT, H., 80, 93
LORIUS, C., *see* BARNOLA, J. M. et al.
LORIUS, C., *see* JOUZEL, J.
LORIUS, CL., JOUZEL, J., RAYNAUD, D., HANSEN, J. and LE TREUT, H., 31, 52, 55, 65
LORIUS, CL., JOUZEL, J., RAYNAUD, D., HANSEN, J. and LE, J., 22
LORIUS, CL., *see* CHAPPELLAZ, J. et al.
LORIUS, CL., *see* GENTHON, C. et al.

LORIUS, CL., *see* LEGRAND, M. R. et al.
LORIUS, CL., *see* RAYNAUD, D. et al.
LOTTI, R., *see* KUKLA, G. et al.
LOUBERE, P., *see* DOWSETT, H. J. and LOUBERE, P.
LOUGH, J. M., 166, 186
LOUGH, J.M., 222, 239
LOUTRE, M. F., 57, 65
LOUTRE, M. F., BERGER, A., BRETAGNON, P. and BLANC, P.-L., 200, 239
LOUTRE, M. F., *see* BERGER, A. and LOUTRE, M. F.
LOUTRE, M. F., *see* BERGER, A. et al.
LOVELAND, T. R., *see* RUNNING, S. W. et al.
LOVELOCK, J. E., 4, 18, 537, 539, 540, 541, 542, 543, 550, 551, 552
LOVELOCK, J. E. and KUMP, L. R., 544, 545, 552
LOVELOCK, J. E. and WHITFIELD, M., 537, 550, 552
LOVELOCK, J. E., *see* CHARLSON, R. J. et al.
LOVELOCK, J. E., *see* WATSON, A. J. and LOVELOCK, J. E.
LOWRIE, W., ALVAREZ, W. and ASARO, F., 101, 142
LOWRY, W. P., 481, 500, 512
LUBIN, D., MITCHELL, B. G., FREDERICK, J. E., ROBERTS, A. D., BOOTH, C. R., LUCAS, T. and NEUSCHULER, D., 425, 431
LUCAS, T., *see* LUBIN, D. et al.
LUCAS, T., *see* STAMNES, K. et al.
LUCKMAN, B. H., 196, 239
LUCKMAN, B. H., *see* OSBORN, G. and LUCKMAN, B. H.
LUDWIG, F. L., *see* SEAMAN, N. L. et al.
LUDWIG, J. H., *see* MCCORMICK, R. A. and LUDWIG, J. H.
LUGINA, K. M., *see* VINNIKOV, K. YA. et al.
LUNDE, A. T., *see* BIRCHFIELD, G. E. et al.
LURIA, M., *see* BOATMAN, J. F. et al.
LUTHER, M. E., O'BRIEN, J. J. and PRELL, W. L., 38, 65
LUYENDYK, B., FORSYTH D. and PHILLIPS, J., 75, 84, 93
LYELL, C., 95, 142
LYJAK, L. V., *see* GILLE, J. C. and LYJAK, L. V.
LYONS, J. B. and OFFICER, C. B., 134, 142
LYONS, J. D., *see* RADKE, L. F. et al.
LYONS, J. H., *see* RADKE, L. F. et al.
LYONS, T. J. and FORGAN, B. W., 486, 512
LYONS, T. J., *see* JOHNSON, G. T. et al.
LYONS, W. B., *see* MAYEWSKI, P. A. et al.
LYV, G., *see* DANARD, M. et al.

MA, S., *see* XU, D. et al.
MAALOE, S., *see* STORETVEDT, K. M. et al.
MAASCH, K. A. and SALTZMAN, B., 42, 43, 65
MAASCH, K. A., *see* GRIMM, E. C. et al.
MAASCH, K. A., *see* SALTZMAN, B. and MAASCH, K. A.
MAASCH, K. A., *see* SALTZMAN, B. et al.
MABBUTT, J. A., 441, 465, 472
MACCRACKEN, M. C. and MOSES, H., 181, 186
MACCRACKEN, M. C., *see* ELSAESSER, H. W. et al.
MACDOUGAL, J. D., 101, 142
MÄCHEL, H., *see* FLOHN, H. et al.
MACHTA, L., *see* MUNN, R. E. and MACHTA, L.
MACINTYRE, A., *see* RIND, D. et al.
MACINTYRE, S., *see* SMITH, R. C. et al.
MACKENSEN, A., 119
MACKENSEN, A. and EHRMANN, W. U., 142
MACKENSEN, A., *see* EHRMANN, W. U. and MACKENSEN, A.
MACLEOD, N. and KELLER, G., 109, 110, 142
MADALA, R. V., *see* HARMS, D. E. et al.
MADDEN, R. A. and JULIAN, P. R., 297, 305, 312
MADDEN, R. A. and RAMANATHAN, V., 181, 186
MADDEN, R. A., SHEA, D. J., BRANSTATOR, G. W., TRIBBIA, J. J. and WEBER, R., 159, 186
MADDISON, B. J., *see* BARATH, F. T. et al.
MADDOX, R. A., *see* NEGRI, A. J. et al.
MADRONICH, S., 421, 422, 423, 426, 431
MADRONICH, S. and DE GRUIJL, F. R., 431
MADRONICH, S. and GRANIER, C., 399, 424, 425, 431
MADRONICH, S., BJÖRN, L. O., ILYAS, M. and CALDWELL, M. M., 420, 424, 431
MAENHAUT, W., *see* ARTAXO, P. et al.
MAENO, N., *see* MOORE, J. C. et al.
MAGARITZ, M., 114, 142
MAGARITZ, M., BAR, R., BAUD, A. and HOLSER, W. T., 120, 142
MAGARITZ, M., HOLSER, W. T. and KIRSCHVINK, J. L., 142
MAGARITZ, M., MOUNT, J. F., DOEHNE, E., SHOWERS, W. and WARD, P., 127
MAGARITZ, M., *see* HOLSER, W. T. and MAGARITZ, M.
MAGARITZ, M., *see* HOLSER, W. T. et al.
MAGIE, G., *see* HAUGLUSTAINE, D. A. et al.
MAHAJAN, I. K., *see* FRAMJI, K. K. and MAHAJAN, I. K.
MAHLMAN, J. D., PINTO, J. P. and UMSCHEID, L. J., 336, 345
MAHONEY, J. J., 131, 142
MAHONEY, L., *see* FRASER, R. S. et al.
MAHONEY, R. L., *see* KAUFMAN, Y. J. et al.
MAIER-REIMER, E. and MIKOLAJEWICZ, U., 16, 18, 530, 534
MAIER-REIMER, E., MIKOLAJEWICZ, U. and CROWLEY, T., 84, 93
MALINGREAU, J. P., STEPHENS, G. and FELLOWS, L., 465, 472
MALKOWSKI, K., *see* GRUSZCZYNSKI, M. et al.
MALMBERG, S.-A., *see* DICKSON, R. R. et al.
MALMGREN, B., *see* WIDMARK, J. G. V. and MALMGREN, B.
MANABE, S., 318, 530

MANABE, S. and BROCCOLI, A. J., 35, 40, 65
MANABE, S. and HAHN, D. G., 35, 36, 65
MANABE, S. and HOLLOWAY JR., J. L., 466, 472
MANABE, S. and STOUFFER, R. J., 33, 34, 37, 40, 58, 65, 88, 93, 226, 232, 239, 534
MANABE, S. and STRICKLER, R. F., 345
MANABE, S., BRYAN, K. and SPELMAN, M. J., 84, 93
MANABE, S., *see* BROCCOLI, A. J. and MANABE, S.
MANABE, S., *see* DELWORTH, T. et al.
MANABE, S., *see* MITCHELL, J. F. B. et al.
MANABE, S., *see* SPELMAN, M. J. and MANABE, S.
MANABE, S., *see* STOUFFER, R. J. et al.
MANABE, S., SPELMAN, M. J. and STOUFFER, R., 327, 331, 345
MANABE, S., STOUFFER, R. J., SPELMAN, M. J. and BRYAN, K., 327, 331, 332, 345
MANGERUD, J., 47, 65
MANKIN, W. G., COFFEY, M. T. and GOLDMAN, A., 413, 431
MANLEY, G., 151, 186
MANNEY, G. L., *see* WATERS, J. W. et al.
MANSELL, J. W., *see* ACKERMAN, B. and MANSELL, J. W.
MANTON, M. J., *see* AYERS, G. P. et al.
MAO, H., *see* SHINE, K. P. et al.
MAO, X., *see* XU, D. et al.
MAPES, B. E., 305, 312
MARCHAND, D. S., *see* STREET-PERROTT, F. A. et al.
MARCUS, S. L., *see* HIDE, R. et al.
MARGOLIS, S. V., CLAEYS, P. and KYTE, F. T., 117, 142
MARGOLIS, S. V., MOUNT, J. F., DOEHNE, E., SHOWERS, W. and WARD, P., 112, 142
MARGOLIS, S. V., *see* CLAEYS, P. et al.
MARIN, V., *see* HUNTLEY, M. E. et al.
MARKGRAF, V., *see* DIAZ, H. F. and MARKGRAF, V.
MARKGRAF, V., *see* MCGLONE, M. S. et al.
MARKHAM, B., *see* GOWARD, S. N. et al.
MARQUES FILHO, A. DE O., *see* SHUTTLEWORTH, W. J. et al.
MARSIAT, I. and BERGER, A., 47, 48, 65
MARSIAT, I., *see* BERGER, A. et al.
MARSIAT, I., *see* GALLÉE, H. et al.
MARSICO, D. C., *see* REYNOLDS, R. W. and MARSICO, D. C.
MARTINSON, D. G., PISIAS, N. G., HAYS, J. D., IMBRIE, J., MOORE, T. C. and SHACKLETON, N. J., 39, 65
MARTINSON, D. G., *see* IMBRIE, J. et al.
MARVIN, U. B., 95, 142
MASAIRE, K. A., *see* DLUGOKENCKY, E. J. et al.
MASKELL, K, *see* ROWELL, D. P. et al..
MASLANIK, J. A., *see* BARRY, R. G. et al.
MASS, C. F. and PORTMAN, D. A., 205, 239
MASUTANI, P. D., *see* HOSKINS, B. J. et al.
MATEER, C. L., 410, 431
MATHEWS, J. T., *see* TURNER II, B. L. et al.
MATLICK, H. A., *see* SMITH, R. C. et al.
MATSON, M., 273, 277
MATTAUCH, R. J., *see* BARATH, F. T. et al.
MATTHAI, S. K., *see* ENGELHARDT, W. V. et al.
MATTHEWS, E., FUNG, I. and LERNER, J., 449, 472
MATTHEWS, R. K. and POORE, R. Z., 81, 93
MATZARAKIS, A. and MAYER, H., 507
MATZARKIS, A. and MAYER, H., 512
MAXWELL, W. D., 120, 123, 142
MAYER, H., *see* MATZARKIS, A. and MAYER, H.
MAYEWSKI, P. A., LYONS, W. B., SPENCER, M. J., TWICKLER, M. S., BUCK, C. F. and WHITLOW, S., 382, 383, 395
MAYEWSKI, P. A., LYONS, W. B., SPENCER, M. J., TWICKLER, M., DANSGAARD, W., KOCI, B., DAVIDSON, C. I. and HONRATH, R. E., 382, 395
MAYEWSKI, P. A., *see* ALLEY, R. B. et al.
MAZUMBER, A., TAYLOR, W. D., MCQUEEN, D. J. and LEAN, D. R. S., 541, 553
MCAVANEY, B. J. and COLMAN, R. A., 292, 312
MCAVANEY, B. J., DAHNI, R. R., COLMAN, R. A., FRASER, J. R. and POWER, S. B., 307, 312
MCAVANEY, B. J., *see* CESS, R. D. et al.
MCAVANEY, B. J., *see* POWER, S. B. et al.
MCBRIDE, J. L., *see* HOLLAND, G. J. et al.
MCCARTNEY, M. S., *see* TALLEY, L. D. and MCCARTNEY, M. S.
MCCAUGHEY, J. H., *see* OKE, T. R. and MCCAUGHEY, J. H.
MCCAULEY, S., *see* BICE, D. M. et al.
MCCLELLAND, L., *see* SIMKIN, T. et al.
MCCORMICK, M. P. and HOOD, L. L., 319, 345
MCCORMICK, M. P., HAMILL, P., PEPIN, T. J., CHU, W. P., SWISSLER, T. J. and MCMASTER, L. R., 411, 432
MCCORMICK, M. P., VEIGA, R. E. and CHU, W. P., 319, 335, 337, 346
MCCORMICK, R. A. and LUDWIG, J. H., 347, 395
MCCRACKEN, A. D., *see* GOODFELLOW, W. D. et al.
MCCUTCHEON, J., *see* ROUSE, W. R. et al.
MCELROY, C. T., *see* KERR, J. B. and MCELROY, C. T.
MCELROY, M. B., *see* SPIVAKOVSKY, C. M. et al.
MCELROY, M. B., *see* WOFSY, S. C. et al.
MCELROY, M., *see* KNOX, F. and MCELROY, M.
MCEVEDY, C. and JONES, R, 472
MCEWAN-MASON, J., *see* RICH, T. H. et al.
MCGANN, R., *see* SALINGER, M. J. et al.
MCGHEE JR, G. R., 124, 142
MCGHEE JR, G. R. et al., 115, 124, 142
MCGLADE, J. M., *see* ADAMS, J. M. et al.
MCGLONE, M. S., KERSHAW, A. P. and MARKGRAF, V., 224, 239
MCGOVERN, T. H., 169, 186

McGowan, J. A., *see* Venrick, E. L. et al.
McGregor, J. L. and Walsh, K., 321, 346
McGuffie, K., Henderson-Sellers, A., Zhang, H., Durbidge, T. B. and Pitman, A. J., 434, 461, 472
McGuffie, K., *see* Henderson-Sellers, A. and McGuffie, K.
McGuffie, K., *see* Henderson-Sellers, A. et al.
McGuffie, K., *see* Henderson-Sellers, B. et al.
McIntyre, A., *see* Imbrie, J. et al.
McIntyre, A., *see* Rind, D. et al.
McIntyre, A., *see* Ruddiman, W. F. and McIntyre, A.
McIntyre, M. E., *see* Hoskins, B. J. et al.
McKay, C. P. and Thomas, G. E., 100, 142
McKay, C. P., *see* Milne, D. H. and McKay, C. P.
McKee, T. B., *see* Jenne, R. L. and McKee, T. B.
McKenna, M. C., 82, 93
McKenzie, J. A., *see* Hsu, K. J. and McKenzie, J. A.
McKenzie, J. A., *see* Hsu, K. J. et al.
McKenzie, D. and Bickle, M. J., 128, 142
McKenzie, D., *see* White, R. and McKenzie, D.
McKenzie, R. L., *see* Seckmeyer, G. and McKenzie, R. L.
McKinlay, A. F. and Diffey, B. L., 420, 432
McKinney, M. L., 113, 142
McKinney, R. P., *see* Barath, F. T. et al.
McLain, D. R., *see* Ebbesmeyer, C. C. et al.
McLaren, D. J., 124, 125, 142
McLaren, D. J. and Goodfellow, W. D., 98, 114, 115, 119, 120, 126, 134, 143
McLaren, D. J., *see* Geldsetzer, H. H. J. et al.
McLaren, D. J., *see* Goodfellow, W. D. et al.
McLean, D. M., 97, 129, 143
McMahon, C. K., *see* Patterson, E. M. and McMahon, C. K.
McManus, J., *see* Bond, G. et al.
McMaster, L. R., *see* McCormick, M. P. et al.
McNab, A. L., *see* Gallo, K. P. et al.
McNally, A. P., *see* Eyre, J. R. et al.
McPeters, R. D., *see* Stolarski, R. S.
McQueen, D. J., *see* Mazumber, A. et al.
McRoberts, C. A., *see* Bice, D. M. et al.
Meaden, G. T., *see* Elsom, D. M. and Meaden, G. T.
Meadows, A., *see* Sellers, A. and Meadows, A.
Mearns, L. O., *see* Giorgi, F. and Mearns, L. O.
Meehl, G. A., 213, 218, 224, 240
Meehl, G. A. and Branstator, G. W., 225, 229, 240
Meehl, G. A., *see* Gates, W. L. et al.
Meehl, G. A., *see* Karoly, D. J. et al.
Meehl, G. A., *see* Washington, W. M. and Meehl, G. A.
Meese, D. A., *see* Alley, R. B. et al.
Megaw, W. J., *see* Flyger, H. et al.
Mégie, G., *see* Hauglustaine, D. A. et al.
Meincke, J., *see* Dickson, R. R. et al.
Meira Filho, L. G., *see* Watson, R. T. et al.
Meira Filho, L. G., *see* Watson, R. W. et al.
Meko, D. M., *see* Mitchell Jr., J. M. et al.
Meko, D. M., *see* Stockton, C. W. and Meko, D. M.
Meko, D. M., *see* Stockton, C. W. et al.
Meleshko, V. P., *see* Cess, R. D. et al.
Meleshko, V. P., *see* Gates, W. L. et al.
Meleshko, V., *see* Mitchell, J. F. B. et al.
Mélice, J. L., *see* Berger, A. et al.
Melosh, H. J., 100, 129, 143
Melosh, H. J., Schneider, N. M., Zahnle, K. J. and Latham, D., 101, 143
Melosh, H. J., *see* Vickery, A. M. and Melosh, H. J.
Mengel, J. G., *see* Crowley, T. J. et al.
Mengel, J. G., *see* North, G. R. et al.
Mengel, J. G., *see* Short, D. A. et al.
Menzel, P., *see* Kaufman, Y. J. et al.
Menzel, P., *see* Rutledge, G. et al.
Menzel, W. P., *see* King, M. D. et al.
Menzies, D., *see* Smith, R. C. et al.
Mercer, J. H., 196, 240
Mergenthaler, J. L., *see* Roche, A. E. et al.
Meroney, R. N., 494, 512
Merrill, J. T., *see* Andreae, M. O. et al.
Merrill, J. T., *see* Berresheim, H. et al.
Merrill, J. T., *see* Betzer, P. R. et al.
Merrill, J. T., *see* Duce, R. A. et al.
Merrill, J. T., *see* Galloway, J. N. et al.
Merrill, J. T., *see* Prospero, J. M. et al.
Merrill, J. T., *see* Savoie, D. L. et al.
Method, T. and Carlson, T. N., 486, 512
Meyer, W. B., *see* Turner II, B. L. et al.
Michaelsen, J. and Long, A., 240
Michaelsen, J. and Thompson, L. G., 222, 240
Michaelsen, J., *see* Graham, N. E. et al.
Michaud, R. and Derome, J., 271, 277
Michel, H. V., *see* Alvarez, L. W. et al.
Michel, H. V., *see* Alvarez, W. et al.
Michel, H. V., *see* Asaro, F. et al.
Middleton, N. and Thomas, D. S. G., 436, 472
Middleton, W. E. K., 152, 186
Mihalopoulos, N., *see* Bingemer, H. G. et al.
Mikolajewicz, U., *see* Maier-Reimer, E. and Mikolajewicz, U.
Mikolajewicz, U., *see* Maier-Reimer, E. et al.
Milankovitch, M. M., 30, 65
Miller, A. J., Nagatani, R. M., Tiao, G. C., Niu, X. F., Reinsel, G. C., Wuebbles, D. and Grant, K., 319, 346
Miller, J. M., *see* Duce, R. A. et al.
Miller, K. G., Fairbanks, R. G. and Mountain, G. S., 118, 119, 143

MILLER, K. G., JANACEK, T. R., KATZ, M. E. and KEIL, D. J., 86, 93
MILLER, M. J., *see* BETTS, A. K. et al.
MILLS, G. M. and ARNFIELD, A. J., 490, 512
MILNE, D. H. and MACKAY, C. P., 98, 143
MINNIS, P., 352
MINNIS, P., HARRISON, E. F., STOWE, L. L., GIBSON, G. G., DENN, F. M., DOELLING, D. R. and SMITH, W. L., 395
MINNIS, P., *see* CESS, R. D. et al.
MINNIS, P., *see* HARRISON, E. F. et al.
MINNIS, P., *see* RAMANATHAN, V. et al.
MINTZ, Y., *see* SHUKLA, J. and MINTZ, Y.
MINTZ, Y., *see* SUD, Y. C. et al.
MITCHELL, B. G., *see* LUBIN, D. et al.
MITCHELL, J. F. B., 33, 34, 37, 38, 65, 347, 366, 367
MITCHELL, J. F. B., GRAHAME, N. S. and NEEDHAM, K. J., 38, 66
MITCHELL, J. F. B., MANABE, S., MELESHKO, V. and TOKIOKA, T., 166, 186
MITCHELL, J. F. B., *see* CAO, H. X. et al.
MITCHELL, J. F. B., *see* CESS, R. D. et al.
MITCHELL, J. F. B., *see* GATES, W. L. et al.
MITCHELL, J. F. B., *see* GREGORY, J. M. and MITCHELL, J. F. B.
MITCHELL, J. F. B., *see* SCHLESINGER, M. E. and MITCHELL, J. F. B.
MITCHELL, J. F. B., *see* SENIOR, C. A. and MITCHELL, J. F. B.
MITCHELL, J. F. B., *see* STREET-PERROTT, F. A. et al.
MITCHELL, J. F. B., SENIOR, C. A. and INGRAM, W. J., 260, 277
MITCHELL, J. G., *see* STORETVEDT, K. M. et al.
MITCHELL, JR., J. M., 156, 186, 240, 395
MITCHELL, JR., J. M., *see* GILMAN, D. L. et al.
MITCHELL, JR., J. M., *see* STOCKTON, C. W. et al.
MITCHELL, JR., J. M., STOCKTON, C. W. and MEKO, D. M., 199, 200, 202, 240
MITCHELL, T. P., 191, 192, 231
MITCHELL, T. P., *see* WRIGHT, P. B. et al.
MITRE, M. E., *see* KELLER, M. et al.
MIX, A. C., 520, 547, 553
MIX, A. C. and FAIRBANKS, R. G., 534
MIX, A. C., *see* IMBRIE, J. et al.
MIX, A. C., *see* SHACKLETON, N. J. et al.
MO, K. and RASMUSSON, E. M., 300, 312
MO, K. C. and LIVEZEY, R. E., 204, 240
MO, T., *see* WANG, W.-C. et al.
MOHNEN, V. A., GOLDSTEIN, W. and WANG, W.-C., 319, 346
MOHNEN, V., *see* PROSPERO, J. M. et al.
MOLFINO, B., *see* IMBRIE, J. et al.
MOLINA, L. T. and MOLINA, M. J., 420, 432
MOLINA, M. J., *see* DEMORE, W. B. et al.
MOLINA, M. J., *see* MOLINA, L. T. and MOLINA, M. J.
MOLION, L. C. B., *see* SHUTTLEWORTH, W. J. et al.
MÖLLER, D., 357, 395
MOLNAR, G., *see* WANG, W.-C. et al.
MONAHAN, E. C., *see* CIPRIANO, R. J. et al.
MONTANARI, A., 115, 118, 143
MONTANARI, A., *see* ALVAREZ, W. et al.
MONTANARI, A., *see* SMIT, J. et al.
MONTIGNY, R., *see* COURTILLOT, V. et al.
MONTIGNY, R., *see* VANDAMME, D. et al.
MOOK, W. G., *see* KEELING, C. D.
MOOLEY, D. A., *see* BHALME, H. N. and MOOLEY, D. A.
MOOR, E., *see* NEFTEL, A. et al.
MOORE, A. M., *see* POWER, S. B. et al.
MOORE, C. J., *see* SHUTTLEWORTH, W. J. et al.
MOORE, J. C., NARITA, H. and MAENO, N., 177, 186
MOORE, M., *see* COLE, J. E. et al.
MOORE, T. C., PISIAS, N. G. and DUNN, D.A., 31, 66
MOORE, T. C., *see* MARTINSON, D. G. et al.
MOORE, T. C., *see* TAYLOR, K. C. et al.
MORCRETTE, J. J., *see* CESS, R. D. et al.
MORCRETTE, J. J., *see* FOUQUART, Y. et al.
MORESHET, S., *see* STANHILL, G. and MORESHET, S.
MORIYAMA, S., 347, 362, 395
MORLEY, J. J., *see* IMBRIE, J. et al.
MORRIS, P., *see* TAYLOR, F. W. et al.
MORRISON, D., *see* CHAPMAN, C. R. and MORRISON, D.
MOSES, H., *see* MACCRACKEN, M. C. and MOSES, H.
MOSLEY-THOMPSON, E., *see* THOMPSON, L. G. et al.
MOUNT, J. F., *see* MARGOLIS, S. V. et al.
MOUNTAIN, G. S., *see* MILLER, K. G. et al.
MOYERS, J., *see* PROSPERO, J. M. et al.
MUELLER, P. A., *see* JONES, D. S. et al.
MULLEN, G., *see* SAGAN, C. and MULLEN, G.
MÜLLER, H., 49, 66
MULLER, J.-P., *see* RUNNING, S. W. et al.
MUNN, R. E. and MACHTA, L., 437, 472
MUNN, R. E., *see* CLARK, W. C. and MUNN, R. E.
MURPHY, J. M., 33, 40, 66
MURPHY, R. E., *see* SELLERS, P. J. et al.
MURRAY, R. J. and SIMMONDS, I., 302, 312
MYERS, N., 435, 436, 438, 444, 445, 464, 472, 473
MYLNE, M. F. and ROWNTREE, P. R., 453, 454, 459, 473
MYSAK, L. A. and LIN, C. A., 226, 240
MYSAK, L. A., *see* STOCKER, T. F. and MYSAK, L. A.
MYSAK, L. A., *see* STOCKER, T. W. et al.

NAGATANI, R. M., *see* MILLER, A. J.
NAKAJIMA, T. and KING, M. D., 258, 277
NAKAJIMA, T., *see* KAUFMAN, Y. J. and NAKAJIMA, T.
NAMIAS, J., 226, 240
NAMIAS, J. and CAYAN, D. R., 226, 240
NAMIAS, J., *see* DOUGLAS, A. V. et al.
NANCE, J. D., *see* RADKE, L. F. et al.
NAPIER, W. M., 95
NAPIER, W. M. and CLUBE, S. V. M., 97, 143
NARITA, H., *see* MOORE, J. C. et al.
NASRALLAH, H. A., BRAZEL, A. J. and BALLING JR., R. C., 501, 512
NATIONAL RESEARCH COUNCIL (NRC), 230, 240, 259, 277
NAUJOKAT, B., 202, 240
NEAL, V. T., *see* QUINN, W. H. et al.
NEEDHAM, K. J., *see* MITCHELL, J. F. B. et al.
NEES, R. T., *see* PROSPERO, J. M. and NEES, R. T.
NEES, R. T., *see* PROSPERO, J. M. et al.
NEFTEL, A., MOOR, E., OESCHGER, H. and STAUFFER, B., 524, 534
NEFTEL, A., OESCHGER, H., SCHWANDER, J., STAUFFER, B. and ZUMBRUNN, R., 524, 534
NEFTEL, A., OESCHGER, H., STAFFELBACH, T. and STAUFFER, B., 525, 526, 534
NEFTEL, A., *see* BARNOLA, J. M. et al.
NEFTEL, A., *see* HAMMER, C. U.
NEGRI, A. J., ADLER, R. F., MADDOX, R. A., HOWARD, K. W. and KEEHN, P. R., 261, 277
NEGRI, A. J., *see* ADLER, R. F. et al.
NEINCKE, J., *see* DICKSON, R. R. et al.
NELLIS, W. J., *see* GRATZ, A. J. et al.
NEUMANN, G. and PIERSON, JR, W. J., 545, 546, 553
NEUSCHULER, D., *see* LUBIN, D. et al.
NEWCOMB, W. W., *see* TUCKER, C. J. et al.
NEWELL, R. E. and KIDSON, J. W., 457, 473
NEWELL, R. E. and WEARE, B. C., 220, 229, 240
NEWELL, R. E., *see* BOTTOMLEY, M. et al.
NEWELL, R. E., *see* HSIUNG, J. and NEWELL, R. E.
NEWHALL, C. G. and SELF, S., 205, 240
NEWHALL, C., *see* SIMKIN, T. et al.
NEWIGER, M., 368, 395
NEWMAN, M. J. and ROOD, R. T., 537, 553
NEWSOME, J., *see* BOULTON, G. S. et al.
NEWTON, C. R., *see* BICE, D. M. et al.
NEZVAL, YE. I., *see* GARADZHA, M. P. and NEZVAL, YE. I.
NGUYEN, B. C., BONSANG, B. and GAUDRY, A., 348, 395
NGUYEN, B. C., *see* BINGEMER, H. G. et al.
NICHOLLS, N., 216, 223, 224, 238, 240, 309, 312
NICHOLLS, N. and KATZ, R. W., 174, 186
NICHOLLS, N. N., *see* HOLLAND, G. J. et al.
NICHOLLS, N., *see* ALLAN, R. J. et al.
NICHOLLS, N., *see* FOLLAND, C. K. et al.
NICHOLLS, N., *see* GLANTZ, M. H. et al.
NICHOLLS, S., 373, 395
NICHOLS, D. J., JARZEN, D. M., ORTH, C. J. and OLIVER, P. Q., 112, 117, 143
NICHOLS, D. J., *see* JOHNSON, K. R. et al.
NICHOLS, F. H., *see* EBBESMEYER, C. C. et al.
NICHOLSON, S. E., 166, 187
NICHOLSON, S. E. and ENTEKHABI, D., 218, 229, 240
NICKESON, J. E., *see* HALL, F. G. et al.
NICOLET, M., *see* BATES, D. R. and NICOLET, M.
NICOLIS, C., 43, 66
NIGAM, S., *see* LINDZEN, R.S. and NIGAM, S.
NIGHTINGALE, T., *see* TAYLOR, F. W. et al.
NISBET, E. G., *see* DLUGOKENCKY, E. J. et al.
NITTA, T. and YAMADA, S., 168, 187
NIU, X. F., *see* MILLER, A. J.
NOAD, D., *see* ROUSE, W. R. et al.
NOBRE, C. A., SELLERS, P. J. and SHUKLA, J., 435, 455, 465, 473
NOBRE, C., *see* SHUKLA, J. et al.
NOBRE, J. C. A., *see* SHUTTLEWORTH, W. J. et al.
NORMAN, J. M., *see* STARKS, P. J. et al.
NORTH, G. R., 66, 94
NORTH, G. R. and CROWLEY, T. J., 43, 66
NORTH, G. R., MENGEL, J. G. and SHORT, D. A., 32, 66
NORTH, G. R., *see* CROWLEY, T. J. and NORTH, G. R.
NORTH, G. R., *see* CROWLEY, T. J. et al.
NORTH, G. R., *see* SHORT, D. A. et al.
NORTH, G. R., *see* SIMPSON, J. et al.
NOVAES, F. C., *see* SALATI, E. et al.
NOWLAN, G. S., *see* GOODFELLOW, W. D. et al.
NRIAGU, J. O., *see* APSIMON, H. et al.
NUCCIARONE, J. J., *see* PRABHAKARA, C. et al.
NUNEZ, M. and OKE, T. R., 490, 492, 512
NYENZI, B. S., *see* FOLLAND, C. K. et al.

O'BRIEN, D. P., *see* CURRIE, R. G. and O'BRIEN, D. P.
O'BRIEN, J. J., *see* LUTHER, M. E. et al.
O'DOWD, C. D. and SMITH, M. H., 351, 382, 395
O'DOWD, C. D., SMITH, M. H. and JENNINGS, S. G., 395
O'KEEFE, J. D. and AHRENS, T. J., 98, 100, 102, 143
O'NEILL, C. A., *see* PENNER, J. E. et al.
O'NIELL, C. E., *see* PENNER, J. E. et al.
OBERHANSLI, H., *see* HSU, K. J. et al.
OECD, 381, 395
OERLEMANS, J., 40, 66
OERLEMANS, J. and VERNEKAR, A. D., 32, 66
OERLEMANS, J., *see* WARRICK, R. A. and OERLEMANS, J.
OESCHGER, H., 67, 534
OESCHGER, H., *see* BARNOLA, J. M. et al.

PITMAN, A. J., *see* HENDERSON-SELLERS, A. and PITMAN, A. J.
PITMAN, A. J., *see* HENDERSON-SELLERS, A. et al.
PITMAN, A. J., *see* MCGUFFIE, K. et al.
PITTOCK, A. B., 175, 179, 187, 199, 202, 241
PIVOVAROVA, Z. I., 177, 187
PLANET, W., *see* OHRING, G. et al.
PLANTICO, M. S., KARL, T. R., KUKLA, G. and GAVIN, J., 232, 241
PLATT, C. M. R., *see* ETHRIDGE, D. et al.
PLATT, T., *see* SATHYENDRANATH, S. et al.
PLAYFORD, P. E., *see* BECKER, R. T. et al.
PLUMMER, N., *see* JONES, P. D. et al.
PLUMMER, N., *see* KARL, T. R. et al.
POLCHER, J. and LAVAL, K., 463, 473
POLLACK, J. B. et al., 98, 99, 143
POLLACK, J. B., *see* KASTING, J. F. et al.
POLLACK, J. B., *see* SAGAN, C. et al.
POLLACK, J. B., *see* TURCO, R. P. et al.
POLLACK, J. B., TOON, O. W., SAGAN, C., SUMMERS, A., BALDWIN, B. and CAMP, W. V., 205, 241
POLLAK, L. D., *see* CONRAD, V. and POLLAK, L. D.
POLLARD, D., 40, 66
POLLARD, D. and THOMPSON, S. L., 76, 93
POLLARD, D., *see* BARRON, E. J. et al.
POLLARD, D., *see* BONAN, G. B. et al.
PONS, A., *see* GUIOT, J. et al.
POORE, R. Z., *see* MATTHEWS, R. K. and POORE, R. Z.
PORCH, W. M., *see* ENSOR, D. S. et al.
PORTER, S. C. and DENTON, G. H., 196, 241
PORTES, J., *see* LE TREUT, H. et al.
PORTMAN, D. A., *see* MASS, C. F. and PORTMAN, D. A.
POTTER, G. L., *see* CESS, R. D. et al.
POTTER, J. F., *see* ROCHE, A. E. et al.
POTTER, K. W., 156, 187
POWER, S. B., COLMAN, R. A., MCAVANEY, B. J., DAHNI, R. R., MOORE, A. M. and SMITH, N. R., 308, 312
POWER, S. B., *see* MCAVANEY, B. J. et al.
PRABHAKARA, C., DALU, G., LIBERTI, G. L., NUCCIARONE, J. J. and SUHASINI, R., 261, 277
PRATHER, M., *see* HANSEN, J. et al.
PREINING, O., 358, 396
PREISENDORFER, R. W. and BARNETT, T. P., 450, 473
PRELL, W. L., 31, 66
PRELL, W. L. and KUTZBACH, J. E., 38, 66
PRELL, W. L., *see* IMBRIE, J. et al.
PRELL, W. L., *see* LUTHER, M. E. et al.
PRELLER, R. H., *see* BARRY, R. G. et al.
PREMOLI-SILVA, I., *see* CORFIELD, R. M. et al.
PRENTICE, K. C., 467, 473
PRENTICE, K. C. and FUNG, I. Y., 466, 473
PREZELIN, B. B., *see* SMITH, R. C. et al.
PRICE, J. C., 502, 513
PRICE, R., *see* APSIMON, H. et al.
PRINN, R. G. and FEGLEY, B., 100, 143
PROBÁLD, F., 486, 513
PROFITT, M. H., *see* ANDERSON, J. G. et al.
PROSPERO, J. M., 350, 396
PROSPERO, J. M. and NEES, R. T., 350, 388, 396
PROSPERO, J. M., CHARLSON, R. J., MOHNEN, V., JAENICKE, R., DELANY, A. C., MOYERS, J., ZOLLER, W. and RAHN, K., 364, 396
PROSPERO, J. M., SAVOIE, D. L., NEES, R. T., DUCE, R. A. and MERRILL, J., 380, 396
PROSPERO, J. M., *see* DUCE, R. A. et al.
PROSPERO, J. M., *see* SAVOIE, D. L. and PROSPERO, J. M.
PROSPERO, J. M., *see* SAVOIE, D. L. et al.
PROSPERO, J. M., *see* UEMATSU, M. et al.
PROTHERO, D. R., *see* SWISHER III, C. C. and PROTHERO, D. R.
PRUPPACHER, H. R. and KLETT, J. D., 372, 396
PSZENNY, A. A. P., *see* GALLOWAY, J. N. et al.
PSZENNY, A. A. P., *see* HANSEN, A. D. A. et al.
PUESCHEL, R. F., *see* ELLIS, H. T. and PUESCHEL, R. F.
PUESCHEL, R. F., VAN VALIN, C. C., CASTILLO, R. C., KADLECEK, J. A. and GANOR, E. J., 372, 388, 396
PULWARTY, R. S., *see* DIAZ, H. F. and PULWARTY, R. S.

QIN-WEN, Z., *see* DAO-YI, X. et al.
QUADRENNIAL OZONE SYMPOSIUM, 339, 340, 346
QUAITE, F. E., 432
QUAITE, F. E., SUTHERLAND, B. M. and SUTHERLAND, J. C., 424
QUAY, P. D., *see* STUIVER, M. and QUAY, P. D.
QUAYLE, R. G., *see* KARL, T. R. et al.
QUINBY-HUNT, M. S., *see* WILDE, P. et al.
QUINN, P. K., BATES, T. S., JOHNSON, J. E., COVERT, D. S. and CHARLSON, R. J., 381, 396
QUINN, P. K., *see* BATES, T. S. et al.
QUINN, W. H., 194, 218, 222, 223, 241
QUINN, W. H., NEAL, V. T. and ANTUNEZ DE MAYOLO, S. E., 222, 241
QUINN, W. H., ZOPF, D. O., SHORT, K. S. and KUO YANG, R. T. W., 222, 241
QUINTANA, L. R., *see* ORTH, C. J. et al.
QUIRK, W. J., *see* CHARNEY, J. G. et al.

RABINOFF, R., *see* HERMAN, B. M. et al.
RADICK, R. R., LOCKWOOD, G. W. and BALIUNAS, S. L., 199, 241
RADKE, L. F., 372, 376, 396
RADKE, L. F., COAKLEY JR., J. A. and KING, M. D., 373, 396
RADKE, L. F., HEGG, D. A., HOBBS, P. V., NANCE, J. D., LYONS, J. H., LAURSEN, K. K., WEISS, R.

E., RIGGAN, P. J. and WARD, D. E., 366, 376, 396
RADKE, L. F., HEGG, D. A., LYONS, J. D., BROCK, C. A., HOBBS, P. V., WEISS, R. and RASMUSSEN, R., 363, 396
RADKE, L. F., *see* HEGG, D. A. et al.
RADKE, L. F., *see* HOBBS, P. V. and RADKE, L. F.
RADKE, L. F., *see* HOBBS, P. V. et al.
RADKE, L. F., *see* KING, M. D. et al.
RADKE, L. F., *see* TWOHY, C. H. et al.
RAEMDONCK, H., *see* ANDREAE, M. O. et al.
RAHN, K., *see* PROSPERO, J. M. et al.
RAISBECK, G., *see* JOUZEL, J. et al.
RAJAGOPALAN, G., *see* SUKUMAR, R. et al.
RAJU, D. S. N., *see* JAIPRAKASH, B. C. et al.
RALSTON, C. W., *see* ARMENTANO, T. V. and RALSTON, C. W.
RAM, M., *see* ALLEY, R. B. et al.
RAMAGE, C. S., 202, 241
RAMAKRISHNAN, P. S., 459, 473
RAMAN, S., *see* HARMS, D. E. et al.
RAMANATHAN, V., 53, 66, 318, 346
RAMANATHAN, V. and COLLINS, W., 228, 241
RAMANATHAN, V., CESS, R. D., HARRISON, E. F., MINNIS, P., BARKSTROM, B. R., AHMAD, E. and HARTMANN, D., 228, 241, 319, 346
RAMANATHAN, V., *see* CESS, R. D. et al.
RAMANATHAN, V., *see* HARRISON, E. F. et al.
RAMANATHAN, V., *see* HARTMANN, D. L. et al.
RAMANATHAN, V., *see* KIEHL, J. T. and RAMANATHAN, V.
RAMANATHAN, V., *see* MADDEN, R. A. and RAMANATHAN, V.
RAMANATHAN, V., *see* OWEN, T. et al.
RAMANKUTTY, N., *see* SCHLESINGER, M. E. and RAMANKUTTY, N.
RAMASWAMY, V., SCHWARZKOPF, M. D. and SHINE, K. P., 319, 320, 336, 346, 419, 432
RAMASWAMY, V., *see* ISAKSEN, I. S. A. et al.
RAMASWAMY, V., *see* SCHWARZKOPF, M. D. and RAMASWAMY, V.
RAMASWAMY, V., *see* SHINE, K. P. et al.
RAMESH, R., *see* SUKUMAR, R. et al.
RAMPINO, M. R., 122, 123, 128, 143
RAMPINO, M. R. and CALDEIRA, K., 97, 128, 143
RAMPINO, M. R. and HAGGERTY, B. M., 114, 116, 143
RAMPINO, M. R. and SELF, S., 98, 130, 144
RAMPINO, M. R. and STOTHERS, R. B., 97, 102, 124, 127, 128, 132, 144
RAMPINO, M. R. and VOLK, T., 106, 144
RAMPINO, M. R., *see* CALDEIRA, K. and RAMPINO, M. R.
RAMPINO, M. R., *see* CALDEIRA, K. et al.
RAMPINO, M. R., *see* LEARY, P. N. and RAMPINO, M. R.
RAMPINO, M. R., *see* SELF, S. et al.
RAMPINO, M. R., *see* STOTHERS, R. B. et al.
RAMPINO, M. R., SELF, S. and STOTHERS, R. B., 129, 131, 132, 144
RANDALL, D. A. and TJEMKES, S. A., 254, 277
RANDALL, D. A., *see* CESS, R. D. et al.
RAPER, S. C. B., *see* JONES, P. D. et al.
RAPER, S. C. B., *see* WIGLEY, T. M. L. and RAPER, S. C. B.
RAPER, S. C. B., CHERRY, B. S. G., GOODESS, C. M. and WIGLEY, T. M. L., 156
RASKIN, P., *see* SUBAK, S. et al.
RASMUSSEN, E. M. and CARPENTER, T. H., 173, 187
RASMUSSEN, R., *see* KAUFMAN, Y. J. et al.
RASMUSSEN, R., *see* RADKE, L. F. et al.
RASMUSSON, E. M., 267, 277
RASMUSSON, E. M. and CARPENTER, T. H., 207, 211, 213, 214, 216
RASMUSSON, E. M. and MO, K., 301
RASMUSSON, E. M., *see* MO, K. and RASMUSSON, E. M.
RASMUSSON, E. M., WONG, X. and ROPELEWSKI, C. E., 213, 241
RASMUSSON, R. M. and CARPENTER, T. H., 241
RASOOL, S. I. and SCHNEIDER, S. H., 347, 396
RAUP, D. M., 96, 114, 118, 120, 144
RAVINDRAN, P., *see* SATHYENDRANATH, S. et al.
RAVISHANKARA, A. R., *see* DEMORE, W. B. et al.
RAY, J. D., *see* BOATMAN, J. F. et al.
RAYMO, M. E. et al., 117, 144
RAYMO, M. E., *see* IMBRIE, J. et al.
RAYMOND, D. J. and JIANG, H., 305, 313
RAYNAUD, D., 524
RAYNAUD, D. and BARNOLA, J. M., 534
RAYNAUD, D., JOUZEL, J., BARNOLA, J. M., CHAPPELLAZ, J., DELMAS, R. J. and LORIUS, CL., 47, 66
RAYNAUD, D., *see* BARNOLA, J. M. et al.
RAYNAUD, D., *see* CHAPPELLAZ, J. et al.
RAYNAUD, D., *see* GENTHON, C. et al.
RAYNAUD, D., *see* JOUZEL, J. et al.
RAYNAUD, D., *see* LORIUS, C. et al.
RAYNAUD, Y., *see* BARNOLA, J. M. et al.
RAZUVAYEV, V. N., *see* KARL, T. R. et al.
REA, D. K., LEINEN, M. and JANACEK, T. R., 84, 86, 93
REA, D. K., *see* JANACEK, T. R. and REA, D. K.
REA, D. K., *see* OWEN, R. M. and REA, D. K.
READ, P. L., *see* ALLEN, M. R. et al.
READ, W. G., *see* WATERS, J. W. et al.
RECKER, E. E., *see* REED, R. J. and RECKER, E. E.
REDMOND, K. T., *see* EBBESMEYER, C. C. et al.
REED, R. J., 202, 241
REED, R. J. and RECKER, E. E., 305, 313
REEH, N., *see* HUYBRECHTS, PH. et al.
REHKOPF, J., 368, 396
REID, G. C., 195, 199, 200, 242

REID, G. C., *see* GAGE, K. S. and REID, G. C.
REILLE, M., *see* GUIOT, J. et al.
REINERS, P. W., *see* BICE, D. M. et al.
REINHARDT, K. H., *see* DUCE, R. A. et al.
REINSEL, G. C., *see* BOJKOV, R. D. et al.
REINSEL, G. C., *see* MILLER, A. J.
REISS, M., *see* RUDOLF, B. et al.
REMER, L. A., *see* SOMERVILLE, R. C. J. and REMER, L. A.
RENNE, P. R. and BASU, A. R., 127, 144
RENNE, P. R. et al., 127, 144
RETALLACK, G. J., LEAHY, G. D. and SPOON, M. D., 111, 144
REUTER, D., *see* SUSSKIND, J. et al.
REYNOLDS, J. F., *see* SCHLESINGER, W. H. et al.
REYNOLDS, R. W., 272, 277
REYNOLDS, R. W. and MARSICO, D. C., 154, 187
RHEIN, M., *see* SCHLOSSER, P. et al.
RHINES, P. B., *see* BREWER, P. G. et al.
RIBEIRO, M. N. G., *see* SHUTTLEWORTH, W. J. et al.
RICH, P. V., *see* RICH, T. H. et al.
RICH, T. H., RICH, P. V., WAGSTAFF, B., MCEWAN-MASON, J., DOUTHITT, C. B., GREGORY, R. T. and FELTON, E. A., 73, 93
RICHARDS, F. and ARKIN, P. A., 261, 278
RICHARDS, G. R., 231, 242
RICHARDS, J. F., 434, 439, 473
RICHARDS, J. F., *see* HOUGHTON, R. A. et al.
RICHARDS, J. F., *see* TURNER II, B. L. et al.
RICHARDSON III, J. B., *see* ROLLINS, H. B. et al.
RICHARDSON, S. M., *see* KASTING, J. F. et al.
RICHARDSON, S. M., *see* KASTING, J. K. and RICHARDSON, S. M.
RICHES, M. R., *see* CESS, R. D. et al.
RICKARDS, R. B., *see* COCKS, L. R. M. and RICKARDS, R. B.
RIEBSAME, W. E., 438, 473
RIGGAN, P. J., *see* RADKE, L. F. et al.
RIKUS, L., *see* CESS, R. D. et al.
RILEY, J. P. and CHESTER, R., 545, 553
RIND, D., 35, 66, 84, 93, 530
RIND, D. and CHANDLER, M., 84, 85, 86, 93
RIND, D. and OVERPECK, J., 200, 242
RIND, D. and PETEET, D., 36, 66, 84, 93
RIND, D., GOLDBERG, R. and RUEDY, R., 164, 187
RIND, D., PETEET, D. and KUKLA, G., 39, 67
RIND, D., PETEET, D. M., BROECKER, W. S., MACINTYRE, A. and RUDDIMAN, W. F., 37, 66, 84, 93, 535
RIND, D., *see* BROECKER, W. S. et al.
RIND, D., *see* CHANDLER, M. et al.
RIND, D., *see* HANSEN, J. et al.
RIND, D., *see* LEAN, J. and RIND, D.
RIND, D., *see* TSELIOUDIS, G. et al.
RISER, J., *see* PETIT-MAIRE, J. R. et al.
RITCHEY, N. A., *see* DARNELL, W. L. et al.
RITCHEY, N. A., *see* WHITLOCK, C. H. et al.
RITCHIE, E. A. and HOLLAND, G. J., 305, 313
RITCHIE, E. A., HOLLAND, G. J. and LANDER, M., 305, 313
RITZ, C., *see* JOUZEL, J. et al.
ROBERT, C. and CHAMLEY, H., 112, 144
ROBERTS, A. D., *see* LUBIN, D. et al.
ROBERTS, J., *see* SHUTTLEWORTH, W. J. et al.
ROBERTSON, A. W., *see* HOSKINS, B. J. et al.
ROBINSON, D. A., 278
ROBINSON, D. A., DEWEY, K. F. and HEIM JR., R. R., 273
ROBINSON, D. A., SERREZE, M. C., BARRY, R. G., SCHARFEN, G. and KUKLA, G., 272, 273, 278
ROBOCK, A., 173, 177, 187
ROBUIN, G., *see* STORETVEDT, K. M. et al.
ROCCHIA, R., BOCLET, D., BONTE, P., BUFFETAUT, E., ORUE-ETXEBARRIA, JAEGER, J.-J. and JEHANNO, C., 144
ROCCHIA, R., BOCLET, D., COURTILLOT, V. and JAEGER, J.-J., 134, 144
ROCHE, A. E., KUMER, J. B., MERGENTHALER, J. L., ELY, G. A., UPLINGER, W. G., POTTER, J. F., JAMES, T. C. and STERRIT, L. W., 411, 432
RODDA, J. C., 155, 187
RODDY, D. J., *see* TINUS, R. W. and RODDY, D. J.
RODGERS, C. D., *see* TAYLOR, F. W. et al.
RODGERS, C. F., *see* HUDSON, J. G. et al.
RODHE, H., *see* CHARLSON, R. J. et al.
RODHE, H., *see* ENGARDT, M. and RODHE, H.
RODHE, H., *see* ISAKSEN, I. S. A. et al.
RODHE, H., *see* LANGNER, J. and RODHE, H.
RODHE, H., *see* LANGNER, J. et al.
RODHE, H., *see* WATSON, R. W. et al.
ROECKNER, E., *see* BARNETT, T. P. et al.
ROECKNER, E., *see* CESS, R. D. et al.
ROELOFFZEN, H., *see* KEELING, C. D.
ROGERS J., *see* VAN LOON, H. and ROGERS, J.
ROGERS, C. F., HUDSON, J. G., ZIELINSKA, B., TANNER, R. L., HALLETT, J. and WATSON, J. G., 376, 396
ROGERS, C. F., *see* HALLETT, J. et al.
ROGERS, J. C., 218, 242
ROLLINS, H. B., RICHARDSON III, J. B. and SANDWEISS, D. H., 224, 242
RONOV, A., *see* BUDYKO, M. and RONOV, A.
ROOD, R. B., *see* SCHUBERT, S. D. et al.
ROOD, R. T., *see* NEWMAN, M. J. and ROOD, R. T.
ROOSEN, G. R., ANGIONNE, R. J. and KLEMCKE, C. H., 379, 396
ROOTH, C., 43, 67
ROOTH, C. G., *see* BREWER, P. G. et al.
ROPELEWSKI, C. E., *see* RASMUSSON, E. M. et al.
ROPELEWSKI, C. F. and HALPERT, M. S., 214, 220, 221, 242
ROPELEWSKI, C. F. and JONES, P. D., 174, 187
ROSEN, R. D., SALSTEIN, D. A. and WOOD, T. M., 291, 292, 313

ROSEN, R. D., *see* HIDE, R. et al.
ROSENFIELD, J., *see* SUSSKIND, J. et al.
ROSS, S. L. and OKE, T. R., 479, 513
ROSSOW, W. B., 255, 278
ROSSOW, W. B. and SCHIFFER, R. A., 255, 278
ROSSOW, W. B., *see* FU, R. et al.
ROSSOW, W. B., *see* TSELIOUDIS, G. et al.
ROSSOW, W., *see* HANSEN, J. et al.
ROTH, M., 496, 513
ROTH, M. and OKE, T. R., 496, 513
ROTH, M., OKE, T. R. and EMERY, W. J., 502, 513
ROTH, M., OKE, T. R. and STEYN, D. G., 496, 513
RÖTHLISBERGER, F. and GEYH, M. A., 196, 242
ROUSE, W. R., NOAD, D. and MCCUTCHEON, J., 486, 487, 513
ROWELL, D. P., FOLLAND, C. K., MASKELL, K, OWEN, J. A. and WARD, M. N., 166, 187
ROWNTREE, P. R., 455, 473
ROWNTREE, P. R., *see* LEAN, J. and ROWNTREE, P. R.
ROWNTREE, P. R., *see* MYLNE, M. F. and ROWNTREE, P. R.
ROWNTREE, P. R., *see* WALKER, J. and ROWNTREE, P. R.
ROY, C. T., GIES, H. P. and GRAEME, E., 425, 432
ROYER, A., *see* PETIT, J.-R. et al.
ROYER, J. F., 38
ROYER, J. F., DEQUE, M. and PESTIAUX, P., 39, 67
ROYER, J. F., *see* CESS, R. D. et al.
RUBENSTONE, J. L., *see* GUILDERSON, TH. P. et al.
RUDDIMAN, W. F. and KUTZBACH, J. E., 34, 67
RUDDIMAN, W. F. and MCINTYRE, A., 24, 40, 67
RUDDIMAN, W. F., KIDD, R. B., THOMAS, E. et al., 31, 67
RUDDIMAN, W. F., *see* RIND, D. et al.
RUDISILL, J. H., *see* PETERSON, J. T. et al.
RUDOLF, B., HAUSCHILD, H., REISS, M. and SCHNEIDER, U., 156, 187
RUDOLPH, J., *see* EHHALT, D. H. et al.
RUEDI, R., *see* HANSEN, J. et al.
RUEDY, R., *see* HANSEN, J. et al.
RUEDY, R., *see* RIND, D. et al.
RUFFELL, A., *see* SIMMS, M. J. and RUFFELL, A.
RUMBAUGH, W. F., 154, 187
RUNNING, S. W., JUSTICE, C. O., SALOMONSON, V., HALL, D., BARKER, J., KAUFMANN, Y. J., STRAHLER, A. H., HUETE, A. R., MULLER, J.-P., VANDERBILT, V., WAN, Z. M., TEILLET, P. and CARNEGGIE, D., 267, 278
RUNNING, S. W., LOVELAND, T. R., PIERCE, L.L. and HUNT JR., E. R., 269, 278
RUSSEL, G., *see* HANSEN, J. et al.
RUSSELL III, J. M., *see* GILLE, J. C. and RUSSELL III, J. M.
RUTHKOSKY, M. S., *see* WOLFF, G. R. et al.
RUTLEDGE, G., LEGG, E. and MENZEL, P., 262, 278
RYAN, B. F., WATTERSON, I. G. and EVANS, J. L., 310, 313
SACHSE, G. W., *see* ANDREAE, M. O. et al.
SADLER, J. C., 305, 313
SAGAN, C. and MULLEN, G., 537, 553
SAGAN, C., *see* POLLACK, J. B. et al.
SAGAN, C., *see* TURCO, R. P. et al.
SAGAN, C., TOON, O. B. and POLLACK, J. B., 231, 242, 435, 440, 452, 474
SAITO, T., YAMANOI, T. and KAIHO, K., 117, 144
SALATI, E., DOUROJEANNI, M. J., NOVAES, F. C., DE OLIVEIRA, A. E., PERRITT, R. W., SCHUBART, H. O. R. and UMANA, J. C., 437, 474
SALINGER, M. J., HAY, J. E., MCGANN, R. and FITZHARRIS, B. B., 164, 187
SALOMONSON, V., *see* RUNNING, S. W. et al.
SALSTEIN, D. A., *see* HIDE, R. et al.
SALSTEIN, D. A., *see* ROSEN, R. D. et al.
SALTZMAN, B., 40, 43, 67
SALTZMAN, B. and MAASCH, K. A., 42, 67
SALTZMAN, B. and SUTERA, A., 43, 67
SALTZMAN, B., HANSEN, A. R. and MAASCH, K. A., 42, 67
SALTZMAN, B., *see* MAASCH, K. A. and SALTZMAN, B.
SALTZMAN, E., *see* BRASS, G. et al.
SALTZMAN, E. S., *see* LEGRAND, M. R. et al.
SALTZMAN, E. S., *see* SAVOIE, D. L. et al.
SANCETTA, C., HEUSSER, L. and HALL, M. A., 117, 144
SANCETTA, C., IMBRIE, J. and KIPP, N. G., 23, 67
SANDBERG, C. A., ZIEGLER, W., DREESEN, R. and BUTLER, J. L., 124, 125, 144
SANDERSON, M., KUMANAN, I., TANGUAY, T. and SCHERTZER, W., 486, 513
SANDWEISS, D. H., 224, 242
SANDWEISS, D. H., *see* ROLLINS, H. B. et al.
SANHUEZA, E., HAO, W. M., SCHARFFE, D., DONESO, L. and CRUTZEN, P. J., 448, 474
SANHUEZA, E., *see* SCHARFFE, D. et al.
SANHUEZA, E., *see* WATSON, R. T. et al.
SANHUEZA, E., *see* WATSON, R. W. et al.
SANTER, B. D., WIGLEY, T. M. L. and JONES, P. D., 181, 188
SANTER, B., BERGER, A., EDDY, J. A., FLOHN, H., IMBRIE, J., LITT, T., SCHNEIDER, S. H., SCHWEINGRUBER, F. H. and STUIVER, M., 39, 67
SANTER, R., *see* FOUQUART, Y. et al.
SARDESHMUKH, P. D. and HOSKINS, B. J., 289, 290, 298, 299, 304, 313
SARDESHMUKH, P. D., *see* HOSKINS, B. J. et al.
SARGENT, N. E., *see* BOER, G. J. and SARGENT, N. E.
SARMIENTO, J. L., 527
SARMIENTO, J. L. and TOGGWEILER, R., 535
SATA, M., *see* HANSEN, J. et al.
SATHYENDRANATH, S., GOUVEIA, A. D., SHETYE, S. R., RAVINDRAN, P. and PLATT, T., 541, 553

SHUGART, H. H., *see* SMITH, T. M. and SHUGART, H. H.
SHUKLA, J., 474
SHUKLA, J. and MINTZ, Y., 260, 278, 435, 474
SHUKLA, J., NOBRE, C. and SELLERS, P. J., 437, 455, 459
SHUKLA, J., *see* NOBRE, C. A. et al.
SHUKLA, J., *see* SUD, Y. C. et al.
SHUKLA, P. N., *see* BHANDARI, N. et al.
SHULKA, J., *see* BENGTSSON, L. and SHULKA, J.
SHUMAN, C. A., *see* ALLEY, R. B. et al.
SHUMWAY, S. E., *see* HOBBS, P. V. et al.
SHUTTLEWORTH, W. J., GASH, J. H. C., LLOYD, J. C. R., MOORE, C. J., ROBERTS, J., MARQUES FILHO, A. DE O., FISCH, G., DE PAULA SILVA FILHO, V., RIBEIRO, M. N. G., MOLION, L. C. B., DE ABREU SÁ, L. D., NOBRE, J. C. A., CABRAL, G. M. R., PATEL, S. R. and DE MORAES, J. C., 453, 474
SIEBERT, L., *see* SIMKIN, T. et al.
SIEBES, G., *see* BARATH, F. T. et al.
SIEGENTHALER, U., 527
SIEGENTHALER, U. and WENK, T., 535
SIEGENTHALER, U., *see* LEUENBERGER, M. and SIEGENTHALER, U.
SIEGENTHALER, U., *see* STAUFFER, B. et al.
SIEGENTHALER, U., *see* WATSON, R. W. et al.
SIEGMUND, P. C., 313
SIELMANN, F., *see* KÖNIG, W. et al.
SIEVERING, H., *see* BOATMAN, J. F. et al.
SIEVERING, H., *see* GALLOWAY, J. N. et al.
SIGNOR III, P. W. and LIPPS, J. H., 114, 145
SIGURDSSON, H., 96, 129, 130, 145, 205, 242
SIGURDSSON, H., D'HONDT, S. and CAREY, S., 102, 145
SIGURDSSON, H., D'HONDT, S., ARTHUR, M. A. et al., 134, 145
SILVA DIAS, P. L., SCHUBERT, W. H. and DEMARIA, M., 290, 313
SILVA DIAS, P. L., *see* ANDREAE, M. O. et al.
SILVA DIAS, P. L., *see* SCHUBERT, W. H. et al.
SIMET, C., *see* BOND, G. et al.
SIMKIN, T., SIEBERT, L., MCCLELLAND, L., BRIDGE, D., NEWHALL, C. and LATTER, J. H., 177, 188
SIMMONDS, I., *see* MURRAY, R. J. and SIMMONDS, I.
SIMMS, M. J., 119, 145
SIMMS, M. J. and RUFFELL, A., 119, 145
SIMMS, M. J., *see* JOHNSON, A. L. A. and SIMMS, M. J.
SIMPSON, J. and WIGGERT, V., 373, 397
SIMPSON, J. S., KEENAN, T. D., FERRIER, B., SIMPSON, R. H. and HOLLAND, G. J., 305, 313
SIMPSON, J., ADLER, R. F. and NORTH, G. R., 261, 278
SIMPSON, R. H., *see* SIMPSON, J. S. et al.
SINGH, I. B., *see* AHARON, P. et al.
SINGH, J., *see* JAIPRAKASH, B. C. et al.
SINHA, S. K., *see* KILADIS, G. N. and SINHA, S. K.
SKINNER, B., *see* APSIMON, H. et al.
SKOLE, D. and TUCKER, C., 246, 278
SKOLE, D. L. and TUCKER, C. J., 444, 474
SKOLE, D. L., *see* HOUGHTON, R. A. and SKOLE, D. L.
SKUMANICH, A., *see* LEAN, J. et al.
SKUMANICH, L., *see* LEAN, J. et al.
SLAPER, H., *see* DE GRUIJL, F. R. et al.
SLINGO, A., 378, 397
SLINGO, A. and SLINGO, J. M., 278
SLINGO, A., *see* CESS, R. D. et al.
SLINGO, A., *see* SHINE, K. P. et al.
SLINGO, J. M. and SLINGO, A., 254
SLINGO, J. M., *see* SLINGO, A. and SLINGO, J. M.
SLINGO, J., *see* TANRE, D. et al.
SLINN, W. G. N., 384, 397
SLOAN, J., *see* BRASS, G. et al.
SLOAN, L. C. and BARRON, E. J., 83, 84, 86, 87, 88, 94
SLUSSER, J., *see* STAMNES, K. et al.
SLUTZ, R. J., *see* WOODRUFF, S. D. et al.
SMIC, 350, 351, 397
SMILEY, C. J., 73, 94
SMIT, J., 96, 103, 107, 110, 111, 112, 113, 145
SMIT, J. and HERTOGEN, J., 95, 145
SMIT, J., MONTANARI, A. et al., 96, 145
SMITH, E. A., COOPER, H. J., CROSSON, W. L. and HENG-YI, W., 270, 278
SMITH, E. A., *see* WILBER, A. C. et al.
SMITH, G. D., *see* BOULTON, G. S. et al.
SMITH, G. L., *see* BARKSTROM, B. R. and SMITH, G. L.
SMITH, L. A., *see* ALLEN, M. R. et al.
SMITH, L. D. and VONDER HAAR, T. H., 257, 278
SMITH, L., *see* FOUQUART, Y. et al.
SMITH, M. H., *see* LATHAM, J. and SMITH, M. H.
SMITH, M. H., *see* O'DOWD, C. D. and SMITH, M. H.
SMITH, M. H., *see* O'DOWD, C. D. et al.
SMITH, N. R., *see* POWER, S. B. et al.
SMITH, R. C., PREZELIN, B. B., BAKER, K. S., BIDIGARE, R. R., BOUCHER, N. P., COLEY, T., KARENTZ, D., MACINTYRE, S., MATLICK, H. A., MENZIES, D., ONDRUSEK, M., WAN, Z. and WATERS, K. J., 425, 432
SMITH, T. M. and SHUGART, H. H., 465, 474
SMITH, W. E., *see* SUD, Y. C. and SMITH, W. E.
SMITH, W. L., *see* MINNIS, P. et al.
SODEN, B. J. and BRETHERTON, F. P., 263, 278
SOKOLOV, A. P., *see* CESS, R. D. et al.
SOLOMON, A., *see* TRENBERTH, K. E. and SOLOMON, A.
SOLOMON, S., *see* BRASSEUR, G. P. and SOLOMON, S.

SOLOW, A., *see* GORNITZ, V. and SOLOW, A.
SOLTIS, F. S., *see* BARATH, F. T. et al.
SOMERVILLE, R. C. J. and REMER, L. A., 260, 278
SOUTHAM, J., *see* BRASS, G. et al.
SOWERS, T., *see* JOUZEL, J. et al.
SPELMAN, M. J. and MANABE, S., 85, 94
SPELMAN, M. J., *see* MANABE, S. et al.
SPENCER, M. J., *see* MAYEWSKI, P. A. et al.
SPENCER, R. W., 156, 188
SPENCER, R. W. and CHRISTY, J. R., 169, 170, 171, 172, 173, 188, 205, 220, 232, 242, 262, 278
SPICER, R. A. and CORFIELD, R. M., 72, 73, 94
SPICER, R. A. and PARRISH, J. T., 73, 94
SPIRO, P. A., JACOB, D. J. and LOGAN, J. A., 357, 397
SPITTLEHOUSE, D. L., *see* KALANDA, B. D. et al.
SPIVAKOVSKY, C. M., YEVICH, R., LOGAN, J. A., WOFSY, S. C. and MCELROY, M. B., 338, 346
SPOON, M. D., *see* RETALLACK, G. J. et al.
SPYERS-DURAN, P. A., *see* FITZGERALD, J. W. and SPYERS-DURAN, P. A.
SQUIRES, P., *see* TWOMEY, S. and SQUIRES, P.
STAEHELIN, J., *see* STOLARSKI, R. S. et al.
STAFFELBACH, T., *see* NEFTEL, A. et al.
STALLARD, R. F., *see* D'HONDT, S. et al.
STALLARD, R. F., *see* KELLER, M. et al.
STAMNES, K., SLUSSER, J., BOWEN, M., BOOTH, C. and LUCAS, T., 425, 432
STANHILL, G. and KALMA, J. D., 486, 508, 513
STANHILL, G. and MORESHET, S., 486, 513
STANLEY, S. M., 97, 122, 123, 145
STANLEY, S. M. and CAMPBELL, L. D., 117, 145
STARKS, P. J., NORMAN, J. M., BLAD, B. L., WALTER-SHEA, E. A. and WALTHALL, C. L., 265, 278
STAUFFER, B., HOFER, H., OESCHGER, H., SCHWANDER, J. and SIEGENTHALER, U., 230, 242, 523, 535
STAUFFER, B., *see* BERNER, W. et al.
STAUFFER, B., *see* JOHNSEN, S. J. et al.
STAUFFER, B., *see* NEFTEL, A. et al.
STAUFFER, R. J., *see* JOHNSEN, S. J.
STAUFFER, R. J., *see* MANABE, S. and STAUFFER, R. J.
STAYLOR, W. F., *see* DARNELL, W. L. et al.
STAYLOR, W. F., *see* WHITLOCK, C. H. et al.
STEDMAN, D. H., *see* WOLFF, G. R. et al.
STEELE, L. P., *see* DLUGOKENCKY, E. J. et al.
STEFFENSEN, J. P., *see* DANSGAARD, W. et al.
STEFFENSEN, J. P., *see* JOHNSEN, S. J. et al.
STEHLI, F. G., *see* SAVIN, S. et al.
STEINNES, E., *see* APSIMON, H. et al.
STEPANOV, V., *see* CAMPBELL, I. H. et al.
STEPHENS, G. L., 370, 397
STEPHENS, G. L. and GREENWALD, T. J., 256, 257, 278
STEPHENS, G. L., *see* TJEMKES, S. A. et al.
STEPHENS, G., *see* MALINGREAU, J. P. et al.
STEPHENSON, D. and HELD, I. M., 313
STERENBORG, H. J. C. M., *see* DE GRUIJL, F. R. et al.
STERRIT, L. W., *see* ROCHE, A. E. et al.
STEURER, P. M., *see* KARL, T. R. and STEURER, P. M.
STEURER, P. M., *see* WOODRUFF, S. D. et al.
STEVENS, C. M., *see* CRAIG, H. et al.
STEVENSON, D. J., *see* WALKER, J. C. G. et al.
STEVENSON, M. P., *see* EMANUEL, W. R. et al.
STEYN, D. G., 496, 513
STEYN, D. G. and OKE, T. R., 488, 513
STEYN, D. G., *see* JOHNSON, G. T. et al.
STEYN, D. G., *see* OKE, T. R. et al.
STEYN, D. G., *see* ROTH, M. et al.
STIEVENARD, M., *see* JOUZEL, J.
STIEVENARD, M., *see* JOUZEL, J. et al.
STOCKER, T. F. and MYSAK, L. A., 192, 195, 226, 242
STOCKER, T. W. and WRIGHT, D. G., 37, 43, 68
STOCKER, T. W., WRIGHT, D. G. and MYSAK, L. A., 43, 68
STOCKTON, C. W. and MEKO, D. M., 199, 242
STOCKTON, C. W., MITCHELL JR., J. M. and MEKO, D. M., 201, 242
STOCKTON, C. W., *see* MITCHELL JR., J. M. et al.
STOFFEL, T. L., *see* PETERSON, J. T. and STOFFEL, T. L.
STOLARSKI, R. S. and CICERONE, R. J., 401, 432
STOLARSKI, R. S., BLOOMFIELD, P., MCPETERS, R. D. and HERMAN, J. R., 335, 337, 346
STOLARSKI, R. S., BOJKOV, R., BISHOP, L., ZEREFOS, C., STAEHELIN, J. and ZAWODNY, J., 421, 425, 432
STONE, P. H., 84, 94
STONE, P., *see* HANSEN, J. et al.
STONE, R. S., *see* KAHL, J. D. W. et al.
STORETVEDT, K. M., MITCHELL, J. G., ABRANCHES, M. C., MAALOE, S. and ROBUIN, G., 127, 145
STORMS, H., *see* ARTAXO, P. et al.
STOTHERS, R. B., 128, 145
STOTHERS, R. B., *see* RAMPINO, M. R. and STOTHERS, R. B.
STOTHERS, R. B., *see* RAMPINO, M. R. et al.
STOTHERS, R. B., WOLFE, J. A., SELF, S. and RAMPINO, M. R., 130, 131
STOTT, L. D. and KENNETT, J. P., 104, 108, 110, 134, 145
STOTT, L. D., *see* ZACHOS, J. C. et al.
STOUFFER, R., 345
STOUFFER, R. J., MANABE, S. and BRYAN, K., 40, 68, 175, 188
STOUFFER, R. J., MANABE, S. and VINNIKOV, K. Y., 34, 68
STOUFFER, R. J., *see* DELWORTH, T. et al.
STOUFFER, R. J., *see* GATES, W. L. et al.

STOUFFER, R. J., *see* KAROLY, D. J. et al.
STOUFFER, R. J., *see* MANABE, S. and STOUFFER, R. J.
STOUFFER, R. J., *see* MANABE, S. et al.
STOWE, L. L., *see* HUSAR, R. B. and STOWE, L. L.
STOWE, L. L., *see* MINNIS, P. et al.
STOWE, L., *see* OHRING, G. et al.
STRAHLER, A. H., *see* RUNNING, S. W. et al.
STRAMSKI, D., *see* BRICAUD, A. and STRAMSKI, D.
STRAPP, J. W., *see* LEAITCH, W. R. et al.
STREBEL, D. E., *see* SELLERS, P. J. et al.
STREET, F. A. and GROVE, A. T., 24, 68
STREET-PERROTT, F. A., 34, 68
STREET-PERROTT, F. A. and HARRISON, S. P., 35, 68
STREET-PERROTT, F. A., MITCHELL, J. F. B., MARCHAND, D. S. and BRUNNER, J. S., 38, 68
STREET-PERROTT, F. A., *see* KUTZBACH, J. E. and STREET-PERROTT, F. A.
STREET-PERROTT, F. A., *see* WEBB, T. et al.
STRETEN, N. A., 208, 243
STRETEN, N. A. and ZILLMAN, J. W., 208, 243
STRICKLER, R. F., *see* MANABE, S. and STRICKLER, R. F.
STROUP, D. P., *see* WOLFF, G. R. et al.
STUIVER, M. and BRAZIUNAS, T. F., 175, 188
STUIVER, M. and QUAY, P. D., 200, 243
STUIVER, M., *see* GROOTES, P. M. et al.
STUIVER, M., *see* SANTER, B. et al.
SUAREZ, M. J. and HELD, I. M., 32, 68
SUBAK, S., RASKIN, P. and HIPPEL, D. V., 322, 323, 346
SUCKLING, P. W., 487, 513
SUD, Y. C. and FENNESSY, M., 456, 458, 474
SUD, Y. C. and SMITH, W. E., 437, 474
SUD, Y. C., CHAO, W. C. and WALKER, G. K., 269, 279
SUD, Y. C., SHUKLA, J. and MINTZ, Y., 455, 474
SUHASINI, R., *see* PRABHAKARA, C. et al.
SUKUMAR, R., RAMESH, R., PANT, G. B. and RAJAGOPALAN, G., 194, 243
SUMAN, D. O., 355, 397
SUMMERS, A., *see* POLLACK, J. B. et al.
SUN, D. Z., 38
SUN, D. Z. and LINDZEN, R. S., 36, 68
SUN, Y., *see* XU, D. et al.
SUSSKIND, J., 262, 271, 279
SUSSKIND, J., ROSENFIELD, J. and REUTER, D., 263, 279
SUSSKIND, J., *see* CHAHINE, M. T. and SUSSKIND, J.
SUTERA, A., *see* SALTZMAN, B. and SUTERA, A.
SUTHERLAND, B. M., *see* QUAITE, F. E. et al.
SUTHERLAND, J. C., *see* QUAITE, F. E. et al.
SUTTER, C. B., *see* DESHLER, T. et al.
SUTTIE, R. A., *see* BARATH, F. T. et al.
SVEINBJORNSDOTTIR, A. E., *see* DANSGAARD, W. et al.
SWAID, H. and HOFFMAN, M. E., 492, 513
SWAIN, A. M., *see* BRYSON, R. A. and SWAIN, A. M.
SWAP, R., GARSTANG, M., GRECO, S., TALBOT, R. and KÅLLBERG, P., 350, 397
SWIFT, J. H., *see* BREWER, P. G. et al.
SWISHER III, C. C. and PROTHERO, D. R., 118, 145
SWISSLER, T. J., *see* MCCORMICK, M. P. et al.
SYTKUS, J. I., *see* FRAKES, L. A. et al.

TAESLER, R., *see* OKE, T. R. et al.
TAKAHASHI, T., *see* BREWER, P. G. et al.
TAKAMURA, T., 488, 513
TALBOT, R. W., ANDREAE, M. O., ANDREAE, T. W. and HARRISS, R. C., 353, 356, 376, 397
TALBOT, R. W., ANDREAE, M. O., BERRESHEIM, H., ARTAXO, P., GARSTANG, M., HARRISS, R. C., BEECHER, K. M. and LI, S. M., 350, 356, 397
TALBOT, R. W., *see* ANDREAE, M. O. et al.
TALBOT, R. W., *see* SWAP, R. et al.
TALLEY, L. D. and MCCARTNEY, M. S., 88, 94
TANAKA, M., 304, 313
TANAKA, M., *see* YAMAMOTO, G. M. et al.
TANG, I. N., 363, 397
TANG, W., *see* LIU, W. T. et al.
TANGUAY, T., *see* SANDERSON, M. et al.
TANIMOTO, T., *see* ANDERSON, D. L. et al.
TANNER, R. L., *see* ROGERS, C. F. et al.
TANRE, D., GELEYN, J. F. and SLINGO, J., 368, 397
TANRE, D., *see* KAUFMAN, Y. J. et al.
TANRE, D., *see* KING, M. D. et al.
TANS, P. P., *see* DLUGOKENCKY, E. J. et al.
TAPPER, N. J., 503, 513
TARPLEY, J. D., *see* GALLO, K. P. et al.
TARPLEY, J. D., *see* OHRING, G. et al.
TARRASON, L. and IVERSEN, T., 379, 397
TARSALA, J. A., *see* BARATH, F. T. et al.
TAYLOR, B. L., GAL-CHEN, T. and SCHNEIDER, S. H., 205, 243
TAYLOR, F. W., RODGERS, C. D., WHITNEY, J. G., WERRETT, S. T., BARNETT, J. J., PESKETT, G. D., VENTERS, P., BALLARD, J., PALMER, C. W. P., KNIGHT, R. J., MORRIS, P., NIGHTINGALE, T. and DUDHA, A., 411, 432
TAYLOR, K. C., HAMMER, C. U., ALLEY, R. B., CLAUSEN, H. B., DAHL-JENSEN, D., GOW, A. J., GUNDESTRUP, N. S., KIPSTUHL, J., MOORE, J. C. and WADDINGTON, E. D., 25, 68
TAYLOR, K. C., *see* ALLEY, R. B. et al.
TAYLOR, K. C., *see* CESS, R. D. et al.
TAYLOR, K. E., *see* CESS, R. D. et al.
TAYLOR, K. E., *see* GHAN, S. J. et al.
TAYLOR, K. E., *see* MAZUMBER, A. et al.
TAYLOR, W. D., *see* GHAN, S. J. et al.
TAYLOR, W. D., *see* MAZUMBER, A. et al.
TEDESCO, K., *see* BOND, G. et al.
TEGART, W. J. MCG., SHELDON, G. W. and GRIFFITHS, D. C., 2, 18

TEILLET, P., *see* RUNNING, S. W. et al.
TELLER, J. T., *see* BROECKER, W. S. et al.
TEN BRINK, H. M., SCHWARTZ, S. E. and DAUM, P. H., 363, 397
THIERSTEIN, H. R., 103, 109, 145, 146
THOMAS, D. S. G., *see* MIDDLETON, N. and THOMAS, D. S. G.
THOMAS, E., 86, 94
THOMAS, E., *see* RUDDIMAN, W. F. et al.
THOMAS, G. E., *see* MCKAY, C. P. and THOMAS, G. E.
THOMAS, R. H., 273, 279
THOMPSON, A. M., 319, 342, 346
THOMPSON, L. G., 243
THOMPSON, L. G., MOSLEY-THOMPSON, E. and ARNAO, B. M., 222, 243
THOMPSON, L. G., MOSLEY-THOMPSON, E. and THOMPSON, P. A., 223, 227, 243
THOMPSON, L. G., *see* MICHAELSEN, J. and THOMPSON, L. G.
THOMPSON, P. A., 223
THOMPSON, P. A., *see* THOMPSON, L. G. et al.
THOMPSON, S. L., *see* BARRON, E. J. et al.
THOMPSON, S. L., *see* BONAN, G. B. et al.
THOMPSON, S. L., *see* COVEY, C. and THOMPSON, S. L.
THOMPSON, S. L., *see* COVEY, C. et al.
THOMPSON, S. L., *see* POLLARD, D. and THOMPSON, S. L.
THOMPSON, S. L., *see* SCHNEIDER, S. H. and THOMPSON, S. L.
THOMPSON, S. L., *see* TRIPOLI, G. J. and THOMPSON, S. L.
THORDARSON, TH. and SELF, S., 129, 130, 146
THORNTON, I., *see* APSIMON, H. et al.
THUNELL, R. C., 119, 146
THUNELL, R. C., *see* GERSTEL, J. et al.
TIAO, G. C., *see* BOJKOV, R. D. et al.
TIAO, G. C., *see* MILLER, A. J.
TIEDTKE, M., *see* HECKLEY, W. A. et al.
TINDALE, N. W., *see* BETZER, P. R. et al.
TINUS, R. W. and RODDY, D. J., 99, 101, 146
TIWARI, R. S. and VIJAYA, 120, 146
TJEMKES, S. A., *see* RANDALL, D. A. and TJEMKES, S. A.
TJEMKES, S. A., STEPHENS, G. L. and JACKSON, D. L., 263, 279
TOGGWEILER, J. R., *see* IMBRIE, J. et al.
TOGGWEILER, R., *see* SARMIENTO, J. L. and TOGGWEILER, R.
TOKIOKA, T., *see* MITCHELL, J. F. B. et al.
TOKOS, J., *see* CHURCH, T. M. et al.
TOMINAGA, K., *see* YOSHIDA, A. et al.
TOON, O. B. et al., 98, 99, 100, 146
TOON, O. B., *see* KASTING, J. F. et al.
TOON, O. B., *see* SAGAN, C. et al.
TOON, O. B., *see* TURCO, R. P. et al.
TOON, O. W., *see* POLLACK, J. B. et al.
TORRES, A. L., *see* ANDREAE, M. O. et al.
TOWNSEND, R. D. and JOHNSON, D. R., 285, 313
TRAINER, M., *see* FEHSENFELD, F. et al.
TRAMONTANO, J. M., *see* CHURCH, T. M. et al.
TRENBERTH, K. E., 33, 68, 156, 167, 168, 169, 188, 208, 209, 211, 213, 227, 228, 243, 268, 279, 283, 284, 286, 302, 303, 313
TRENBERTH, K. E. and OLSON, J. G., 170, 188, 283, 313
TRENBERTH, K. E. and PAOLINO, D. A., 156, 188
TRENBERTH, K. E. and SHEA, D. J., 213, 243
TRENBERTH, K. E. and SOLOMON, A., 286, 293, 294, 295, 313
TRENBERTH, K. E., BERRY, J. C. and BUJA, L. E., 286, 313
TRENBERTH, K. E., BRANSTATOR, G. W. and ARKIN, P. A., 262, 279
TRENBERTH, K. E., CHRISTY, J. R. and HURRELL, J. W., 159, 170, 188
TRENBERTH, K. E., *see* HURRELL, J. W. and TRENBERTH, K. E.
TRIBBIA, J. J., *see* MADDEN, R. A. et al.
TRICOT, C., 52, 53, 68
TRICOT, C. and BERGER, A., 30, 68
TRICOT, C., *see* BERGER, A. et al.
TRICOT, C., *see* GALLÉE, H. et al.
TRIPOLI, G. J. and THOMPSON, S. L., 130, 146
TROTTER, D. E., *see* EDDY, J. A. et al.
TRUMBORE, S., *see* BROECKER, W. S. et al.
TSELIOUDIS, G., ROSSOW, W. B. and RIND, D., 256, 259, 260, 279
TSONIS, A. A., *see* ELSNER, J. B. and TSONIS, A. A.
TUCKER, C. J., DREGNE, H. E. and NEWCOMB, W. W., 436, 441, 443, 452, 474
TUCKER, C. J., *see* KAUFMAN, Y. J. et al.
TUCKER, C. J., *see* SKOLE, D. L. and TUCKER, C. J.
TUEDY, R., *see* CHANDLER, M. et al.
TUNCEL, G., ARAS, N. K. and ZOLLER, W. H., 380, 397
TURCO, R. P., TOON, O. B., ACKERMAN, T. P., POLLACK, J. B. and SAGAN, C., 100, 101, 135, 146
TUREKIAN, K. K., *see* SAVOIE, D. L. et al.
TURNER II, B. L., CLARK, W. C., KATES, R. W., RICHARDS, J. F., MATHEWS, J. T. and MEYER, W. B., 437, 474
TURVEY, D. E., *see* AYERS, G. P. et al.
TUSHINGHAM, A. M. and PELTIER, W. R., 24, 68
TWICKLER, M. S., *see* MAYEWSKI, P. A. et al.
TWIST, D., *see* ELSTON, W. E. and TWIST, D.
TWOHY, C. H., CLARKE, A. D., WARREN, S. G., RADKE, L. F. and CHARLSON, R. J., 377, 397
TWOMEY, S., 348, 373, 377, 378, 387, 397, 398
TWOMEY, S. A., PIEPGRASS, M. and WOLFE, T. L., 377, 382, 398
TWOMEY, S. and SQUIRES, P., 372, 398

TWOMEY, S. and WARNER, J., 372, 398
TWOMEY, S., DAVIDSON, K. A. and SETON, J., 372, 380, 398
TWOMEY, S., *see* WARNER, J. and TWOMEY, S.

UDO, R. K., AREOLA, O. O., AYOADE, J. O. and AFOLAYAN, A. A., 438, 474
UEMATSU, M., DUCE, R. A. and PROSPERO, J. M., 350, 398
UEMATSU, M., *see* BETZER, P. R. et al.
UHL, C. and BUSCHBACHER, R., 465, 474
UMANA, J. C., *see* SALATI, E. et al.
UMSCHEID, L. J., *see* MAHLMAN, J. D. et al.
UNCOD, 439, 441, 474
UNITED NATIONS, 435, 441, 474, 475, 480, 513
UNITED NATIONS ENVIRONMENT PROGRAM, 420, 432, 433, 434, 436, 441, 442, 474, 475, 480, 481, 513
UPCHURCH JR., G. R., 111, 112, 146
UPCHURCH JR., G. R., *see* WOLFE, J. A. and UPCHURCH JR., G. R.
UPLINGER, W. G., *see* ROCHE, A. E. et al.
UPPALA, S., *see* HOLLINGSWORTH, A. et al.
URBACH, F., BERGER, D. and DAVIES, R. E., 420, 432
URBACH, F., *see* SCOTTO, J. et al.
UREY, H. C., 95, 97, 146
US NATIONAL ACADEMY of SCIENCES, 5, 18

VALDES, P., 90, 94
VALDES, P. J., *see* HALL, N. M. J. et al.
VALDES, P. J., *see* HOSKINS, B. J. and VALDES, P. J.
VALENTINE, J. W., 120, 122, 146
VAN DER LEUN, J. C., *see* DE GRUIJL, F. R. et al.
VAN DER MERSCH, I., *see* PESTIAUX, P. et al.
VAN DORLAND, R., *see* SHINE, K. P. et al.
VAN GRIEKEN, R., *see* ARTAXO, P. et al.
VAN LOON, H., 156, 188
VAN LOON, H. and LABITZKE, K., 175, 179, 188, 202, 203, 204, 243
VAN LOON, H. and ROGERS, J., 167, 188, 204, 243
VAN LOON, H. and SHEA, D. J., 174, 188, 213, 243
VAN LOON, H., *see* CHEN, W. Y. et al.
VAN LOON, H., *see* KILADIS, G. N. and VAN LOON, H.
VAN LOON, H., *see* LABITZKE, K. and VAN LOON, H.
VAN LOON, H., *see* WILLIAMS, J. and VAN LOON, H.
VAN VALIN, C. C., *see* BOATMAN, J. F. et al.
VAN VALIN, C. C., *see* PUESCHEL, R. F. et al.
VAN WEELDEN, H., *see* DE GRUIJL, F. R. et al.
VAN WEERING, T. C. E., *see* DUPLESSY, J. CL. et al.
VAN YPERSELE, J. P., *see* ADEM, J. et al.
VAN YPERSELE, J. P., *see* BERGER, A. et al.
VAN YPERSELE, J. P., *see* GALLÉE, H. et al.
VANDAMME, D., COURTILLOT, V., BESSE, J. and MONTIGNY, R., 127, 146
VANDAMME, D., *see* COURTILLOT, V. et al.
VANDERBILT, V., *see* RUNNING, S. W. et al.
VARNEY, S. K., *see* HOUGHTON, J. T. et al.
VAUTARD, R. and GHIL, M., 201, 243
VAUTARD, R., *see* GHIL, M. and VAUTARD, R.
VEHIL, R., *see* ESTOURNEL, C. et al.
VEIGA, R. E., *see* MCCORMICK, M. P. et al.
VEIZER, J., *see* GRUSZCZYNSKI, M. et al.
VENKATESAN, M. I. and DAHL, J., 102, 146
VENRICK, E. L., MCGOWAN, J. A., CAYAN, D. R. and HAYWARD, T. L., 227, 243
VENTERS, P., *see* TAYLOR, F. W. et al.
VERNEKAR, A. D., *see* OERLEMANS, J. and VERNEKAR, A. D.
VERSTRAETE, M. M., 446, 465, 475
VERSTRAETE, M. M. and SCHWARTZ, S. A., 441, 475
VERSTRAETE, M. M., *see* DICKINSON, R. E. et al.
VICKERY, A. M. and MELOSH, H. J., 115, 146
VIJAYA, *see* TIWARI, R. S. and VIJAYA
VILLALBA, R., 194, 243
VINCENT, D. G., 208, 243
VINNIKOV, K. YA., GROISMAN, P. YA. and LUGINA, K. M., 158, 188
VINNIKOV, K. YA., *see* FOLLAND, C. K. et al.
VINNIKOV, K. YA., *see* STOUFFER, R. J. et al.
VIRGINIA, R. A., *see* SCHLESINGER, W. H. et al.
VISSCHER, H. and BRUGMAN, W. A., 117, 120, 146
VITERBO, P., *see* BETTS, A. K. et al.
VOGEL, J. L., *see* CHANGNON JR., S. A. et al.
VOLK, T., *see* CALDEIRA, K. et al.
VOLK, T., *see* KUMP, L. R. and VOLK, T.
VOLK, T., *see* RAMPINO, M. R. and VOLK, T.
VOLK, T., *see* SCHWARTZMAN, D. W. and VOLK, T.
VON RUDLOFF, H., 152, 188
VONDER HAAR, T. H., *see* CARISSIMO, B. C. et al.
VONDER HAAR, T. H., *see* SMITH, L. D. and VONDER HAAR, T. H.
VOOGT, J. A. and OKE, T. R., 499, 514
VOOGT, J. A., *see* JOHNSON, G. T. et al.
VUKOVICH, F. M., 502, 514
VULIS, I. L., *see* CESS, R. D. and VULIS, I. L.

WADDINGTON, E. D., *see* ALLEY, R. B. et al.
WADDINGTON, E. D., *see* TAYLOR, K. C. et al.
WAGGONER, A. P., *see* COVERT, D. S. et al.
WAGGONER, A. P., WEISS, A. P., AHLQUIST, N. C., COVERT, D. S. and CHARLSON, R. J., 363, 398
WAGSTAFF, B., *see* RICH, T. H. et al.
WALKER, G. K., *see* SUD, Y. C. et al.
WALKER, G. T., 214, 243
WALKER, J. and ROWNTREE, P. R., 456, 475
WALKER, J. C. G., 84, 94, 539, 553
WALKER, J. C. G. and KASTING, J. F., 549, 553
WALKER, J. C. G., HAYS, P. B. and KASTING, J. F., 537, 538, 547, 553
WALKER, J. C. G., KLEIN, C., SCHIDLOWSKI, M., STEVENSON, D. J. and WALTER, M. R., 543, 553

WALKER, J. C. G., *see* KASTING, J. F. et al.
WALLACE, J. M., 228, 243
WALLACE, J. M. and GUTZLER, D. S., 204, 243, 297, 313
WALLACE, J. M., LIM, G.-H. and BLACKMON, M. L., 302, 313
WALLACE, J. M., *see* DESER, C. and WALLACE, J. M.
WALLACE, J. M., *see* HOREL, J. D. and WALLACE, J. M.
WALLACE, J. M., *see* WRIGHT, P. B. et al.
WALSH, J. E., *see* CHAPMAN, W. L. and WALSH, J. E.
WALSH, K., *see* MCGREGOR, J. L. and WALSH, K.
WALTER, L. S., *see* BLUTH, G. J. S. et al.
WALTER, M. R., *see* WALKER, J. C. G. et al.
WALTER, R. C., *see* EBINGER, C. J. et al.
WALTERS, S., *see* BRASSEUR, G. P. et al.
WALTER-SHEA, E. A., *see* STARKS, P. J. et al.
WALTHALL, C. L., *see* STARKS, P. J. et al.
WALTHER, E. G., *see* FLYGER, H. et al.
WALTON, J. J., *see* COVEY, C. et al.
WALTON, J. J., *see* ELSAESSER, H. W. et al.
WALZEBUCK, J., *see* ENGELHARDT, W. V. et al.
WAN, Z. M., *see* RUNNING, S. W. et al.
WAN, Z., *see* SMITH, R. C. et al.
WANDIGA, S. O., *see* ZIMMERMAN, P. R. et al.
WANG, C. and LIU, J., 486, 514
WANG, J-X., CHAI, Z-F. et al., 115, 116, 146
WANG, K., 115, 146
WANG, K., ORTH, C. J., ATTREP JR, M., CHATTERTON, B. D. E., HOU, H. and GELDSETZER, H. H. J., 115, 124, 125, 146
WANG, K., ORTH, C. J., ATTREP JR, M., CHATTERTON, B. D. E., WANG, X. and LI, J-J., 146
WANG, W.-C. and ISAKSEN, I. S. A., 344, 346
WANG, W.-C., DUDEK, M. P. and LIANG, X., 327, 328, 331, 346
WANG, W.-C., DUDEK, M. P., LIANG, X.-Z. and KIEHL, J. T., 328, 346
WANG, W.-C., PINTO, J. P. and YUNG, Y. L., 319, 336, 339, 346
WANG, W.-C., *see* DUDEK, M. P. et al.
WANG, W.-C., *see* JONES, P. D. et al.
WANG, W.-C., *see* MOHNEN, V. A. et al.
WANG, W.-C., *see* SHINE, K. P. et al.
WANG, W.-C., WUEBBLES, D. J., WASHINGTON, W. M., ISAACS, R. G. and MOLNAR, G., 318, 346
WANG, W.-C., YUNG, Y. L., LACIS, A. A., MO, T. and HANSEN, J. E., 318, 346
WANG, W.-C., ZHUANG, Y.-Z. and BOJKOV, R. D., 319, 320, 336, 340, 343, 346
WANG, X., *see* WANG, K. et al.
WANG, Y. and HOLLAND, G. J., 305, 313
WARD, D. E., 353, 398
WARD, D. E. and HARDY, C. C., 353, 398
WARD, D. E., *see* EINFELD, W. et al.
WARD, D. E., *see* KAUFMAN, Y. J. et al.
WARD, D. E., *see* RADKE, L. F. et al.
WARD, M. N., *see* ROWELL, D. P. et al.
WARD, P., 114
WARD, P., *see* MARGOLIS, S. V. et al.
WARNECK, P., 352, 353, 356, 358, 363, 364, 380, 398
WARNER, C. W., *see* CALDWELL, M. M. et al.
WARNER, J. and TWOMEY, S., 372, 376, 380, 398
WARNER, J., *see* TWOMEY, S. and WARNER, J.
WARNER, T. T., *see* SEAMAN, N. L. et al.
WARREN, S. G. and CLARKE, A. D., 382, 398
WARREN, S. G., HAHN, C. J., LONDON, J., CHERVIN, R. M. and JENNE, R. L., 362, 398
WARREN, S. G., *see* ANDERSON, T. L. et al.
WARREN, S. G., *see* CHARLSON, R. J. et al.
WARREN, S. G., *see* TWOHY, C. H. et al.
WARRICK, R. A. and OERLEMANS, J., 231, 243
WARRICK, R. A., BARROW, E. M. and WIGLEY, T. M. L., 57, 68
WARRILOW, D. A., *see* LEAN, J. and WARRILOW, D. A.
WASHINGTON, W. M., 345, 552
WASHINGTON, W. M. and MEEHL, G. A., 40, 68, 224, 243
WASHINGTON, W. M. and PARKINSON, C. L., 33, 68
WASHINGTON, W. M., *see* BARRON, E. J. and WASHINGTON, W. M.
WASHINGTON, W. M., *see* CESS, R. D. et al.
WASHINGTON, W. M., *see* WANG, W.-C. et al.
WASHINGTON, W. M., *see* WILLIAMS, J. et al.
WASSON, J. T., *see* KYTE, F. T. and WASSON, J. T.
WASSON, J. T., *see* KYTE, F. T. et al.
WATATANI, S., *see* YOSHIDA, A. et al.
WATERHOUSE, J. B., 146
WATERMAN, L. S., *see* KEELING, C. D. et al.
WATERS, J. W., FROIDEVAUX, L., READ, W. G., MANNEY, G. L., ELSON, L. S., FLOWER, D. A., JARNOT, R. F. and HARWOOD, R. S., 407, 432
WATERS, J. W., *see* BARATH, F. T. et al.
WATERS, K. J., *see* SMITH, R. C. et al.
WATSON, A. J. and LOVELOCK, J. E., 539, 540, 553
WATSON, I. D., *see* JOHNSON, G. T. et al.
WATSON, I. D., *see* OKE, T. R. et al.
WATSON, J. G., *see* ROGERS, C. F. et al.
WATSON, R. T., MEIRA FILHO, L. G., SANHUEZA, E. and JANETOS, A., 177, 189, 446, 448, 449, 450, 451, 475
WATSON, R. W., RODHE, H., OESCHGER, H. and SIEGENTHALER, U., 177, 188
WATTERSON, I. G., *see* FARRELL, B. and WATTERSON, I. G.
WATTERSON, I. G., *see* RYAN, B. F. et al.
WATTS, M., 441, 475
WATTS, R. G., 62
WATTS, R. G. and HAYDER, M. E., 41, 69

WATTS, W. A., *see* GRIMM, E. C. et al.
WEARE, B. C., 210, 244
WEARE, B. C., *see* NEWELL, R. E. and WEARE, B. C.
WEBB, P. N., *see* BARRERA, E. et al.
WEBB, R. S. *see* ANDERSON, D. M. and WEBB, R. S.
WEBB, T., STREET-PERROTT, F. A. and KUTZBACH, J. E., 37, 69
WEBER, R., *see* MADDEN, R. A. et al.
WEBSTER, P. J., 209, 210, 226, 244, 289, 290, 313
WEBSTER, P. J. and CHANG, H.-R., 298, 305, 314
WEBSTER, P. J. and HOLTON, J. R., 298, 314
WEBSTER, P. J. and YANG, S., 214
WEBSTER, P. J., *see* CHANG, H.-R. and WEBSTER, P. J.
WEBSTER, P. J., *see* ZHANG, C. and WEBSTER, P. J.
WEEMS, R. E., 119, 146
WEERTMAN, J., *see* BIRCHFIELD, G. E. and WEERTMAN, J.
WEERTMAN, J., *see* BIRCHFIELD, G. E. et al.
WEFERS, M., 350, 398
WEICKMANN, K. M., *see* KILADIS, G. N. and WEICKMANN, K. M.
WEISS, A. P., *see* WAGGONER, A. P. et al.
WEISS, R. E., *see* COVERT, D. S. et al.
WEISS, R. E., *see* RADKE, L. F. et al.
WEISSERT, H., *see* HSU, K. J. et al.
WEISSMAN, P. R., *see* COVEY, C. et al.
WELCH, R. M., *see* WISCOMBE, W. J. et al.
WELHAN, J. A., *see* CRAIG, H. et al.
WELLMAN, D. L., *see* BOATMAN, J. F. et al.
WELLS, L. E., 224, 244
WENDEL, G. J., *see* WOLFF, G. R. et al.
WENDLAND, W. M., 244
WENDLAND, W. M. and BRYSON, R. A., 194
WENK, T., *see* SIEGENTHALER, U. and WENK, T.
WENTZ, F. J., 272, 279
WENTZ, F. J., *see* LIU, W. T. et al.
WERRETT, S. T., *see* TAYLOR, F. W. et al.
WESTBERG, H., *see* FEHSENFELD, F. et al.
WETHERALD, R. T., *see* CESS, R. D. et al.
WETHERALD, R. T., *see* KAROLY, D. J. et al.
WEZEL, F. C., *see* CROCKET, J. H. et al.
WHELPDALE, D. M., KEENE, W. C., HANSEN, A. D. A. and BOATMAN, J., 381, 398
WHELPDALE, D. M., *see* CHURCH, T. M. et al.
WHELPDALE, D. M., *see* GALLOWAY, J. N. et al.
WHISTON, W., 95, 147
WHITE, G. H., *see* HOSKINS, B. J. et al.
WHITE, J. M., EATON, F. D. and AUER JR., A. H., 487, 488, 514
WHITE, J. W. C., *see* ALLEY, R. B. et al.
WHITE, J. W. C., *see* GROOTES, P. M. et al.
WHITE, O., *see* LEAN, J. et al.
WHITE, R. and MCKENZIE, D., 128, 147
WHITE, W. H., 363, 398
WHITEAND, J. W. C., *see* DANSGAARD, W. et al.
WHITFIELD, M., *see* LOVELOCK, J. E. and WHITFIELD, M.
WHITFORD, W. G., *see* SCHLESINGER, W. H. et al.
WHITLOCK, C. H., CHARLOCK, T. P., STAYLOR, W. F., PINKER, R. T., LASZLO, I., DIPASQUALE, R. C. and RITCHEY, N. A., 265, 279
WHITLOW, S., *see* MAYEWSKI, P. A. et al.
WHITMORE, T. C., 547, 553
WHITNEY, J. G., *see* TAYLOR, F. W. et al.
WHORF, T. P., *see* KEELING, C. D.
WHORF, T. P., *see* KEELING, C. D. et al.
WIDMARK, J. G. V. and MALMGREN, B., 108, 147
WIEBE, H. A., *see* LEAITCH, W. R. et al.
WIGGERT, V., *see* SIMPSON, J. and WIGGERT, V.
WIGLEY, T. M. L., 17, 18, 40, 69, 158, 326, 348, 383, 385, 387, 398, 545, 553
WIGLEY, T. M. L. and BARNETT, T. P., 178, 189
WIGLEY, T. M. L. and JONES, P. D., 166, 189
WIGLEY, T. M. L. and RAPER, S. C. B., 175, 178, 189, 346
WIGLEY, T. M. L., ANGELL, J. K. and JONES, P. D., 170, 189
WIGLEY, T. M. L., JONES, P. D. and KELLY, P. M., 170, 189
WIGLEY, T. M. L., *see* ISAKSEN, I. S. A. et al.
WIGLEY, T. M. L., *see* JONES, P. D. and WIGLEY, T. M. L.
WIGLEY, T. M. L., *see* JONES, P. D. et al.
WIGLEY, T. M. L., *see* KELLY, P. M. and WIGLEY, T. M. L.
WIGLEY, T. M. L., *see* SANTER, B. D. et al.
WIGLEY, T. M. L., *see* WARRICK, R. A. et al.
WIGLEY, T. M. L., *see* WILLIAMS, J. and WIGLEY, T. M. L.
WIGLEY, T. M. W., *see* JONES, P. D. et al.
WILBER, A. C., *see* DARNELL, W. L. et al.
WILBER, A. C., *see* GUPTA, S. K. et al.
WILBER, A. C., SMITH, E. A. and COOPER, H. J., 276
WILDE, P., BERRY, W. B. N. and QUINBY-HUNT, M. S., 147
WILDE, P., BERRY, W. B. N., QUINBY-HUNT, M. S. et al., 125
WILHEIT, T. J., CHANG, A. T. C. and CHIU, L. C., 261, 279
WILHEIT, T. T., *see* CHANG, A. T. C. and WILHEIT, T. T.
WILKINSON, B. H., *see* OPDYKE, B. N. and WILKINSON, B. H.
WILKINSON, S. K., *see* BOATMAN, J. F. et al.
WILLIAMS JR., C. N., *see* KARL, T. R. and WILLIAMS JR., C. N.
WILLIAMS JR., C. N., *see* KARL, T. R. et al.
WILLIAMS R. T., *see* BROECKER, W. S. et al.
WILLIAMS, J. and VAN LOON, H., 156, 189
WILLIAMS, J. and WIGLEY, T. M. L., 168, 189

WILLIAMS, J., BARRY, R. G. and WASHINGTON, W. M., 35, 69
WILLIAMS, M., 443, 444, 475
WILLIAMS, R. T., *see* BREWER, P. G. et al.
WILLIAMS, R. T., *see* HALTINER, G. J. and WILLIAMS, R. T.
WILLMOTT, C. J., *see* LEGATES, D. R. and WILLMOTT, C. J.
WILLSON, R. C., 244, 245, 253, 279
WILLSON, R. C. and HUDSON, H. S., 175, 189, 199, 253, 279
WILSON, C., *see* FALKOWSKI, P. G. et al.
WILSON, H., *see* HANSEN, J. et al.
WILSON, M. F., *see* DICKINSON, R. E. et al.
WILSON, M. F., *see* GRAETZ, D. et al.
WILSON, M. F., *see* HENDERSON-SELLERS, A. and WILSON, M. F.
WILSON, W. E., *see* FALKOWSKI, P. G. et al.
WILSON, W. E., *see* HUSAR, R. B. and WILSON JR., W. E.
WILSON, W. E., *see* HUSAR, R. B. and WILSON, W. E.
WILSON, W. J., *see* BARATH, F. T. et al.
WINCHESTER, J. W., *see* LAWSON, D. R. and WINCHESTER, J. W.
WING, S. L. and GREENWOOD, D. R., 83, 94
WIRICK, C., *see* FALKOWSKI, P. G. et al.
WISCOMBE, W. J., WELCH, R. M. and HALL, W. D., 375, 398
WISE, K. P. and SCHOPF, T. J. M., 114, 147
WISE, S. W., *see* ZACHOS, J. C. et al.

WOFSY, S. C., MCELROY, M. B. and YUNG, Y. L., 401, 432
WOFSY, S. C., *see* ANDREAE, M. O. et al.
WOFSY, S. C., *see* SPIVAKOVSKY, C. M. et al.
WOLBACH, W. S., GILMOUR, I., ANDERS, E., ORTH, C. J. and BROOKS, R. R., 101, 110, 147
WOLBACH, W. S., LEWIS, R. S. and ANDERS, E., 101, 147
WOLBACH, W. S., *see* ANDERS, E. et al.
WOLBACH, W. S., *see* GILMOUR, I. et al.
WOLDEGABRIEL, G., *see* EBINGER, C. J. et al.
WOLFE, G. V., *see* ANDERSON, T. L. et al.
WOLFE, J. A., 82, 94, 111, 147
WOLFE, J. A. and UPCHURCH JR., G. R., 74, 94, 111, 147
WOLFE, J. A., *see* STOTHERS, R. B. et al.
WOLFE, R. F., *see* SHOEMAKER, E. M. et al.
WOLFE, T. L., *see* TWOMEY, S. A. et al.
WOLFF, G. R., RUTHKOSKY, M. S., STROUP, D. P., KORSOG, P. E., FERMAN, M. A., WENDEL, G. J. and STEDMAN, D. H., 380, 398
WOLFF, J. A., *see* DE SILVA, S. L. et al.
WOLFI, W., *see* BROECKER, W. S. et al.
WOLTER, K., 229, 244
WONG, X., *see* RASMUSSON, E. M. et al.
WOOD, T. M., *see* ROSEN, R. D. et al.
WOODRUFF, S. D., SLUTZ, R. J., JENNE, R. J. and STEURER, P. M., 152, 189
WOODS, P., 465, 475
WOODWARD, F. I., *see* ADAMS, J. M. et al.
WOOLF, D. K., *see* CIPRIANO, R. J. et al.
WORLD CLIMATE RESEARCH PROGRAMME (WCRP), 156, 189
WORLD METEOROLOGICAL ORGANIZATION (WMO), 156, 189, 317, 319, 320, 322, 323, 325, 335, 340, 346, 399, 406, 409, 410, 420, 425, 432, 448, 449, 475
WORLD METEOROLOGICAL ORGANIZATION (WMO) and UNITED NATIONS ENVIRONMENT PROGRAMME (UNEP), 160, 164, 189
WORLD RESOURCES INSTITUTE, 435, 440, 444, 475
WRIGHT JR., H. E., *see* KUTZBACH, J. E. and WRIGHT JR., H. E.
WRIGHT, D. G., *see* STOCKER, T. W. and WRIGHT, D. G.
WRIGHT, D. G., *see* STOCKER, T. W. et al.
WRIGHT, H. E., 196, 244
WRIGHT, P. B., *see* JONES, P. D. et al.
WRIGHT, P. B., WALLACE, J. M., MITCHELL, T. P. and DESER, C., 221, 244
WU, G., *see* BERGER, W. H. et al.
WU, K.-D., *see* CHEN, W. Y. et al.
WU, M. L. C. and CHANG, L. A., 265, 279
WUEBBLES, D. J., *see* LACIS, A. A. et al.
WUEBBLES, D. J., *see* MILLER, A. J.
WUEBBLES, D. J., *see* WANG, W.-C. et al.
WYRTKI, K., 210, 212, 244

XU, D., MA, S., CHAI, Z, MAO, X., SUN, Y., ZHANG, Q. and YANG, Z., 120
XU, D., MA, S., CHAI, Z., MAO, X., SUN, Y., ZHANG, Q. and YANG, Z., 147

YABUSHITA, S. and ALLEN, A. J., 102, 147
YAGAI, I., *see* CESS, R. D. et al.
YAMADA, S., *see* NITTA, T. and YAMADA, S.
YAMAMOTO, G. M., TANAKA, M. and ARAO, K., 379, 398
YAMANOI, T., *see* SAITO, T. et al.
YAMASAKI, M., 305, 314
YAMASHITA, S., 486, 514
YAMASHITA, S. and SEKINE, K., 479, 514
YAN, Z., *see* DAO-YI, X. et al.
YANG, J., *see* GOWARD, S. N. et al.
YANG, R., *see* FAN, D. et al.
YANG, S., 244
YANG, Z., *see* XU, D. et al.
YAO, J., *see* LI., Z. et al.
YAOQI, Z., *see* CHIFANG, C. et al.
YAP, D., 491, 514
YASUNARI, T., 218, 244
YEMANE, T., *see* EBINGER, C. J. et al.

YEN, M.-C., *see* CHEN, W. Y. et al.
YERSEL, M. and GOBLE, R., 496, 514
YEVICH, R., *see* SPIVAKOVSKY, C. M. et al.
YIM, W., *see* APSIMON, H. et al.
YIOU, F., *see* JOUZEL, J.
YIOU, F., *see* JOUZEL, J. et al.
YIOU, P., *see* JOUZEL, J. et al.
YI-YIN, S., *see* DAO-YI, X. et al.
YOSHIDA, A., 487, 514
YOSHIDA, A., TOMINAGA, K. and WATATANI, S., 490, 514
YOSHIKADO, H., 505, 514
YOUNG, K. C., 156, 189
YOUNG, K., *see* GALL, R. et al.
YOUNG, P. J., *see* KARL, T. R. et al.
YOUNG, R. W., *see* BETZER, P. R. et al.
YUNG, Y. L., *see* WANG, W.-C. et al.
YUNG, Y. L., *see* WOFSY, S. C. et al.
YUREVICH, F. B., *see* COAKLEY JR., J. A. et al.

ZACHOS, J. C. and ARTHUR, M. A., 103, 104, 106, 108, 110, 147
ZACHOS, J. C., ARTHUR, M. A. and DEAN, W. E., 108, 111, 147
ZACHOS, J. C., BREZA, J. R. and WISE, S. W., 118, 119, 147
ZACHOS, J. C., *see* CALDEIRA, K. et al.
ZACHOS, J. C., *see* GERSTEL, J. et al.
ZACHOS, J. C., STOTT, L. D. and LOHMANN, K. C., 82, 83, 94
ZAHNLE, K., 101
ZAHNLE, K. and GRINSPOON, D., 102, 147
ZAHNLE, K. J., 147
ZAHNLE, K. J., *see* KASTING, J. F. et al.
ZAHNLE, K., *see* MELOSH, H. J. et al.
ZARDECKI, A., *see* GERSTL, S. A. and ZARDECKI, A.
ZAWODNY, J., *see* STOLARSKI, R. S. et al.
ZEBIAK, S. E., 211, 244
ZEBIAK, S. E., *see* CANE, M. A. and ZEBIAK, S. E.
ZEBIAK, S. E., *see* GORDON, A. L. et al.
ZEIGLER, A. M. *see* KUTZBACH, J. E. and ZEIGLER, A. M.
ZEREFOS, C., *see* STOLARSKI, R. S. et al.
ZETTERBERG, P., *see* BRIFFA, K. R. et al.
ZEUNER, G., *see* OKE, T. R. et al.
ZHAN, L., *see* LI., Z. et al.
ZHANG, C. and WEBSTER, P. J., 298, 314
ZHANG, H., *see* MCGUFFIE, K. et al.
ZHANG, M. H., *see* CESS, R. D. et al.
ZHANG, Q., *see* XU, D. et al.
ZHANG, S.-H., SHAW, M., SEINFELD, J. H. and FLAGAN, R. C., 356, 398
ZHANG, Y.-S., *see* ANDERSON, D. L. et al.
ZHI-FANG, C., *see* DAO-YI, X. et al.
ZHOU, L. and KYTE, F. T., 120, 147
ZHOU, Y., *see* LI., Z. et al.
ZHOU, Z., *see* KYTE, F. T. et al.
ZHUANG, Y.-Z., *see* WANG, W.-C. et al.
ZIEGLER, W., *see* SANDBERG, C. A. et al.
ZIELINSKA, B., *see* ROGERS, C. F. et al.
ZIELINSKI, G. A., *see* ALLEY, R. B. et al.
ZILLMAN, J. W., *see* STRETEN, N. A. and ZILLMAN, J. W.
ZIMMERMAN, P. R., GREENBERG, J. P., WANDIGA, S. O. and CRUTZEN, P. J., 449, 475
ZIMMERMAN, P., *see* FEHSENFELD, F. et al.
ZIMMERMANN, P., *see* LANGNER, J. et al.
ZINDLER, A., *see* BARD, E. et al.
ZOBRIST, J., *see* DREVER, J. I. and ZOBRIST, J.
ZOLLER, W. H., *see* PROSPERO, J. M. et al.
ZOLLER, W. H., *see* TUNCEL, G. et al.
ZOPF, D. O., *see* QUINN, W. H. et al.
ZUMBRUNN, R., *see* NEFTEL, A.
ZUMBUHL, H. J., 169, 189

Geographical Index

Adelaide, 493
Africa, 295, 323, 376, 388, 440, 457–458, 461–463, 480
Agulhas Plateau, 109
Alaska, 79, 161, 168, 221, 225
Alps, 120
Amazon, 209, 219, 246, 356, 377, 381, 438, 461–462
Antarctic, 75, 78, 108, 119, 162, 291, 302, 318, 408, 428, 520, 524, 547
Arctic, 202
Atlantic, 107, 166, 193, 195, 461–462, 529–532, 545
Australia, 124–125, 162, 216–217, 295, 331, 440, 480
Austria, 119
Azores, 167

Berlin, 491
Bermuda, 380
Bolivia, 445
Brazil, 122, 445
Budapest, 486, 491
Burma, 445
Byrd Station, Antarctica, 524

Cairo, 194, 501
Cameroon, 445
Camp Century, Greenland, 522, 524
Canada, 116, 161, 194, 221, 225, 425, 440, 543
Cape Grim, Tasmania, 364, 373, 380
Central America, 445
Chicago, 504
China, 124–125, 197, 323, 331, 369, 388
Christmas Island, 214
Colombia, 445
Congo, 209, 445
Copenhagen, 427
Coronation Gulf, 129

Darwin, 210
Detroit, 486

Ecuador, 445
El Chicón, 171, 176, 207, 351
England, 119
Ethiopia, 218
Europe, 77, 117, 124, 191, 323, 480

Fairbanks, 491, 497
Falkland Plateau, 122–123, 127
Fiji, 214
Former Soviet Union, 164, 323
France, 110

Gabon, 445
Galapagos Islands, 208
Germany, 110
Great Plains, 111, 201
Greenland, 82, 162, 167–168, 193–194, 272, 382–383, 463, 524, 531
Guadalajara, 500
Guam, 380
Guyanas, 445

Hamilton, 486
Hangchow, 486
Hong Kong, 491

Iceland, 176, 193
India, 445, 497
Indian Ocean, 119, 221, 295
Indonesia, 445
Iowa, 194
Isua, 4
Ivory Coast, 445

Jakarta, 481
Japan, 323

Kalahari, 209
Kampuchea, 445
Karachi, 481, 501
Kilauea volcano, 133
Korea 507
Krakatoa, 176, 205
Krakow, 486
Kuwait, 501

Laki, 129, 205, 351
Laos, 445
London, 504
Los Angeles, 486, 491

Madagascar, 445
Malaysia, 445
Manhattan, 491
Marshall Islands, 214
Maud Rise, 110, 119

Mauna Loa, 382, 539
Mexico, 219, 445, 480, 492, 497, 501
Mexico City, 481, 491–492, 493
Mississippi, 43, 504
Montreal, 491
Moscow, 491
Mount Pinatubo, 171, 176, 207, 227, 352, 413
– St Helens, 176, 351
Myanma, 445

Nairobi, 501
Nevada, 125
New York, 505
New Zealand, 110, 480
Nigeria, 438, 441, 445, 497, 501
Nile River, 194, 218, 222–223
Nordeste, Brazil, 218
North Africa, 126
– America, 77, 83, 109
– Atlantic, 37, 43
Norway, 193
Norwegian Sea, 523

Pacific, 110, 173, 208, 210, 295, 463, 529–531
Pangea, 122
Papua New Guinea, 445
Paraná River, 219
Patagonia, 194
Peru, 207–209, 214, 224, 445
Philippines, 445
Phoenix, 501, 504

Quebec, 125

Rocky Mountains, 111

Sacramento, 493, 500, 501
Sahara, 367, 441, 443
Sahel, 165–166, 218, 229, 331, 441, 443, 456–459, 465
São Paulo, 481
Scandinavia, 169, 194
Seychelles, 218
Shanghai, 486, 507
Shatsky Rise, 110
Shetland Islands, 82
Siberia, 79, 162, 461–462
Singapore, 481
South Africa, 129
– East Asia, 436, 461–463, 497
Spain, 107, 109
Sri Lanka, 218
St Louis, 486, 491, 493, 504, 506
Sweden, 425
Switzerland, 169
Sydney, 491

Tahiti, 210
Thailand, 445, 480, 507
Tibetan plateau, 218
Tokyo, 486, 504–505
Toulouse, 486
Tucson, 491–492, 493, 498

Uruguay, 219
USA, 130, 154, 323, 369, 384, 425, 438, 440, 507

Vancouver, 486, 491, 493
Venezuela, 445
Vietnam, 445
Vostok, 522

Walvis Ridge, 108
Weddell Sea, 110

Yukon, 116, 168

Zäire, 445
Zimbabwe, 218

Subject Index

Absorption bands, 317–318, 359
Acid rain, 100–101, 112, 135
Advanced Earth Observing System (ADEOS), 250
–, Very High Resolution Radiometer (AVHRR), 246, 248, 269–270
Aeolian dust, 83–85, 86, 349–351, 364, 367, 434, 532
Aerosol cooling, 347–365, 382–384
–, trends, 379–382
–, warming, 365–368
Aerosols, 16–17, 50, 130, 131, 176–177, 207, 231, 253, 258, 347–389, 433, 437, 467, 485–486
Africa, 295, 323, 376, 388, 440, 457–458, 461–463, 480
Agricultural trends, 438–439, 448
Airborne Antarctic Ozone Experiment, 416
Air pollution, 480
Alaska, 77, 161, 168, 221, 225
Amazon, 209, 219, 246, 356, 377, 381, 438, 461–462
Anasazi, 194
Anoxic ocean, 125
Antarctic, 75, 78, 108, 119, 162, 291, 302, 318, 408, 428, 520, 524, 547
–, Bottom Water (AABW), 528
–, ice core, 350
Anticyclones, 301–304
Archaean, 4, 542–543
Arctic, 202
Areal runoff, 297
Asteroids, 95, 97–98, 128, 135, 136, 348
Atlantic, 107, 166, 193, 195, 461–462, 529–531, 532, 545
Atmospheric aerosols, 451–452
–, turbidity, 360
–, waves, 288–291
Atmospheric Model Intercomparison Project (AMIP), 272, 292–293, 297, 300–301
Attribution of climate change, 180

Benguela Current, 209
Biological pump (oceans), 523
Biomass burning, 349, 353–355, 364, 366, 376, 450–451
Biophysical feedback, 456
Biosphere-Atmosphere Transfer Scheme (BATS), 457, 460
Bolide, 116
Bowen ratio, 492
Bureau of Meteorology Research Centre (BMRC), 292, 297, 299–301, 307–308

C3/C4 pathways (in plants), 550
Cape Grim, Tasmania, 364, 373, 380
Carbon budget, 324, 446–448, 526, 549
–, silicate weathering, 547–548
Charney mechanism, 456, 458
Chlorine, 402–403
Chlorofluorocarbons (CFCs), 177, 317–334, 338, 340, 400, 404, 416–418, 425
CLAW hypothesis, 348, 375–376, 539, 544–546
CLIMAP, 24, 34, 36, 526, 545
Climate analysis, 283–288
–, models, 10, 21, 32, 282–283
–, regulation, 538–541
–, variability, 225–230
Climatic Research Unit (CRU), 157–158
–, effects of atmospheric aerosols, 358–379
Cloud condensation nuclei (CCN), 54, 106, 258–260, 348, 351, 370–378, 387–388, 452, 504, 539, 541, 549
–, feedbacks, 84, 326, 329, 389, 453, 461
–, forcing, 268
–, variations, 232, 379
CO_2 concentration, 35, 37, 42, 50, 52, 75, 78, 81, 84, 89, 102–105, 126, 133–134, 136, 166, 177, 181, 224–225, 230, 259, 306–307, 317–334, 340, 343, 385–386, 400, 418, 446–451, 464–467, 508, 517–518, 520–522, 524–529, 538–539, 542–543, 549–551
–, fertilization, 465
COHMAP, 24, 36
Comets, 97, 102, 115–116, 128, 135, 136, 348
Comprehensive Ocean-Atmosphere Data Set (COADS), 152–153
Cool Sun paradox, 537
Cretaceous, 71–81, 85, 89, 104, 108–110, 133
–, -Tertiary boundary, 100–112, 114–115, 125–128, 134–136
Cryosphere, 272–274
Cyclones, 301–304

Daisyworld, 537, 539–541
Deep Sea Drilling Project (DSDP), 108–110, 112
Deforestation, 436, 441–446
Demographic change, 324
Desertification, 441–442, 445, 454
Devonian, 115

Dimethyl sulphide (DMS), 54, 106, 136, 348, 357–358, 374, 388, 539, 541, 544–546
Diurnal temperature, 163–164, 384, 502
DNA, 399, 421–422
Dust cloud, 97–98
–, veil index, 177

Earth Observing System (EOS), 250–251, 267, 271, 412
Earthquakes, 128
Earth Radiation Budget (ERB), 254–255, 318
– – –, Experiment (ERBE), 254–255, 257, 258
East African jet, 457
Eccentricity (of Earth orbit), 26–29
Eemian interglacial, 22, 38
El Chicón, 171, 176, 207, 351
Electromagnetic radiation, 245, 359, 487, 492
El Niño, 261, 310
El Niño-Southern Oscillation (ENSO), 173–174, 191, 193, 199, 207–225, 229, 262, 282, 297
–, variability, 222–223
Energy balance model (EBM), 10, 75–79
–, consumption, 509
–, use, 437, 501
Enteric fermentation, 450
Eocene, 5, 71–72, 81–88, 89, 115, 118–119
Equilibrium greenhouse warming, 327–331, 466
Erosion, 434
European Centre for Medium-range Weather Forecasting (ECMWF), 156, 263, 268, 303

Fingerprinting of climate change, 181, 274
First GARP Global Experiment (FGGE), 268, 292, 295
First ISLSCP Field Experiment (FIFE), 270
Flood basalts, 123, 127–128, 130–133
Flux correction, 33, 271
Foraminifera, 73, 81, 107, 226, 518, 526–527
Fossil fuel use, 322–324, 386, 450, 549
Framework Convention on Climate Change, 2
Frasnian-Famennian extinctions, 124–125

GARP Atlantic Tropical Experiment (GATE), 261
GENESIS 76–78, 80, 85
Geophysiology, 537–551
Glaciers, 5, 273
Global Atmospheric Research Programme (GARP), 261, 268
–, climate model (GCM), 10, 12, 32, 35–36, 73, 75–79, 84–85, 89, 99, 155, 159, 163, 166, 178, 180–182, 224–225, 228, 231, 250–252, 255, 272, 282–283, 285, 292, 321, 323, 327, 334, 342, 369, 378, 453, 456–459, 465, 477, 479, 482
–, Precipitation Climatology Project (GPCP), 155–156, 261
Gravity wave drag, 288, 290

Great Plains, 111, 201
Greenhouse effect, 7, 177–178, 180, 230–232, 307–309, 317–344, 352, 365, 385–387, 437, 509, 549
–, gas concentrations, 324–325, 446–451, 538–539
– –, emissions, 322–324
Greenland, 82, 162, 167–168, 193–194, 272, 382–383, 463, 524, 531
–, ice core, 129, 517, 522–523
– – –, project (GRIP), 58, 226, 389, 520

Hadley cell, 8, 10, 45, 300, 306, 455, 457
Historical record, 157–158
History of meteorological observations, 151–152
Holocene, 25, 37, 56, 195, 224, 519, 524
Homogeneity of meteorological record, 152–156
Homoeostatis, 537
Hurricanes – see Tropical cyclones

Ice ages, 22
Ice-albedo feedback, 51, 330, 467
Ice sheet simulation, 46–54, 56
Impact energy, 97
–, extinction, 114
–, volcanism, 127–129
Indian Ocean, 119, 221, 295
Intergovernmental Panel on Climate Change (IPCC), 2–3, 10, 15, 72, 80, 89, 164, 179–180, 230, 321, 324–325, 355, 358, 444, 446, 508
International Oceanographic Commission (IOC), 230
–, Council of Scientific Unions (ICSU), 230
–, Satellite Cloud Climatology Project (ISCCP), 255–256, 265–266, 269
– –, Landsurface Climatology Project (ISLSCP), 270
Interstellar cloud, 102
Intertropical Convergence Zone (ITCZ), 208, 213, 218, 295, 299
Iridium, 115–117, 120, 123–124, 126, 136
Irrigation, 438
Isua, 4

Jurassic, 115–117

Kalahari, 209
Kellwasser event, 124
Kelvin waves, 210, 288–289
Kepler's second law, 29
Kilauea volcano, 133
Krakatoa, 176, 205
KT boundary – see Cretaceous-Tertiary boundary

Lake desiccation, 24
Laki eruption, 129, 205, 351
Landsat, 246–247, 270
Landsurface, 8, 264, 269
Land use change, 246, 331, 433–468
La Niña, 174, 213, 222

Largescale circulation, 8
Last glacial maximum (LGM), 5, 16, 22–23, 34–36, 52, 54
Laurentide ice sheet, 25, 35, 37
Leaf area index, 269
Little ice age, 168, 175, 195–199, 222, 253
Los Angeles Basin, 356

Macrofossils, 82, 86
Marine mass extinctions, 113
–, stratocumulus, 258
Mass extinctions, 97, 114, 134
Matuyama-Brunhes boundary, 22
Maud Rise, 110, 119
Mauna Loa, 382, 539
Maunder minimum, 175, 200
Maximum temperature, 154, 163, 497–498
Mediaeval warm period, 168, 193–195, 224
Megacities, 480
Mesospheric ice clouds, 100
Meteorites, 95
Meteorological data assimilation, 263–264, 283
Methane concentration, 177, 230, 317–334, 340, 400, 404, 446–451, 464, 520–522, 543
METROMEX, 504
Microtektites, 96, 114,–115, 123, 136
Microwave sounding unit (MSU), 169–173, 176, 205–206, 232, 248, 262–263
Milankovitch theory, 30–31, 40, 44, 231, 547
Minimum temperature, 154, 163, 497–498
Miocene, 5
Model hierarchy, 10, 21, 32
Monsoon, 38, 194, 209, 214, 216–217, 304–306
–, gyre, 305
Montreal protocol, 2, 324, 426–427
Mount Pinatubo, 171, 176, 207, 227, 352, 413
–, St Helens, 176, 351

National Aeronautics and Space Administration (NASA), 246–247, 412, 416
Nile River, 194, 218, 222–223
Nitrate aerosols, 356–357
Nitrogen fertilizer, 449
Non-methane hydrocarbons, 356
Nordeste, Brazil, 218
North Atlantic, 37, 43
– –, deep water, 178, 226, 520, 524, 528–532, 547
– –, Oscillation (NAO), 167
NO_x, 100–101, 103, 135, 317–335, 340, 400–401, 404–407, 425, 446–451, 464
Nuclear winter, 135–136, 347
Numerical weather prediction, 263–264

Obliquity (of Earth orbit), 26–29, 51
Ocean alkalinity, 104–106, 527–528
Oceanic conveyor belt, 16, 529–531
–, heat transport, 84–85, 88, 326, 331

Oligocene, 5, 115, 118
Orbital insolation, 29, 34
–, parameters (of the Earth), 25–29, 200
Ordovician, 125–126
Oxygen isotope, 22–23, 41–42, 47, 49, 73, 80–82, 108–112, 117, 122, 125, 127, 195, 226, 230, 517–525
Ozone concentration (stratosphere), 100, 334–338, 343, 399–429
–, "hole", 406–408, 416, 428
–, photochemistry, 400–406
Pacific, 110, 173, 208, 210, 295, 463, 529–531
–, North America pattern (PNA), 297, 304
Palaeoclimate, 22, 32
Palmer Drought Area Index (PDAI), 201–202
– –, Severity Index (PDSI), 200–201
Pangaea, 122
Paraná River, 219
Permian, 124
– –, Triassic boundary, 120–124
Phanerozoic, 113
Phase locking (of climate), 41
Pleistocene, 43, 100, 114, 191, 230
Pliocene, 89, 100, 116–117
Polar stratospheric clouds (PSCs), 406, 415
Policy development, 2, 22, 180, 425–428
Population, 435, 438, 468, 477–478, 507
Precambrian, 116, 124–125
Precession (of Earth orbit), 26–29, 51
–, of the perihelion of the axis, 28
Precipitation from satellites, 260–264
–, records, 164–167, 217

Quasi-biennial oscillation (QBO), 179, 199, 202–205, 213, 288
Quaternary, 5, 55, 59, 539, 544–548

Radiative convective model, 52–53, 98
–, forcing, 320
Reanalysis project, 302
Red noise, 192
Reforestation, 464
Regional climate change, 321
Reservoir construction, 438–440
Rice cultivation, 449–450
Rossby radius of deformation, 290
–, waves, 288–290, 299–301
Roughness length, 454–455, 478–479, 484, 494
Runaway greenhouse, 6, 59

Sahara, 367, 441, 443
Sahel, 165–166, 218, 229, 331, 441, 443, 456–459, 465
Satellites, 245–274, 335–337, 411, 492, 502, see also EOS, Landsat, MSU, SPOT
Sea breeze, 505
–, ice, 273

–, salt, 351
Sea-level change, 526
Severe weather events, 309
Skin cancer, 420–423
Snow albedo, 46, 51
Soil erosion, 436, 440
Solar insolation, 101, 175, 206, 221, 252–253, 293
–, variability, 199–205
South Pacific Convergence Zone (SPCZ), 208, 218
Southern Oscillation Index (SOI), 174, 207, 210–211, 213
SPECMAP, 31, 42, 49
Spectral analysis, 47
SPOT, 247
Statistical-dynamical models, 10, 14, 44–54, 339
Storm tracks, 301–304
Stratospheric cooling, 328
–, ice clouds 406–408
–, ozone – see Ozone concentration (stratosphere)
–, temperatures, 203
Stromatolites, 117
Sulphate aerosols, 130, 164, 173, 177, 231, 347–389, 412–415, 452
Sunspots, 175, 179, 199, 201–202
Surface albedo, 437, 452, 453, 459, 461, 488–489, 540
–, Heat Island Model (SHIM), 499
–, pressure records, 167–168, 203–204
–, Radiation Budget (SRB), 265
Surprises (climate catastrophes), 14, 117, 193, 226, 389, 517–533, 550

Teleconnections, 214, 224, 281–282
Temperature records, 158–159, 172, 196–198, 382–384
Termites, 450
Terpenes, 355–356
Tertiary Strangelove Ocean, 103, 106, 127
Thermohaline circulation, 85, 88, 226, 529–531
Thermometers, 153
Tibetan plateau, 218
Time series analysis, 192
Tissue damage due to UV increase, 424
TOPEX-POSEIDON, 272
Total Ozone Mapping Spectrometer (TOMS), 409–410, 412, 421–422

Transient response to greenhouse warming, 331–334
Tree rings, 168, 196, 201–222
Triassic, 119
Tropical cyclones (typhoons, hurricanes), 209, 304–305, 309–310
–, deforestation, 322, 435, 444–445, 455, 459–463
–, Rainfall Measuring Mission (TRMM), 250, 261–262
Tropospheric ozone, 317–335, 338–344, 487
Tunguska, 101
Typhoons – see Tropical cyclones

UK Meteorological Office GCM, 334
Ultraviolet radiation, 399, 405, 420, 425, 427, 429, 487, 506
Upper air data, 169–172
Urban boundary layer (UBL), 483, 486–487, 494–496
–, canopy layer (UCL), 482–483, 486, 490–491, 494–496
–, heat island, 155, 197, 478, 498–500
–, humidity, 502–503
–, plume, 505
–, rain islands, 504

Vegetation change, 466
–, indices (VI), 269–270
Volcanic eruptions, 131, 171, 176, 205–207, 347, 349, 351–352, 537
–, explosivity index, 205
Vostok, 522

Walker circulation, 214, 306, 455, 463
–, Gilbert, 214
Walvis Ridge, 108
Warm epochs, 71–88
Water vapour feedback, 52–54, 99–100, 228
Weddell Sea, 110
– –, Bottom Water (WSBW), 520
Wildfires, 101
World Climate Research Programme (WCRP), 155–156, 265, 272
–, Meteorological Organization (WMO), 152, 154, 159, 164, 230, 324

Younger Dryas, 37, 43, 517, 523–524, 531